CONTAMINATED LAND
Problems and Solutions

CONTAMINATED LAND
Problems and Solutions

Edited by

T. CAIRNEY
Consultant
W.A. Fairhurst Environment Division
Newcastle upon Tyne

BLACKIE ACADEMIC & PROFESSIONAL

An Imprint of Chapman & Hall

London · Glasgow · New York · Tokyo · Melbourne · Madras
Published in the USA and Canada by Lewis Publishers
Boca Raton · Ann Arbor · London · Tokyo

**Published by Blackie Academic & Professional, an imprint of Chapman
& Hall, Wester Cleddens Road, Bishopbriggs, Glasgow G64 2NZ, UK**

Chapman & Hall, 2–6 Boundary Row, London SE1 8HN, UK

Blackie Academic & Professional, Wester Cleddens Road, Bishopbriggs,
Glasgow G64 2NZ, UK

Chapman & Hall Japan, Thomson Publishing Japan, Hirakawacho Nemoto
Building, 6F, 1–7–11 Hirakawa-cho, Chiyoda-ku, Tokyo 102, Japan

DA Book (Aust.) Pty Ltd, 648 Whitehorse Road, Mitcham 3132, Victoria,
Australia

Chapman & Hall India, R. Seshadri, 32 Second Main Road, CIT East,
Madras 600 035, India

First edition 1993

© 1993 Chapman and Hall

Published in the USA and Canada by Lewis Publishers, 2000 Corporate Blvd,
N.W., Boca Raton, FL 33431

Typeset in 10/12pt Times by Pure Tech Corporation, India

Printed in Great Britain by St Edmundsbury Press Ltd, Bury St Edmunds, Suffolk

ISBN 0 7514 0065 3 (HB)   0 87371 870 4 (USA)

A catalogue record for this book is available from the British Library

Library of Congress Cataloging-in-Publication data

Contaminated land: problems and solutions/edited by T. Cairney
    p.   cm.
Includes bibliographical references and index.
ISBN 0–87371–870–4
1. Soil pollution.   I. Cairney, T. (Thomas)
TD878.C65   1993
628.5′5—dc20   92–16710
                                                   CIP

Preface

A decade and a half of reclaiming contaminated land for beneficial reuse has yielded a large amount of useful experience. For many aspects of the reclamation process, it is possible to specify and advocate good practice, in particular for devising cost-effective site investigations; specifying and installing clean covers and in-ground barriers; quantifying the potential for buried materials to undergo subterranean heating; selecting methods to prevent fire hazards; revegetating reclaimed sites; and ensuring that the required reclamation quality is provably attained. For other aspects such certainty has yet to be attained, either because a particular problem is especially complex (as is the case with some landfill gas situations) or because practical experience is still too limited. With more research and experience, these less certain aspects will become better understood. The fundamental message of this book is that at the present time it is possible to identify the basic factors that are likely to be critical in ensuring appropriately safe reclamations.

Variations in national legislation and controls do affect the reclamation options that can be employed, and the clean-up standards (and costs) that are necessary. These variations have tended to obscure the essential commonality of land contamination as a problem for all industrialized and developed societies. Fortunately national emphases are gradually changing, and a consensus appreciation of the problem now looks more likely to evolve. Thus, the information on legislation, controls and practice that forms the core of this book has a global relevance.

Reactions to the presence of land contamination in sites earmarked for redevelopment have sometimes been overly concerned. This book argues that much of the problem of land contamination is relatively easily resolved by the application of quite simple but logically planned solutions, and that perceived effects on human health and the wider environment are still largely potential rather than provable hazards.

T.C.

Contributors

Mrs G.S. Beauchamp Senior Landscape Architect, Environment Division, W.A. Fairhurst & Partners, No. 1 Arngrove Court, Barrack Road, Newcastle upon Tyne NE4 6DB

Mr M.J. Beckett Consultant — land assessment and reclamation, Formerly Secretary to the ICRCL, 1 Falcon Court, Alton, Hampshire GU34 2LP

Dr T. Cairney Consultant, Environment Division, W.A. Fairhurst & Partners, Mast House, Derby Road, Bootle, Merseyside L20 1EA

Dr R.H. Clucas School of Built Environment, Liverpool Polytechnic, Clarence Street, Liverpool L3 5UG

Mr D.M. Hobson Divisional Director, Environment Division, W.A. Fairhurst & Partners, No. 1 Arngrove Court, Barrack Road, Newcastle upon Tyne NE4 6DB

Dr S.A. Jefferis Golder Associates (UK) Ltd, Consulting Engineers, 54 Moorbridge Road, Maidenhead, Berkshire SL6 8BN and Technical Co-ordinator, European Centre for Pollution Research, Queen Mary College, University of London, E1 4NS

Mrs W.K. Lewis H.A.S.T.A.M., Aston Science Park, Love Lane, Birmingham B7 4BJ

Mr G.J. Longbottom Principal Engineer, Environment Division, W.A. Fairhurst & Partners, No. 1 Arngrove Court, Barrack Road, Newcastle upon Tyne NE4 6DB

Mr M.K. Meenan Halliburton NUS Environmental Ltd, Thorncroft Manor, Dorking Road, Leatherhead, Surrey KT22 8JB

Mr T. Sharrock Consultant, Environment Division, W.A. Fairhurst & Partners, Mast House, Derby Road, Bootle, Merseyside L20 1EA

Dr S.A. Simmons   H.A.S.T.A.M., Aston Science Park, Love Lane, Birmingham B7 4BJ

Mr M.V. Smith   L.T.G. Environmental Services, Page Street, Mill Hill, London NW7 2ER

Contents

7 Reclaiming potentially combustible sites   141

R.H. CLUCAS and T. CAIRNEY

8 Landfill gases   160

M.V. SMITH

9 Establishing new landscapes   191

G.S. BEAUCHAMP

10 Quality assurance   212

T. CAIRNEY

11 UK legal framework   228

G.J. LONGBOTTOM

12 Introduction to US waste management approach   263

M.K. MEENAN

13 Health and safety 276

S.A. SIMMONS and W.K. LEWIS

Appendix I: Soil and water quality criteria 305

Appendix II: Waste disposal regulations 334

References 340

Index 349

1 International responses

T. CAIRNEY

1.1 Introduction

When land contamination was first identified in the late 1970s, in a few of the industrialized nations, almost nothing was known of the hazards this could pose. The governments concerned, in the United States, The Netherlands, West Germany and the United Kingdom, thus had to react in a state of partial information. Other industrialized countries, which had not had the problem thrust upon them, found the partial evidence of environmental hazards unconvincing and chose not to treat land contamination as a problem of real concern. Some still persist in this view.

This situation is in sharp contrast to the universally recognized concerns over air and water pollution, whose effects have been obvious even to casual observers for centuries. Indeed, water pollution worries were recorded as far back as the time of imperial Rome, when the effects of the city's untreated sewage on the River Tiber were recognized as dangerous to human health. Likewise, the identification of air pollution as a concern has a long history. The smoke clouds that covered Elizabethan London were clearly related to various human ailments, and quite serious (though unsuccessful) efforts were then made to ban the use of coal for domestic fuel [1].

The growth in industrialization in Western Europe, in the 18th and 19th centuries, gave even more convincing proof of the obvious links between air and water pollution events and such declines in the quality of life as the loss of fish in some rivers adjacent to industrial centres, the destruction of trees and crops lying downwind of smelters and industrial chimneys, and the reduced life expectancies likely to occur in many industrial employments [2]. These persuaded the more enlightened administrations of the day into enacting increasingly effective controls over air and water pollution. As this process was gradual, neighbouring nations had the opportunity to absorb the lessons that had been learned and to gain access to the relevant scientific and engineering information.

When the influential environmental movements appeared in the 1960s (largely as a result of Rachael Carson's [3] apocalyptic predictions of the world our children could inherit if controls were not enforced), a considerable technical background existed on which to base more effective environmental conservation measures. Thus, whilst national differences still persist on some aspects of air and water quality protection, the overall need for such conservation measures is not subject to dispute.

In contrast, little or no concern over land contamination and the fate of spillages and wastes on industrialized land appeared in earlier years. To a large extent, this is still the case and, whilst fears over soil contamination can feature locally and briefly in the press, other concerns, such as the ozone layer depletions or the decline in animal populations in the North Sea, make a much more obvious impact on public perceptions and attract more national and international controls.

In this situation of limited technical guidance and of generally low key public pressures, it is not surprising that different national administrations adopted very different responses to the identification of land contamination.

1.2 Differences in national responses to land contamination

1.2.1 *Various national viewpoints*

That potential hazards can arise from the addition of chemical substances to soils is probably not a matter of dispute. There is, after all, a widespread knowledge that altering the chemical condition of a soil will affect plant life and that substances deposited on land can later migrate to streams and rivers, and then enter the human food chain.

The problem comes when it is necessary to prove that actual hazards have occurred, or are likely to take place and so justify particular control measures.

Some nations have found no difficulty in being absolute in their stance on this, and senior Dutch spokesmen have expressed such definite views as: "soil pollution is a gigantic problem with serious consequences for man, plants (including crops), animals and the abiotic environment" [4].

In contrast, other nations like the United Kingdom have been less certain of the scale of the likely hazards and have chosen to discuss land contamination only in the context of the safe redevelopment of land, e.g.

> Contaminated land is that land which, because of its former uses, now contains substances that give rise to the principal hazards *likely* to affect the proposed form of development and which requires an assessment to decide whether the chosen development may proceed safely, or whether it requires some form of remedial action, which may include changing the layout and form of development [5].

Yet other industrialized nations, such as France, Italy and Spain, apparently have chosen not to see land contamination as an issue of significance and so have avoided taking any particular stance.

The reason for this variation in possible viewpoints is easy to identify.

1.2.2 *The complexity of soil systems*

Soil is an extremely complex medium and different soil factions and constituents have very variable degrees of reactivity to any introduced compounds. Thus the simplicity of the 'cause and effect' situations that can be obvious in air and water pollution incidents is usually unattainable. The complex behaviour of pollutants introduced into soils that support vegetation are well described by Failey and Bell [6], and even when soils contain such extremely high concentrations of a contaminant that vegetables grown in them have at least 10 times the contaminant level of normal urban soils, it can be impossible to demonstrate any hazard at all to consumers of the crops [7].

Thus it is not surprising that demonstrable cases of hazards from contaminated soil are almost totally absent in the technical literature. For example, Dutch studies [8] on a population exposed to chlorinated benzenes and phenols from a household garbage dump failed to prove that any individuals had in fact suffered, despite long-term exposure to what are obviously dangerous substances.

This lack of actual proof that hazards have occurred does not, of course, mean that they are non-existent or negligible, but merely indicates that they are difficult to identify and probably very slow to appear.

However, national administrations, forced on by local public pressures, cannot delay decisions for decades until hazards have actually appeared. Thus intuitive and emotional responses have tended to typify the various national reactions.

1.2.3 *Identification of land contamination and necessary clean-up standards*

In Europe, the Dutch have been most emphatic in seeing land contamination as a major environmental concern. Other countries (Germany, Denmark, Norway and Austria) have in part followed the Dutch lead, though with less apparent emphasis and commitment. The United Kingdom (as indicated in detail in chapter 4) initially chose quite a different approach, but now is moving, as did the Dutch, to compiling registers of contaminated sites and identifying priority cases for clean-up and remediation [9].

The most recent tabulation (Table 1.1) of the scale of clean-up necessary [10] in different countries, makes clear the obvious differences in emphasis between those nations which have succeeded in devising registers of contaminated land. For a small country of the size of Holland to have identified more than twice as many sites requiring clean-up as its much larger and more heavily industrialized German neighbour is obviously not credible, unless — as is the case — the Dutch have chosen to apply a more rigorous definition of land contamination and have had the national will to accept the financial and other consequences. Also, the remarkably small number of US sites requiring remediation is obviously indicative of a very different national emphasis from that in The Netherlands.

Table 1.1 Estimated scale of the necessary clean-up of contaminated land (after [10])

Country	Likely number of suspect sites	Likely number of contaminated sites	Estimated clean-up costs (US$ × 10⁹)
Netherlands	600 000	110 000	25[a]
Germany	200 000	50 000	10–60
United States (Superfund)	35 000	1200	30
Denmark	6000–10 000	Not known	Not known

[a] The Netherlands clean-up costs have also been estimated at $27 billion, or $270 million per annum for the next 100 years [11].

The Dutch lead position arises largely from its adoption of the most severe clean-up standards of any that have so far been defined. These 'A-B-C' standards (Appendix I) have the logical basis of setting a standard for uncontaminated soil (the background level A), a trigger value standard (level B) — which indicates the need for further evaluation and testing to identify the problems that may need to be resolved — and a contaminated standard (level C) above which clean-up is mandatory. Where the Dutch standards are illogical, however, is in their adoption of single concentration values (for each contaminant) at each of the three levels, irrespective of the soil types which were analysed, or the geographical/geological setting of the site in question. This is a surprising situation, since Dutch research [12], prior to the establishment of the national clean-up standards, had clearly shown that concentrations of elements in soils gave significant linear correlations with the type of soil containing the contaminants and, to a lesser extent, with the percentage of organic matter in the soils (Table 1.2). This is precisely what would be expected and it would seem more reasonable to have defined standards that took appropriate note of the soil types involved.

Table 1.2 Comparison of some background values in nature reserve top-soils and the ABC guidelines

Element	Concentration range (mg / kg) in			A value in guidelines (mg / kg)
	Sandy soils	Loamy soils	Clay soils	
Chromium	11–43	33–75	99–117	100
Cobalt	0.3–2.0	2.4–10	14–16	20
Zinc	6.4–43	28–150	135–153	200
Arsenic	1.4–18	5.1–13	19–21	20
Barium	80–266	168–416	466–525	200
Bromium	2–18	5.7–42	14–16	20

It is also apparent from Table 1.2 that the Dutch have chosen to accept the costs and difficulties of cleaning up old industrial land to a condition similar to that of a nature reserve. No other country has so far been willing to accept the financial burdens and technical difficulties that this choice presents.

The standards adopted by other countries are equally easy to criticize. In the United Kingdom, the pitfall of ascribing a single contaminant concentration that makes clean up mandatory has been avoided, but at the cost of leaving the interpretation of the guidelines open to a large element of professional judgment. This can give rise to very different clean-up standards being used on adjacent sites and makes it difficult to justify any claim of a national standard.

In those countries where federal and state authorities coexist, the situation can be more complex. In the United States, the Environmental Protection Agency has developed a comprehensive and evolving set of clean-up criteria, yet a state like New Jersey (with a particularly rigorous view on environmental clean-up) makes use of its own informal guidelines for the soil and water concentrations that must trigger site clean-up activity [13]. Likewise in Germany, the federal authorities have established a national strategy to remove contaminated land hazards, but the implementation is necessarily left to the various Laender authorities, which still lack any uniform and agreed evaluation system on which contamination hazards are judged [14].

Given the fact that provable land contamination hazards have yet to be widely demonstrated, the complexity and variability of different soil systems and the very real difficulty in identifying what are acceptable daily intakes for the vast range of possible contaminant compounds, it is not at all surprising that this variation in defined clean-up standards and their applications has arisen. When technical information is absent or partial, political judgments inevitably tend to supplant scientific data.

1.2.4 *Variation in legislation and controls*

The differences in national emphases obviously have governed the legislation and controls that are seen as necessary.

The Dutch view of soil multifunctionality logically requires contamination clean-up to a very high standard. This, in turn, has led to The Netherlands Interim Soil Clean-up Act (1983), the Soil Protection Act (1986) and the recent Nuisance Act, which permits local authorities to set constraints on the environmental impacts of industry. Without doubt, this is a particularly comprehensive set of land contamination controls and is supported by a national programme of research and development to prove the effectiveness of innovative clean-up technologies. A recent effect of these policies is that the state has found it impossible to organize, fund and execute all the clean-ups that are seen as necessary [11] and efforts are in progress to encourage industry voluntarily to clean up some sites. Also, technical difficulties of attaining the required clean-up level, in some cases, have led to criticisms of the national standards, and the recent modifications of the Dutch Building Act (which permit local authorities to include soil contamination as a factor governing whether or not a building permit is granted) look likely to harm the necessary inner city rebuilding programme, since a great many sites have been levelled with dredging spoils that are more contaminated than the Dutch A standard. Thus the Dutch position looks to be one where a laudable emphasis on environmental conservation is increasingly coming into conflict with other equally desirable national aims.

The United States situation is, in some ways, analagous to that in The Netherlands and environmental controls have been enforced to a level that is surprising, given the usual US emphasis on constitutional freedoms. The Comprehensive Environmental Response, Compensation and Liability Act of 1980 and the Superfund Amendments and Reauthorization Act of 1986 give the federal Environmental Protection Agency (EPA) very considerable powers. Funding to support technology trials and demonstrations is also provided on a scale much greater than in most other countries. The EPA can force abatements of environmental threats and damage and can compel responsible persons to pay the costs of rectifying environmental degradation. Since these powers can be applied retro-actively (unlike those of UK or EC legislation), many commentators [15] have seen the system as unfair, more likely to involve expensive litigation than land clean-up, and a direct cause of the near impossibility of insuring against environmental liabilities. Other writers [13] have noted that the system has led to an increasing number of derelict (and possibly contaminated) sites, which have been abandoned by bankrupt companies. Thus, the required environmental protection might actually be being reduced by the severity of the legislative controls.

Other European countries have avoided the severity of the Dutch and US legislation, though most have gone some way to imposing environmental restraints on land contamination, and on the use of such land.

What has been learnt in the past decade seems to be that land contamination is simply one factor that should be of concern in a modern industrialized society and that an overly rigorous legislative emphasis on it can, in itself, adversely affect other desirable environmental objectives.

1.3 Signs of an international consensus

Even if some existing legislation is amended, there are good reasons for differences to persist in national policies. A country like Holland, with dominantly permeable soils and a high groundwater table widely used for drinking supplies, has obviously more reason to be concerned over the slow leaching of contaminants from industrial fills. Other countries, with greater safeguards provided by a boulder clay land surfacing can, in contrast, afford to take a less concerned view of this sort of degradation.

The costs of contaminated land reclamation, the large numbers of sites that have been identified in several countries, and the general failure of the apparently simple policy of 'making the polluter pay' have encouraged an increasing consensus that:

(a) registers of contaminated sites are an essential first step;

(b) consistent methodologies are needed to identify the more dangerous sites, from those on the registers, so that scarce national resources can be better targeted;

(c) on-going interchange of technical information such as the NATO/CCMS study [16] is useful and saves every nation having to undertake all the research that is needed;

(d) integrated pollution control is the logical long-term answer, as this should prevent more contaminated sites being created;

(e) fuller research is especially essential on the rankings of the various contamination hazards in the different possible soil and environmental conditions.

Thus, the next decade offers the hope that the distinctions between national attitudes may reduce, that a sounder basis for establishing the actual hazards from land contamination may appear, and that misapplication of scarce national resources can be minimized.

Whilst the bulk of this book is based on UK information and methods, there is, in fact, no reason to see this as a parochial attitude. Contamination of soils from past industrial uses does not differ technically from one country to another; all that changes are the national policies.

2 Land contamination

M.J. BECKETT

2.1 Introduction

The problems of contaminated land can be divided very simply into two broad categories: (i) direct, i.e. technical problems caused by the presence of contaminants and the hazards created by them; and (ii) indirect, i.e. those caused by the ways in which existing legislative and administrative systems handle the direct problems.

Problems of a technical nature can be solved by technical means such as those described in later chapters of this book. There will inevitably be a price to pay for these solutions; the more elaborate the solution the higher the cost will be. Given adequate resources, however, technical problems can be solved. For indirect problems, this may not be the case. Most indirect problems can be regarded as self-imposed constraints. Why do people purchase land without first finding out if it is suitable for their needs? How is it that planning applications and decisions do not always take contamination into account properly? It is not for the want of advice on such matters, but because so many other issues become involved, including questions of legal liability and economic viability, that the indirect problems as a set can appear so intricate that complete solution is virtually impossible. In such circumstances the possibility of solving the technical problems then seems more attractive: these can usually be solved at a price. Thus, even though the technical problems may not be the most important they will — despite the inherent costs — receive attention, while the indirect ones remain unaddressed.

Solutions for the technical problems of contaminated sites represent a challenge, which has been met in many different ways. Solving the indirect problems requires just as much ingenuity from experts in the complex legal and administrative fields, though as yet relatively little attention has been given to such matters. By the combined efforts of all the relevant disciplines, a complete solution to any contaminated land problem should be possible in the future.

2.2 The problem of definition: what is contaminated land?

The difficulty in reaching agreement on how this apparently simple question should be answered is central to the development of policies and approaches for dealing with contamination. There are many possible answers, and as many different approaches can be found in countries where the problems of contaminated land have been recognized.

In the United Kingdom, the Royal Commission on Environmental Pollution (RCEP) [1] drew the following distinction between 'contamination' and 'pollution':

> *Pollution* can be defined as the introduction by man into the environment of substances or energy liable to cause hazards to human health, harm to living resources and ecological systems, damage to structures or amenity, or interference with legitimate uses of the environment. Substances introduced into the environment become pollutants only when their distribution, concentration or physical behaviour are such as to have undesirable or deleterious consequences.
>
> For comparison, *contamination* can be defined as the introduction or presence in the environment of alien substances or energy, on which we do not wish or are unable to pass judgement on whether they cause, or are liable to cause, damage or harm. *Contamination* is therefore a necessary, but not sufficient, condition for *pollution*.

According to the above definitions, most attitudes and approaches to contaminated land seem to imply that the land is polluted rather than merely contaminated. Because it is harder to exercise the 'judgement' inherent in the RCEP definition of contamination, the public, local authorities, and others prefer to assume the definition of pollution, which they see as more positive. In fact this appearance of greater certainty is deceptive. The condition of the land in question seldom justifies the assumption of pollution. This often causes the significance of contamination to be overestimated, with unfortunate consequences for assessment and treatment of the land. When remedial action is recommended, designed and implemented on a contaminated site, it usually has to be based on the principle that it is prudent to avoid problems whenever possible, rather than to allow them to occur and force later emergency action. Even so, the decision to take avoiding action is seldom justified by conclusive evidence that a problem has occurred, or might occur in the future. Such positive evidence is extremely difficult to obtain. Before deciding on what action to take when contaminants have been discovered on a site, it is therefore important to avoid hasty overreaction. This is particularly relevant where hazards to health are suggested as a reason for taking action, or where the commissioning of elaborate programmes intended for monitoring purposes is contemplated.

Neither of the RCEP definitions is an ideal way of describing the types of land and the individual sites with which this book is concerned. Before examining why this difficulty arises, some possible alternative ways of classifying land need to be considered.

One way of classifying land is by its use or location, for example:

By use	By location
Agricultural	Coastlines
Industrial	Moorlands
Residential	Uplands, i.e. hills and
Recreational	mountains

Each category has particular characteristics, which are quite distinctive and easy to recognize. Taken to an individual site in any of these categories for the first time, a visitor would have no difficulty in recognizing the type of site and describing the surroundings.

Another possible basis for classification uses the condition or status of land, for example:

By condition	By status
Polluted	Vacant
Damaged	Under-used
Contaminated	Derelict
	Waste

Each of these categories has at one time or another been cited in official UK statistics or policy statements. Clearly, therefore, they have some significance. Yet none of these terms has distinctive characteristics that uniquely identify sites, thus the first-time visitor test mentioned above will usually fail. How does vacant land differ from land which is under-used? What distinguishes waste land from land which is derelict? Derelict sites are not necessarily contaminated, nor are contaminated sites always derelict: many are in continuous regular use, and seem not to be hazardous to their users. All of the above terms are vague and indistinct; their characteristics overlap and make precise definition and description difficult if not impossible. Mainly for this reason, there are no reliable estimates of the amount of contaminated land in the United Kingdom. There is, in practice, no simple definition of contaminated land that could be used to make such an estimate. It follows that in countries where estimates have been published and lists of individual sites compiled, their accuracy and validity can be no better than those of the definition used as the basis for identification. Matters of definition, identification and estimation of contaminated land are inseparable, and are strongly linked to national policies and approaches to the problems of such land.

2.3 Estimating the amount of contaminated land in the United Kingdom

Because of the inherent difficulty in defining contaminated land, it follows that no precise statistics of the amount of such land exist and such estimates as are available are difficult to interpret and compare. The methods that have been applied to the estimation of contaminated land in the United Kingdom are of two main types:

- those based on surrogates for contamination, e.g. dereliction;
- those derived from prospective surveys or registers of potentially contaminated sites

Using these methods a number of attempts have been made to estimate the amount of contaminated land in England and Wales. Although consistently indicating a significant area of land, the individual estimates varied widely and cannot be regarded as reliable. The most recent figures suggest that between 50 000 and 100 000 individual sites might be contaminated. The geographical distribution of these sites is primarily related to the pattern of industrial activity, although there are other factors. However, only a small proportion of these sites are likely to pose an immediate threat to public health or to the environment. For these reasons, it is not at present possible to quote a representative value.

In responding to the Select Committee report on contaminated land, the UK Government recognized the need for better information on the location and extent of contaminated land [2, 3]. Acquisition of this information is one of the objectives of the registers to be prepared by local authorities under S. 143 of the Environmental Protection Act 1990 [4].

2.4 Origins of contaminated land

However it is defined, contaminated land consists of many individual sites where the contamination occurred either in the past, or is still occurring. The number of such sites, and hence the area of contaminated land, should diminish with time as these sites are recognized, treated and returned to use. It is of course also possible that some land that has not yet been contaminated will become so at some time in the future. With increased awareness and recognition of the problems, and the introduction of stricter environmental controls, future additions to the stock of contaminated land should be small both in number and area.

2.4.1 *Historical contamination*

A wide range of contaminated sites exists under this description. In many cases these sites became contaminated through their use by industry and by processes and practices, which by current environmental standards would be judged inadequate.

Although much of the contamination caused by these methods occurred since the beginning of the Industrial Revolution, some examples of much older contamination are known. These include the sites of copper and lead workings dating back to Roman times. In some places, for example around Shipham in Somerset, there was a continuous history of metal mining and processing for many centuries. When contaminants accumulate over so long a period at a relatively slow rate, their effects tend to become subsumed into the general environmental changes that take place over time in any area. As such, the contamination may after many years appear to have little discernible impact on the local environment, which has become adapted to the presence of contaminants. Evidence of direct effects, whether adverse or otherwise, due to the presence of the contaminants may be hard to obtain in such circumstances.

The impact of derelict and contaminated sites used by industries that operated until recently is much more readily apparent. Much of the land in the Lower Swansea Valley in South Wales was severely damaged by the fumes discharged into the atmosphere by the local metal smelting industries. Considerable amounts of solid process wastes containing metals in high concentrations were left behind when the smelters closed down. As a result, a large area of land became derelict, with direct consequences for the local environment and its population. The assessment, restoration and reuse of this land was prompted by the need to regenerate the local economy, and represented one of the first attempts in the United Kingdom to deal with contaminated sites on a large scale.

On a more widespread scale in terms of the number of sites affected, land formerly occupied by town gas works, steelworks and foundries, railway goods yards and similar operations represents a large proportion of the total stock of contaminated land in the United Kingdom. A recent DOE consultation paper [5] on the compilation of S.143 Registers lists 16 categories of contaminative uses, most categories containing several commonly occurring industrial and commercial operations. Moreover, because of the tendency for some of these operations to be located close to residential areas, their sites are more readily obvious to the local community and, in addition, more likely to be required for reclamation or redevelopment. For this reason it is much more important to assess and deal with any contamination adequately.

2.4.2 Current contamination

A number of current industrial and commercial operations not only have the potential to contaminate land but actually do so, even though they may possess the necessary licences or other permissions from the appropriate regulatory authorities. Examples of such activities include waste disposal sites (landfills) and scrap yards. Although such sites are now subject to varying degrees of control intended to reduce or eliminate environmental nuisance, this has not always been the case. However, the controls are sometimes less effective than they could be. This is because it is often difficult to equip older sites and activities with the protective measures needed to reach the high degree of security possible on a more recently designed site. Another type of difficulty can arise from the way in which the responsibilities for licensing, supervision and enforcement are allocated. In the United Kingdom, they tend to be distributed between a number of diverse agencies and authorities, none with complete environmental responsibility for the site as a whole. Other concerns are inherent in the activities themselves; for example the generation of gas within a waste disposal site may not only affect the area of the landfill site itself but also, if the gas migrates beyond the site boundaries, adjoining land in different ownership. This can raise very complex issues of responsibility and legal liability.

In such circumstances, while the sites remain in use, there may only be scope for limiting further damage caused by the contamination. Application of the statutory nuisance provisions in current legislation may serve to contain problems, but comprehensive remedial action to improve the local environment may have to be deferred until the activities that cause the contamination have ceased. At this stage, another difficulty may arise: UK legislation does not always enable former landowners to be held responsible for the long-term consequences of their occupation of the land, or the costs of rectifying any damage. Nor are means and resources always available for the regulatory bodies to deal with the legacy of contamination. Despite the much improved effectiveness of recent environmental legislation at the point of control, there remain matters such as these where closer integration of the diverse interests seems desirable. Even in legislative areas where some integration has been achieved, for example control of water pollution, it is not unusual for inconsistencies to occur in implementation between different parts of the regulatory body, as can be the case with the National Rivers Authority.

The above discussion serves to illustrate the complexity of issues related to the handling of contaminated land. It is in these respects that the main difficulties and costs arise, not in connection with the technicalities of the remediation process. The amount of time and attention devoted to specific contamination problems has been considerable: better progress in solving the problem might have been made had the real reasons why contaminated sites continue to exist in the United Kingdom and elsewhere been identified first.

2.4.3 *Recognition of contaminated land as a problem*

In the years following the Aberfan Disaster (1966), a very significant increase in Government support for land reclamation took place. Although not directly prompted by the tragedy, the aim of the programme was to deal with land that had been damaged or rendered derelict by previous uses. This support has been continued and enlarged through a wide range of schemes intended to provide direct financial help for the restoration and reuse of land, with additional incentives and policies on land use as it affects the location of and suitability for development. Much land was successfully restored or reclaimed through this programme, but the resulting environmental gain was broad and not intended to address specific issues such as contamination. It was not until the mid-1970s that contamination, rather than dereliction or other problems, was recognized as a primary constraint on many sites and attention began to be directed towards different categories of land.

The principal aims of Government policies for reclamation of land have been to create jobs, increase investment potential, provide or improve housing, and encourage better use of amenity and recreational land. Where the achievement of these aims has coincided with a need to deal with pollution or contamination, grant support can also include the costs of additional treatment needed to render the land safe for further use, or to improve the environment. In the main, however, environmental objectives were regarded as of secondary importance compared to the more immediate priorities mentioned above. Contamination was seldom the factor that decided whether a grant was provided or not.

This separation of contamination problems from those associated more generally with returning land to use has influenced the approach to contaminated land issues in the United Kingdom for many years. The inability to estimate accurately the amount of contaminated land in the United Kingdom also results from this separation of basic policy issues. Since a well-established programme on derelict land exists, relevant statistics that enable its performance to be measured are regularly compiled and published. Because there has been no comparable programme on land that is contaminated rather than derelict, no reliable figures are available. Only recently have signs of a reappraisal of priorities begun to emerge in UK policies.

2.5 Legislation and policies in the United Kingdom

The principal Government programme on reclamation is primarily targeted at derelict land, defined as land that has been so damaged by industrial or other use that it is now incapable of beneficial use without treatment. This is the *Derelict Land Grant* (DLG) scheme. As previously stated, not all derelict land is contaminated nor are all contaminated sites derelict. Other Government initiatives, such as the Urban Programme, City Grant, and Urban Development Corporations, have funded reclamation schemes that included action to remedy contamination problems, where this was consistent with the main objectives of the grant regime concerned.

Although the DLG was the only land reclamation scheme covering the whole of England, the rate at which grant was provided varied according to location, with the greater proportion of available funds and the highest rate of grant (100%) allocated to those areas that had the largest areas of derelict land. Within this programme, the fact that land was derelict was not in itself sufficient reason to incur expenditure on its reclamation. Other justifications, especially the likelihood that after reclamation the site would attract development that met the Government's preferences for employment, investment or housing, were normally required before a grant was provided.

Thus, contamination, even though it might be the main reason why the land became and remained derelict, could well remain outside the scope of the DLG scheme if the development potential of the site was low. An internal review of derelict land policy carried out by the Department of the Environment in 1989 [6] examined these constraints and identified several ways in which more effective use of the available funds was possible. One of these was to assign greater priority to contaminated sites so that they might become eligible for DLG even where no development after-use was likely. This could enable schemes whose aim was environmental gain rather than development potential to receive a grant.

The main effect of UK environmental legislation since 1970 has been to control polluting activities ever more strictly and, through Integrated Pollution Control, to eliminate them where possible. Sites operating to these newer controls are unlikely to cause problems in the future.

Much less attention has been given to sites where the pollution or damage is of historic origin. Moreover, the principle of making new legislation retrospective in effect is unusual in the United Kingdom. What this has meant in practice is that contamination, however it is defined and measured, has not been regarded as an issue requiring legislation in its own right. It has instead been dealt with by means of legislation on other subjects. The most important of these other areas of interest is the subject of town and country planning, which is concerned with how land uses are allocated and achieved. A succession of *Town and Country Planning Acts* has been introduced since 1948. For development that falls within the scope of these Acts, which in practice is the majority, this legislation provides a comprehensive and systematic means for controlling land use allocation and regulating development. Both these functions are directly relevant to the use, reclamation and redevelopment of contaminated sites.

There are, however, important limitations to the ability of town and country planning legislation to deal with the problems of contaminated land. First, the Acts are founded on a presumption in favour of development: their main function is to control development, not prevent it. Some 500 000 planning applications are made each year in the United Kingdom; after taking into account those that are determined by appeal or called in by the Secretary of State for the Environment, over 94% of all applications are granted. This demonstrates that the legislation achieves its main objective. Second, responsibility for processing and determining applications for planning consent is exercised by individual local planning authorities, of which there are more than 350 in England and Wales. When contaminated land problems began to increase in number and importance, many of these authorities had neither the experience nor the resources to evaluate properly the complex inter-disciplinary technical issues of contamination. For some, this may still be the case. Given the legislative bias in favour of development, it is likely that some inherently flawed planning applications on contaminated sites were granted rather than rejected.

Other relevant fields of legislation are those concerning public health and safety, building control, and environmental health. Within these subject areas, local authorities have various powers to deal with contamination and similar nuisances or threats. Evolution and application of these powers over a long period of time have resulted in a great deal of valuable practical experience being available within the local authority departments concerned. In many cases, however, these departments do not have the primary responsibility for deciding whether development should be allowed: often they can only deal with problems brought to their notice after development has taken place. Where the problems could have been avoided by not granting permission for development, trying to deal with them later is a decidedly inferior option.

It should be apparent from the above discussion that the system for dealing with contaminated land issues in the United Kingdom is very wide-ranging and comprehensive, but in some respects less effective than it could be because of the segregation of functions and lack of overall responsibility at any one level of authority.

2.6 Contaminants, hazards and targets

2.6.1 *Contamination as a concept*

Contaminants can be of natural origin, such as emissions of radon and methane or enhanced concentrations of metals in soils. The true significance of naturally occurring contaminants in the contexts of human health and development remains to be established and, if necessary, included as part of an overall strategy for dealing with contaminated land. More generally, contamination is the result of man's exploitation of land, and is the existence on a site — usually in the soil or groundwater — of potentially harmful substances, which may occur as solids, liquids and gases, or combinations of these.

But what is it that these substances are potentially harmful to, and in what circumstances? There is no single, or simple, answer: much depends on which hazards may occur, how the risks may arise, and how those risks manifest themselves. There is, in fact, a chain of interlinked relationships:

Previous
and/or   ⟶ Contamination ⟶ Hazard ⟶ Target ⟶ Protection
present use

Each of these relationships can be examined individually in order to build up a good understanding of how they affect views about contaminated land. Table 2.1 illustrates the most important contaminant/hazard links, while in Table 2.2 those between land use and hazards are listed.

Table 2.1 Contaminant / hazard links

Main contaminants	Hazards
Heavy metals (plant uptake) Heavy metals (grazing animals) Organics (PCB, etc.)	Ingestion
Asbestos Metal dusts Toxic gases	Inhalation
Acids and alkalis Organics, e.g. phenols Some metal salts, e.g. chromates	Skin contact
Zn/Cu/Ni Sulphates Landfill gas	Phytotoxicity
Sulphate/Sulphide Organics, e.g. oils, tars, phenols Acidity	Building material degradation
Carbonaceous matter Sulphur wastes (spent oxide) Gases	Fires
Landfill gas	Asphyxiation
Landfill gas	Explosion
Biological organisms Chemical warfare agents Biocides	Other (health)
Smell/odour/taint	Other (nuisance)
Biodegradation Fires	Instability

Table 2.2 Land use / hazards links

Proposed use	Principal hazards that may pose a threat
Residential with gardens	All
Residential without gardens (i.e. flats, etc.)	All except phytotoxicity or ingestion
Allotments/market gardens	Phytotoxicity Skin contact
Agriculture : 　Arable 　Grazing	 Phytotoxicity Phototoxicity and ingestion
POS[a]/amenity/recreational	Phytotoxicity Skin contact
Commercial (e.g. offices, retail)	Building material degradation Fires, explosion

Table 2.2 *(Cont.)*

Proposed use	Principal hazards that may pose a threat
Light industry (e.g. warehouses, factory units)	Building material degradation Fires, explosion
Heavy industry	Building material degradation Fires

[a] POS, public open space

The more common 'targets' and the routes by which they may become exposed to the hazards of contamination are as follows:

- People   by direct or indirect ingestion or inhalation of contaminants
- Animals   from grazing on contaminated land
- Plants   growth for agriculture, horticulture and domestic use in contaminated soil
- Buildings   by chemical attack of contaminants on foundations and services installed in aggressive ground

Which of the possible targets is likely to be exposed on any given site depends on the use to which that site is put. If there is no exposure path by which the contaminants present can affect the targets associated with the selected use, then the importance of any hazards from those contaminants is reduced. Because the link between uses and targets is one in a sequence of relationships, it follows that the significance of contaminants is related to the choice of land uses. The UK approach to contaminated land problems was based on this principle. On this basis, contamination only needs to be considered significant if the site contains contaminants in concentrations high enough to pose hazards to the intended use of the land. For a different land use in which exposure paths are eliminated or blocked, there would be less risk from the same contaminants. The importance of the contamination would therefore be lower and more tolerable. This sets an upper limit on the amount of land that needs to be regarded as 'contaminated'. The practical and financial implications of not setting such a limit, i.e. regarding all contaminants as of equal importance wherever they occur, are far-reaching.

For example, on many industrial sites, such as scrapyards, the soil is likely to contain metals such as cadmium, lead, copper and zinc in high concentrations, thereby rendering plant growth impossible or at least of poor quality. For land uses in which good plant growth is essential, such as allotments and gardens, the presence of the metals in the soil would represent a hazard for which remedial action would be essential. However, if instead the intention was to construct a warehouse, office block, hotel or other similar building on the same site, the presence of the same metals in the same concentrations would be of little importance and would not preclude successful use of the site. In these circumstances, it would not be necessary to take remedial action in respect of the metals. In evaluating the site for such uses, other issues, such as the implications for groundwater quality of allowing the contaminants to remain, could then be taken into account.

No simple formal definition can incorporate all the diverse aspects of the above terms. Contamination is therefore best regarded as a concept, not as something capable of precise definition and measurement. The underlying difficulty is that for any given target, or in any specified circumstances, not all of the possible hazards will arise. Even those which do arise will differ in importance depending on the individual circumstances. This is the fundamental reason why no universal definition of contamination is possible, and why no generally applicable criteria for assessing contamination have yet been devised.

2.6.2 *Types of contaminants and their effects*

A large number of potentially harmful substances may be present on a contaminated site, though in most cases their concentrations are low. Examples of such substances include:

- Heavy metals cadmium, lead, zinc, copper, nickel
- Inorganics   sulphate, asbestos
- Organics   oils, tar, chlorinated hydrocarbons, PCBs, dioxins
- Gases   landfill gas

Contaminants present in any of these forms, particularly gases or liquids, can be mobile. If they migrate beyond the boundaries of the contaminated site itself, the contamination can spread to surrounding land. This may have particularly damaging consequences if the underlying groundwater is affected, or if adjoining buildings and structures are put at risk. In many cases, however, contaminants remain on or close to the places where they were formerly used, processed, stored or deposited when the site was in active use. The relationship between site history and the pattern of contamination can prove useful in the investigation and assessment of contaminated land.

2.6.3 *Hazards, risks and their perception*

In this context, *hazard* means an undesirable and usually malign consequence of the presence on a site of contaminants. Examples of possible adverse effects include:

- Uptake of contaminants by food crops
- Ingestion and/or inhalation of harmful substances
- Skin contact with contaminants

- Prevention or inhibition of plant growth (phytotoxicity)
- Contamination of water resources and supplies
- Chemical attack on building materials and services
- Fire and explosion

Risk is the statistical probability that a hazard will occur. For events that occur relatively frequently, it is possible to calculate numerical estimates of this probability in order to compare their importance. For example, some numerical estimates of risk of death from various forms of travel (numbers are deaths per 10^9 km travelled) are shown below:

- To passengers from rail travel   0.45
- To passengers on public road services   1.20
- To passengers on scheduled air services   1.40
- To passengers in a car or taxi   7.00
- To passengers on a two-wheeled motor vehicle   359.00

These values could be used to decide which means of transport gives the lowest risks. Public transport, being by far the safest form of travel, could then be encouraged and private vehicles, especially two-wheeled ones, discouraged. If all decisions and actions could be based primarily on such numerical estimates of risk, all would be well. That this is not done is obvious and needs little further explanation: the assessment of risk cannot be based solely on calculated probabilities without taking into account public attitudes to the acceptance of risk. When allowance is made for the latter factor, the risk accepted becomes a *perceived risk* rather than an actual one and cannot easily be calculated for the purposes of evaluation. In assessing the hazards of contaminated land, it is the perceived risks, rather than the actual ones, which prompt reactions from the public and from local authorities. Perceived risks are extremely difficult to quantify, and are frequently overexaggerated. Hazards from metals in the soil tend to be overestimated, while those of mobile contaminants such as landfill gas or leachate tend to be underestimated. The difficulty in quantifying these risks occurs either because the hazards occur only very rarely, or because their effects are very slow-acting and hence hard to measure. For example, risks involving toxic effects on human beings can only be estimated objectively by studying data on human mortality. This is not readily available, nor can it be obtained experimentally. In these circumstances, the view that it is better to err on the side of caution is understandable. A different view is sometimes taken when considering other hazards. The risks associated with development on or near land affected by migration of landfill gas appear frequently to have been accepted both by prospective developers, and by local planning authorities. Acceptance of risks, followed by efforts to mitigate them, usually incurs higher costs than if a risk avoidance approach was adopted instead. A possible explanation for this anomaly is that risks are

not always recognized during the planning process. Improved ways of quantifying the risks of contaminated land would undoubtedly help to ensure that
sites in need of remedial treatment receive it, while those where the risks
are negligible do not consume valuable resources that could be put to better
use elsewhere. This is an important point: where it has not been done, for
example in The Netherlands, the costs of generally removing or reducing
contamination hazards have proved inordinately high (they can be counted in
billions rather than millions). Expenditure on this scale may be justified when
the risks are significant, and can be quantified. Such justification, however,
was not needed in The Netherlands, where a statutory policy of total clean-up
is in operation. The importance of risk assessment is discussed further in
chapter 3.

2.7 The systematic approach to assessment of contaminated land

During the author's time as secretary of the Interdepartmental Committee on the
Redevelopment of Contaminated Land (ICRCL) (1980–1990), the principles set
out in earlier sections of this chapter were used to develop a systematic approach
to the assessment of contamination and contaminated sites. Although the reasons
why contaminated sites are reclaimed for further use have now been extended
beyond the limits set by earlier policies, the systematic approach continues to
provide a sound basis for understanding and discussing contamination problems.
The approach is based on the principle that contamination cannot be considered
separately from questions of land use (see section 2.6.1). It follows from this that
the significance of contamination on any given site can best be determined in two
stages. The first stage is to identify the principal hazards likely to affect the
intended use of the land, and to investigate the site for the contaminants that give
rise to those hazards. The second stage is to assess the investigation findings to
decide if the site is suitable for the selected use. If it is not suitable, then either
remedial action is necessary or a different use should be selected. By repeating
these two stages for each of the possible land uses under consideration, the most
appropriate use for any given site can be identified. This involves posing and
answering the following sequence of questions.

2.7.1 What is the history of the site?

Site history indicates the likelihood of finding contamination that might affect
decisions on land use. Information on the history of most sites can readily be
obtained at low cost [7] and is a prerequisite for effective ground investigation
(chapter 3). The registers to be compiled under S.143 of the Environmental
Protection Act 1990, together with their supporting documents, will represent
a good starting point for acquiring information on the history of a given site.

2.7.2 What is the intended use of the site?

This provides the opportunity for an initial check on the possible significance of contamination. Different forms of development vary in their sensitivity to the presence of contaminants. If an obviously contaminated site is required to be developed for a particularly sensitive use, then the costs of the necessary remedial action will inevitably be higher than if a less sensitive form of development was chosen. Prospective purchasers and developers of land are thus able to take contamination into account before committing themselves irrevocably to difficult and expensive treatment of their land.

2.7.3 Are contaminants present? If so, in what concentrations and with what distribution?

This is the purpose of site investigation. If investigations are designed on the principles set out in chapter 3 then these questions can be answered. The aim should be to enable the suitability of the land for the proposed development to be assessed with sufficient confidence, without acquiring an excess of expensive and irrelevant data.

2.7.4 Are there likely to be any hazards from the contaminants actually present on the site? If there are, how may those hazards be removed, or at least reduced?

These are perhaps the most crucial questions in the whole sequence, since it is in answering them that incorrect conclusions are most likely to be drawn, often through inexperience and lack of common sense. The assessment of data from site investigations is not simple, and tends to be complicated by the expectation that the detailed groundwork and analysis should provide a single unambiguous result. This is seldom possible. Sites which are so badly contaminated that remedial action would be needed for any use can usually be recognized easily, but few sites are so badly contaminated. The majority of sites require careful interpretation of the ground investigation data to decide whether their condition is suitable for the desired use. The interpretation should follow sound and relevant scientific principles, using sensible criteria (see section 2.8) and taking into account the limitations of sampling practices and analytical methods. Good solutions are unlikely to be achieved when these principles are ignored.

2.7.5 Could the hazards be lessened or eliminated by choosing a different use for the site?

The best way to progress an application for planning consent to develop a contaminated site is to take the possibility of contamination into account at the earliest possible stage, i.e. before buying the land.

If the fact that the site is contaminated is only discovered after the land has been purchased with a particular use in mind and a planning application has been submitted for that use, delay is likely to occur and the costs of treating the land to make it fit for the chosen use may render the proposed development financially unviable. It is simpler to choose a land use that fits the site's condition, than to adjust the site conditions to suit a particularly sensitive reuse. The most cost-effective and practical use will always be that which can best tolerate the existing site conditions.

2.7.6 *What remedial treatment is possible and what monitoring is needed to enable the site to be used for the chosen purpose?*

The range of options available for treating a contaminated site is still limited, although more and better techniques are being developed. The main methods in use at present are:

- Removal of material from the site for disposal elsewhere
- Retention and isolation of material on-site using an appropriate form of cover, barrier or encapsulation system
- Physical, chemical or biological treatment to eliminate or immobilize the contaminants
- Lowering the concentrations by diluting the contaminated material with clean material

Although relatively simple, the above methods can be used on the majority of sites where remedial treatment is needed. Chapter 4 provides additional detail on the remediation options.

After the remedial treatment has been completed some provision for monitoring is desirable:

- to obtain information that can be used for future designs and prediction of performance; or
- to check that the protective measures installed continue to work properly.

In many fields of civil engineering, monitoring is carried out as normal practice. When developing a contaminated site, the pressures of time and cost currently preclude the type of monitoring required to meet the first of these objectives. The second objective may have to be adopted instead, to obtain information that can provide an early indication of possible failure. For this purpose, simpler and more cost-effective methods are likely to be preferred.

2.8 The assessment of data on contamination

The systematic approach provides a good basis for making sound decisions, but chemical data obtained from site investigations must be used carefully when assessing the risks of a given site for a specified purpose. The key questions that need to be answered are:

(a)   Are the contaminants present hazardous for the intended use?
(b)   How can the significance of their concentrations be decided to determine the actual degree of risk?

To assist in answering these key questions, two important sets of criteria have been developed in the United Kingdom and in The Netherlands. Although both systems employ empirical rather than scientifically verifiable data they differ in mode of application, the UK criteria being intended for use as advisory guidelines and the Dutch having a statutory legislative basis.

2.8.1 *ICRCL trigger concentrations (UK)*

The difficulty of obtaining reliable data on which meaningful assessment criteria could be based was identified at an early stage in UK policy development. The objectives set for Government land reclamation programmes implied, however, that some way of assessing sites was needed. The concept of trigger concentrations was specifically devised by ICRCL as a means of evaluating site investigation data, and forms an essential part of the UK approach to contaminated land. From the outset, it was intended to provide a basis for relating hazards (as deduced from ground investigation data) to land uses, thereby differing from other assessment systems based on concentration limits.

Two significant concentrations are generally identified for each relevant contaminant: (i) a lower, or *threshold value*, below which the site can be *regarded as uncontaminated*; and (ii) an upper or *action value*, which, if reached or exceeded, means that the site has to be *regarded as contaminated*. Both these values depend on the use of the site because the various possible uses (actual or proposed) differ in their sensitivity to the hazards that the contaminants produce. The relationship between the two trigger values and the significance of the hazards associated with them is shown in Figure 2.1.

In specifying the two trigger concentrations, contaminants can be assigned to one or other of the following categories:

(a)   Those for which no threshold value can be set because in any concentration, however small, they present some risk. Examples include asbestos and dioxin. For such contaminants, the aim should be to keep the risks to as low as is reasonably possible. This, however, does *not* mean that they must be reduced to zero.
(b)   Those for which both the trigger values can be based on a measurable relationship between the concentration of the contaminant and the size

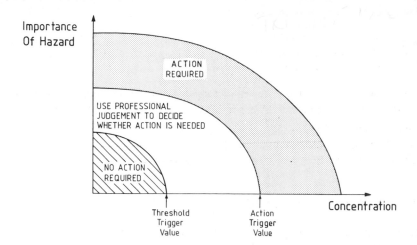

Importance
Of Hazard

ACTION
REQUIRED

USE PROFESSIONAL
JUDGEMENT TO DECIDE
WHETHER ACTION IS NEEDED

NO ACTION
REQUIRED

Threshold
Trigger
Value

Action
Trigger
Value

Concentration

Figure 2.1   Trigger concentration zones.

of the effect it produces. Examples of such relationships that exist and
can be studied are the attack of sulphate on concrete and the phytotoxicity of zinc to plants.

(c)   Those for which trigger values can only be set on the basis of limited
scientific data, because their effects are very difficult to assess, for
example ingestion of plants which have taken up cadmium or lead by
growing in contaminated soils.

When used for the interpretation of site investigation data, the threshold and
action trigger concentrations define three concentration zones, into one or
other of which the analytical results obtained will fall. The first zone is that
below the threshold trigger value, in which a contaminant is present only at
relatively low concentrations and can normally be disregarded. As the concentration increases, a value is reached at which the risks can no longer be
regarded as insignificant: the presence of the contaminant at such concentrations means that the hazards *may* occur. This value is the *threshold trigger
concentration* for that contaminant. That concentration is therefore the value
below which the site can be regarded as uncontaminated for the specified end
use, even though the concentrations found on the site may be greater than
those typical of the normal background values for that area.

The second, or intermediate, zone is that in which the concentration of the
contaminant is above the threshold value but below the action value. Even
though the local background values and the threshold trigger concentration are
exceeded, this does not mean that the risk is sufficient to require action to be
taken automatically: it merely indicates that there is a need to *consider*

whether action is justified. If the risk is judged sufficient to justify action, then it should be taken. In many instances the judgement will have to be made subjectively. The third zone is that in which the concentration of the contaminant exceeds the action trigger value. When this is the case, the risks of the hazards occurring then usually become so high that the presence of the contaminant has to be considered undesirable or even unacceptable. The site must then be regarded as contaminated and action of some kind, ranging from carrying out appropriate remedial treatment to choosing an entirely different use for the site, is then essential.

2.8.2 Netherlands 'A-B-C' values

To implement its statutory legislation on cleaning up contaminated soils, the Dutch Government has developed a comprehensive system of quality criteria for assessing the concentrations of contaminants. These criteria, known as the 'A-B-C' values, are in fact a series of reference values against which the concentrations actually measured on contaminated sites can be compared. In order of increasing concentration the values are as follows:

(a)   The A values represent concentrations typical of clean uncontaminated soils.
(b)   The B values define soils whose higher concentrations indicate that further investigation is needed.
(c)   The C values are defined as concentrations at which risks are deemed to be so high that action must be taken to reduce them.

Sites are evaluated by assessing the investigation data against these values. If and when any of the three reference levels is exceeded, the type of action specified in the above descriptions must be taken. For sites that yield contaminant values in the C category, this usually means that the action required to be taken needs to be capable of reducing the concentrations present to those indicated by the A values, i.e. a high degree of clean-up is required. The A-B-C values ignore the type of land reuse and assume that all land uses are at equal risk. This is one way of stating the principle of *multifunctionality*: the soil should at all times be in a condition such that it can be used for any purpose, no matter how sensitive. In this respect, the Dutch approach differs markedly from that of the United Kingdom, where the concept of multifunctionality has been rejected, and fitness for purpose adopted instead. The A-B-C system also differs from the UK trigger concentrations in that it applies to a much wider range of contaminants, includes concentrations for groundwaters as well as soils, and is presented as single concentration limits rather than ranges of values. There are special geological and hydrogeological conditions, which justify the rigorous approach taken on the presence of contaminants in soils in The Netherlands. It is not obvious that such conditions exist elsewhere where criteria based on similar principles have been advocated.

2.8.3 *Use of assessment criteria in practice*

There are several pitfalls to be avoided when criteria such as trigger concentrations are used for assessing site investigation data:

(a) Trigger concentrations are not substitutes for soil quality standards and should not be interpreted or used as such. Much further research on environmental risk assessment is needed to develop methods from which standards could be derived. Few of the present values have been derived properly, but any extension of the concept would need to consider contaminants not yet included in the present system, and be applicable to other possible targets (such as groundwater). This will be a major task and the Department of the Environment has established a study group to consider the whole issue of risk assessment for contaminated land; the intention being to derive site-specific standards using a systematic approach. Such standards, if they can be produced, will be better suited to the needs than any existing criteria. In the meantime, the present trigger values should be used with great care.

(b) The data being assessed were obtained by analysing soil samples collected during the site investigation. Even when the sampling density corresponds to that recommended in BSI DD 175/88, the amount of material actually collected and examined represents only a tiny proportion of the volume present on the site. It follows that any errors associated with the analysis of the samples will be insignificant compared to those incurred in acquiring the samples. This means that over-rigorous interpretation of the results obtained is therefore unadvisable.

(c) It is only possible to use ICRCL trigger values to assess the suitability of a site for a particular form of development before that development actually begins. The values cannot be used in the same way after development has commenced because other factors, in particular economic considerations and the implications for existing sites already in similar use, are then relevant and must be taken into account. The abandonment of a site which has already been developed and appears to be performing satisfactorily, on the grounds that *after* development it was found that the trigger values were exceeded, might well be judged impractical and economically unacceptable. The question of whether that use would have commenced if it had been known *before* development that the trigger values were exceeded, is another matter entirely. Decisions on the condition and suitability of sites should always be based on adequate information obtained before any commitments are made, not afterwards.

(d) No statutory significance attaches to the ICRCL trigger values. There is no UK legislation that requires compliance with the specified trigger values, nor are there at present any proposals for statutory quality objectives for soil. The concept is an aid to the use of professional judgement, not a substitute for it. Evidence from many sources suggests that these important limitations are frequently overlooked: this illustrates the need for better training and improved guidance for those who need to use the concept.

(e) The typical local background values for contaminants included in the trigger concentrations need to be known so that they can be taken into account in the interpretation process. This is necessary because the distribution and concentrations of some common contaminants, particularly metals, can depend markedly on the local geology and geochemistry. In many mineralized areas such as parts of Cornwall and the Pennines, metals are found in appreciable concentrations even on sites where no contaminating activity has taken place. If such concentrations are compared to the ICRCL values, without making allowance for the locally high background levels, contamination may be inferred and action taken unnecessarily. Conversely, the existence of vegetation on a site does not necessarily indicate that the concentrations of metals are tolerable. Plants can become adapted to the presence of metals, although the diversity of flora may be restricted and the growth of some species affected.

Acknowledgements

The Department of the Environment's permission to publish this chapter is gratefully acknowledged, as is the assistance of former colleagues in the Department during its preparation.

3 Rational site investigations

D.M. HOBSON

3.1 Introduction

Site investigation is a widely used term and is often taken to mean physical exploration on site, such as the excavation of trial pits or the sinking of boreholes. It is, perhaps, for this reason that all too often investigators charge onto site with no real understanding of what they are looking for or why. Such investigations can often end up as nothing more than the examination of a number of unrelated points of detail with no coherent strategy or means of interpretation.

For an investigation to achieve its purpose, a rational approach must be taken to define its objectives, determine the level of information required, and properly plan the various stages necessary. In other words, an investigation must be designed to meet the specific needs of the project to which it relates. This requirement is of greater importance when applied to contaminated land, which rarely follows the comparatively logical sequences of geological soils. The British Standards Draft Code of Practice DD175 [1] defines site investigations as follows:

> The planned and managed sequence of activities carried out to determine the nature and distribution of contaminants on and below the surface of a site that has been identified as being potentially contaminated. These activities comprise identification of the principal hazards; design of sampling and analysis programmes; collection and analysis of samples; and reporting of results for further assessment.

Throughout this chapter the process is described by reference to an example project, which is, in reality, a composite of a number of actual schemes chosen to illustrate the various procedures. A plan of the site is shown in Figure 3.1. The site chosen comprises a riverside industrial area on which is situated an engineering works, now derelict, with associated coal storage and waste tip. An operational oil storage terminal lies to the east and residential housing and industrial warehousing to the south.

3.2 Purpose of site investigation

ICRCL Guidance Note 59/83 [2] defines the main objectives of investigations.

(i) To identify the various buildings or other structures present on the site;

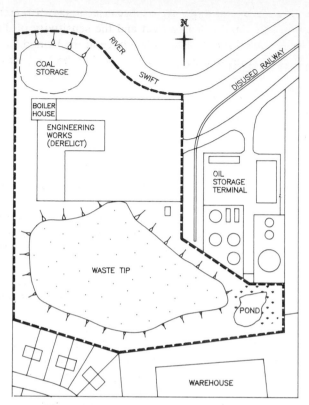

Figure 3.1 Example site. (– – –) Site boundary.

(ii) To identify the contaminants present; and
(iii) To ascertain their distribution over the site area and their concentrations both on and below the surface.

In most cases the investigation of contamination is necessary in order to permit consideration of the effects of its presence. It is rarely carried out simply for its own sake. It is therefore important to consider what these effects might be.

3.2.1 *Effects on users*

The most widely considered effect relates to the hazards and constraints applied to users of the site. In most cases, when a site is being investigated it is because new uses are proposed on land that has been contaminated by previous uses. BSI DD175 [1] defines the principal hazards as:

(a) those that may affect the integrity of buildings, building materials and site services; and
(b) those that may adversely affect the health of the site users or occupiers, or that of plants or animals.

Hazards are affected by the type of use because of the different routes by which contamination can reach its target and the sensitivity of different targets. The ICRCL guidelines [2,3] provide advice in the United.Kingdom on a limited range of contaminants. Hazards are related to the following categories of uses:

- Domestic gardens and allotments
- Parks, playing fields, open space
- Landscaped areas
- Buildings
- Hardcover

In The Netherlands, guidance on a much wider range of contaminants is available, but these are not related to end use and they tend to reflect a more demanding approach than is presently accepted in the United Kingdom.

When using hazard guidelines, it is very important to ensure that they are relevant to the targets and contamination routes being assessed. This is more fully discussed in section 3.8.5.

The investigation will therefore need to identify the presence and nature of contaminants within the surface layers and to a depth consistent with its accessibility to future targets. If major earthworks or deep foundations are envisaged, information at greater depth will be required than if only superficial disturbance of the ground is anticipated. It will be necessary to determine the nature of the ground and define its ability to transmit contaminants laterally to other parts of the site or upward by rising groundwater or the capillary movement of soil moisture. Possible hazards during construction work or during the investigation itself must also be considered.

3.2.2 *Effects on the environment*

Of equal importance is the potential effect of contamination on the environment. This requires completely different considerations since contamination at depth may represent a considerable threat to underlying aquifers, but have little consequence for subsequent users of the site.

Owners of land that causes environmental damage may be liable for its consequences [4] and, if redevelopment is being considered, remedial action is likely to be easier and more economic before construction work commences.

It will therefore be necessary to identify the spatial distribution of contaminants, determine containing strata and barriers, and define permeabilities of the ground to water and, if present, to gases. The hydrogeology of the locality may need to be investigated to an appropriate degree. It will also be necessary to identify the primary sources of contamination that may lead to the release of contaminants into the environment, and to evaluate the ability of the site to produce noxious liquids or gases.

BSI DD175 provides a decision tree illustrating a stepwise approach to determine whether an investigation is necessary. This has been redrawn in Figure 3.2 with the consideration of environmental effects added.

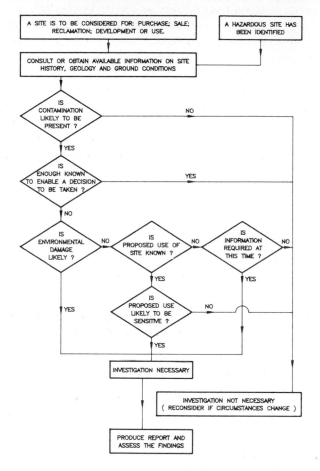

Figure 3.2 Determination of the need for site investigation: decision tree.

3.2.3 *Level of detail*

Investigations are required for a variety of reasons and these will dictate the level of information needed, or indeed, the amount of work that can be afforded. The depth of knowledge that would be required to assist in decisions on whether to acquire land, for instance, may be provided by a fairly general assessment. However, this must be sufficient to identify previous uses, main potential problems, outline of the superficial geology, general soil contamination levels, and surface and groundwater quality. In our example project, if acquisition of the derelict engineering works and waste tip was being considered, it would be necessary to research the original topography before filling, any former uses, the materials handled and what might have been placed in the waste tip. A limited ground investigation might be carried out to search for obvious contamination and landfill gases. The quality of surface or groundwater discharge into the adjacent river would also need to be checked.

The information necessary to support detail design for restoration and re-development would obviously be much greater, but would be directed at more specific requirements.

In reality, investigation of contaminated land is rarely completed in a single stage, but usually occurs in several phases during which the results of previous work are revised to produce an increasingly refined 'model' of the site. This is of particular importance in development projects where expenditure must reflect the degree of certainty that the project can proceed and the amount of time that the costs will be outstanding [5]. Intermediate stages that might be necessary include:

- Acquisition
- Planning
- Preliminary design
- Detail design
- Delineation of 'hot spots'
- During construction (e.g. after demolition)
- On completion (proof testing)

In defining the purpose of any investigation, therefore, the relevant stage must be identified in order to decide on the level of detail necessary.

3.2.4 *Topics of interest*

Having established that an investigation is necessary, its purpose and the level of detail required, it is possible to identify the specific items which are of interest:

(1) Define the physical site condition
(2) Identify likely contaminants
(3) Quantify the extent and severity of contamination
(4) Assess the effects and constraints on future users
(5) Assess the potential for environmental harm
(6) Assess the hazards for restoration or construction workers

A summary of topics is listed in Table 3.1

3.3 Site investigation strategy

In order to get the most out of an investigation, it must be approached in a systematic way, always bearing in mind its purpose, level of detail required

Table 3.1 Topics of interest for site investigation of contaminated land

Physical site conditions
    Natural geology and topography
    Soil types and physical properties
    Disturbances and alterations caused by man's activities
    Structures, buildings and underground features
    Surface water drainage
    Groundwater regimes, locations, depths, directions of flow

Likely contaminants
    Previous uses and contamination
    Activities on and adjacent to the site
    Background contaminants

Extent and severity of contamination
    Chemical concentration in soils and distribution by area, depth and soil type
    Chemical concentration and distribution of surface and groundwaters
    Background quality of soils and water entering the site
    Presence, nature and concentration of gases on the site
    Temperatures within the ground

Effects on users
    Nature and level of soil contamination with reference to user threshold values within the
       depth likely to be of relevance (see section 3.8.5)
    Contamination levels in soils, which might be removed, to allow assessment of waste
       disposal arrangements
    Surface water quality
    Potential for gas emissions
    Potential for combustion of soils

Potential for environmental harm
    Nature and level of soil contamination to the maximum depth of penetration for identifica-
       tion of primary contamination sources
    Presence of contained contaminants subject to future leakages (e.g. drums, tanks)
    Quality of groundwaters with reference to relevant guidelines and background levels
    Identification of water contamination phases (free, suspended, dissolved)
    Definition of aquifers
    Permeability of soils and identification of containing strata and barriers
    Discontinuities and migration pathways for water and gases
    Quality of surface water
    Soil contamination in surface layers likely to become airborne as dust

Hazards during construction
    Contamination of soils and waters with reference to exposure limits
    Presence of drums, tanks or buried containers
    Unstable conditions

and the topics of interest. At the outset the following questions should be addressed [6]:

    (a) What is known about the site?
    (b) What is not known about the site?
    (c) What needs to be known?

The process essentially involves the construction of a theoretical 'model', which can be used to assess the condition and behaviour of the ground, and the mechanisms and processes that lead to hazards and other effects. An initial model must first be constructed using whatever is already known from existing records, previous investigation work (if any), and from what is visibly evident. An investigative programme must then be designed to refine the model to the extent that is necessary and provide specific information required by the project. The design must take into account physical constraints and other limitations, such as cost, access, etc. It must, above all, lead to the collection of data that are capable of interpretation in a logical manner.

Physical exploration work on the site and subsequent laboratory testing must be implemented to a designed plan, with sufficient flexibility to allow adjustments to be made in reaction to conditions encountered.

Finally, the interpretation of the results must follow a rational basis, which can be clearly understood. The results should produce an adjusted and more clearly defined model of the site, an assessment of its implications, and information on the various topics of interest.

The basic strategy that should be adopted is summarized in Table 3.2.

Table 3.2 Site investigation strategy

Preparation	Design	Implementation	Interpretation
Define objectives: level of detail, topics of interest	Identify information required to refine model	Exploratory work on site	Data presentation
Collect and analyse current knowledge	Identify constraints and limitations (e.g. acess, presence of services, financial limitations)	In situ testing	Logical analysis
Visual inspection	Define sampling and interpretation strategy	Sampling	Refine 'model'
Construct theoretical 'model' of site	Determine exploratory techniques and testing programme	Record of investigation logs, photographs, sample details Laboratory work	Identify implications Report

3.4 Preparation

The first stage in any site investigation requires the establishment of what is already known, what can be observed, what can be deduced and what conditions can reasonably be expected. This process is described by BSI DD175 as the *Preliminary Investigation*.

It will, in the first instance, provide the means of constructing a preliminary theoretical 'model' upon which is based the first assessment of site conditions. The Preliminary Investigation is in two parts: (i) study of documentary material; and (ii) site reconnaissance.

3.4.1 *Data collection*

This is the collection and assessment of data that already exist, are available and are relevant to the investigations. The information should be grouped into categories and it is most important that data are collated in a logical manner. To this end it is useful to refer to topics of interest listed in section 3.2.4 and Table 3.1.

Sources of information are considerable and a comprehensive listing is provided in the BSI DD175. However it is not necessary, or desirable, that all such sources are used. Once the accessible and most promising sources have been consulted, the likely value of further data that might be obtained from more obscure sources must be weighed against the effort involved in obtaining them. Since the site investigation under consideration may be a later stage of the phased process described in section 3.2.3, much of the original data collection may already be available in a previous report. This should be reviewed to establish whether more detailed data are now justified as a result of the more advanced stage of the project. The main sources that should normally be consulted are listed in Table 3.3.

Table 3.3   Main sources of information

Local library	Maps
	Book, journals
	Newspaper records
Ordnance survey	Current and superseded maps
National map libraries	Various maps
British Geological Survey	Geological maps and memoirs
	Well and exploration records
	Hydrogeological records
British Coal	Mining records
Minerals Planning Authority	Mineral extraction records
Waste Regulation Authority	Licensed waste disposal activities
Public utilities	Location of services
Present and previous owners, occupiers and users	Details of activities and processes carried out
	Plans and photographs
National Rivers Authority and Water Undertakings	Surface water run-off
	Outfall details
	Abstraction points
	River details
Drainage authorities	Surface water drainage
Aerial photographs	Historical and modern photography

At the start of the data collection, a preliminary inspection of the site should be carried out. This need not be in detail, but should be sufficient to allow the investigator to become familiar with the general topography, conditions, main features of the site and the surrounding land. General panoramic photographs should be taken at this stage. A more detailed inspection will be undertaken after the collected data have been analysed.

The natural topography should be established from superseded maps, which should preferably trace the site back to 'green field' conditions. Examination of subsequent editions of maps will provide a series of snapshot views of the site at specific points in its development. From these it will be possible to deduce any previous modifications to the natural ground structure, topography and drainage. Unfortunately, few historical maps contain accurate ground levels. Examination of smaller scale Ordnance Survey plans, such as 1:10 000 scale maps, for which superseded editions are available, will give coarse details of ground elevations. Old ground surveys provided by site owners are invaluable in providing more detailed information. When using this information it must be remembered that the Ordnance Survey datum was changed in 1940, and the Ordnance Survey must be consulted to obtain the relevant conversion.

Examination of a series of old maps will also provide an indication of the history of the activities that have taken place on the site, and names and details, which can be further researched in library records.

Older maps are not always readily referenced to the modern site and this makes assessment difficult and often leads to misinterpretation. When comparing such maps, therefore, it is useful to produce a composite plan based on the most modern map showing the main points of historical detail. To do this it may be necessary to produce a series of intermediate plans, starting with the second earliest, onto which details of the oldest map are plotted. These details should then be transferred to the following plan in the series and the sequence continued up to the latest. Common points of detail that occur on several maps should also be cross checked to obtain the most accurate fit.

A composite plan for an example site is shown in Figure 3.3. From this it can be seen that the river channel was realigned after 1850 and that the Engineering Works was formerly a brickworks. The modern waste tip has clearly been placed into an old quarry, and the area used for coal storage was previously an ash tip, presumably from the adjacent boiler house. Railway links into the waste tip indicate the possibility of imported wastes. The source of these may be determined by examination of adjacent maps. The oil storage terminal appears to have undergone at least one stage of redevelopment and the names of the quarry and the brickworks are provided.

The names of features and places obtained from maps or other sources can easily be researched in local libraries and in other archives. This will often provide references in historical books, journals and newspaper cuttings, which give useful data about the site. The depth of worthwhile research will depend on individual circumstances and will be a matter for judgement.

Aerial photographs are another useful source of historical information. When available, as stereoscopic pairs, they can be used to obtain an indication of ground levels. This will be particularly useful in determining the depth of the filled quarry in the example site. They might also give an indication of the type of fill being placed and the method of filling. Surface photography will also assist in such details.

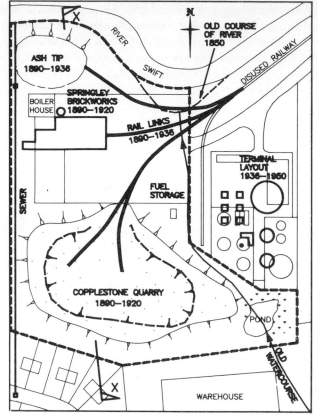

Figure 3.3   Example site: composite plan. (– – –) Site boundary.

3.4.2 *Recorded physical conditions*

The natural geology of the site may be established from geological maps and memoirs, and from previous investigations. This process is well documented elsewhere and forms a normal part of any geotechnical investigation. Its importance to contaminated land is that it establishes the framework onto which the details of the disturbed site can be constructed. In particular, details of the depth to rockhead, drift materials and aquifer locations are of interest. Useful information may also be available on the hydrogeology of the area. More detailed information will be obtained if previous site investigation records can be consulted.

Details obtained from historical maps and other records will indicate modifications that have occurred due to man's activities. For instance, on the example site (Figure 3.3), the waste tip area has been both excavated and then filled. It must, of course, be borne in mind that, since historical records are not a continuous record, they may not show all modifications.

Information obtained from British Coal and the Minerals Planning Authority will give details about mining and mineral activities, and Waste Disposal or

Regulation Authorities can provide information on licensed landfill operations. These all represent modifications to the natural conditions.

The evaluation of the drainage system on the site is of particular importance since original drainage channels and water courses which have been piped, often remain as pathways even after filling. A separate drainage plan should be prepared which indicates historical as well as modern drainage. This will be made up of information from old maps and from information provided by drainage and river authorities.

3.4.3 *Previous uses and contamination potential*

Knowledge of previous activities is essential to the understanding of contamination potential. This information is normally obtained firstly from records and secondly from the detailed questioning of previous users. The importance of the latter depends very much on whether the previous user carried out the most contaminating activities. However, it is quite probable that the more recent uses were the most intensive. Once a list of industrial uses has been established from historical records, possible contaminants must be determined by further research of the industries concerned. The government in the United Kingdom is in the process of preparing profiles of processes and activities with the potential to contaminate land. These are intended for use in the identification of sites for inclusion in contaminated land registers [7] and should be consulted during the preliminary investigations.

BSI DD175 [1] provides general advice on the range of contaminants related to various classes of industry. ICRCL Guidance Notes [1, 2–3, 8–11] provide greater details related to a small number of specific uses.

To make a site specific assessment, it is necessary to consider all the possible ways in which contamination may be brought onto the site and the mechanisms that could result in their deposition within the ground. Operational practices for the relevant industries and, where possible, for the specific plant concerned, must therefore be researched. Tables 3.4 and 3.5 summarize the main sources of contaminating materials and processes. By careful analysis of a particular industrial process, it is often possible to predict with some accuracy both the contaminants that will be present and their most likely locations. This, of course, is rarely comprehensive, since during the life of any industrial site there will be many thousands of unrecorded activities which might lead to contamination.

Information from previous users, if available and willing, will provide details of the activities that occurred during their occupation of the site and permit a much better assessment of contamination potential. Initially a questionnaire should be provided detailing all the information required. This should be followed by an interview in which the operation of the site should be discussed in detail. The most productive approach is to trace the progress of all raw materials entering the site, from delivery and storage to despatch as finished products or wastes.

Table 3.4 Sources of contaminating materials

Raw materials
Products and by-products
Rejects
Transitional products (materials part way through the manufacturing process)
Wastes
Ashes
Lubricants and coolants
Fuels
Building materials (e.g. asbestos, paints)
Vehicle and plant consumables
Cleaning materials
Laboratory chemicals

Table 3.5 Contaminating processes

Storage methods
Leakage and spillage
Transport
Processes
Disposal of wastes
Water treatment
Maintenance
Research and testing
Demolitions and redevelopment

In the example site, such enquiries may reveal that the engineering works had involved the production of metal components, including processes such as machining, metal hardening and finishing. Contaminants such as heavy metals, cyanides and mineral oils may, therefore, be expected.

3.4.4 *Preliminary theoretical model*

Having completed the research of the known site history, the first theoretical model should be produced. This should be represented as plans, indicative sections and descriptive text summarizing what is known. The model should define:

- Natural geology and topography
- Modifications, mining and other alterations
- Filled and disturbed areas
- Locations of potentially contaminating activities
- Historical and modern drainage paths
- Services and other constraints

The main elements for the example site would compromise the composite plan in Figure 3.3, deduced sections, such as shown in Figure 3.4, and drainage and service layouts. The natural topography of the example is a gently sloping valley side falling northwards to the River Swift, but also eastwards towards an old watercourse, which originally ran northwards through, what is now, the oil terminal. Natural geology comprises alluvial sands and gravels adjacent to the river, overlying a succession of laminated and stiff boulder clays and weathered shales. Modifications identified are:

- Filling of old river channel
- Tipping of ash
- Excavation of quarry
- Filling of waste tip
- Regrading of brickworks/engineering works and oil terminal

Contaminating activities identified are:

- Disposal of ash — coal storage area
- Coal storage — coal storage area
- Chimney — brickworks
- Metal machining, hardening and finishing — engineering works

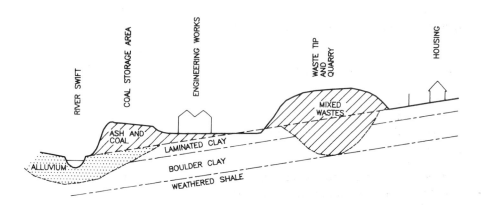

Figure 3.4 Deduced section X–X for the example site.

- Vehicle maintenance — south end of engineering works
- Fuel storage — southeast corner of hardstanding
- Wastes from engineering works — waste tip
- Imported wastes (foundry ash, possible chromium wastes from nearby chemical industry, domestic refuse) — waste tip
- Oil contamination — adjacent to oil storage terminal

The site should be divided into areas of similar type, background or use. This will allow further analysis, observation and investigations to be taken in a logical manner.

Subdivision of the example site (Figure 3.5) might comprise:

(1) Coal storage area
(2) Main engineering works buildings
(3) Boiler house
(4) Vehicle maintenance area
(5) Hardstanding area
(6) Area to west of engineering works
(7) Waste tip
(8) Unfilled land around the waste tip
(9) Ponded area

These areas may be further subdivided to take account of additional details such as the filled river channel, the quarry, the adjacent oil terminal and other features. Further observations and quantitative data obtained in subsequent stages of investigation may require amendment of the subdivision of the site.

3.4.5 *Site reconnaissance*

The purpose of a site reconnaissance is to check the information obtained from documentary evidence and to add further detail. The inspection should therefore be structured to ensure that all aspects of the theoretical model and topics of interest are addressed, and it should be carried out by reference to the areas into which the site has been subdivided.

The reconnaissance should, wherever possible, be conducted on foot and it is usually best to walk around the perimeter of the site first, before inspecting the central area and points of detail. This gives an understanding of the overall scale of the site and allows landmarks to be easily located.

A reasonably large-scale plan of the site, with points of interest marked, should be used to guide the inspector and to allow additional points to be recorded. The recording of detail as numbered points on the plan with descriptive text entered into a notebook is the best approach. Photographs should be taken of points of detail to supplement the more general photography taken during the preliminary inspection.

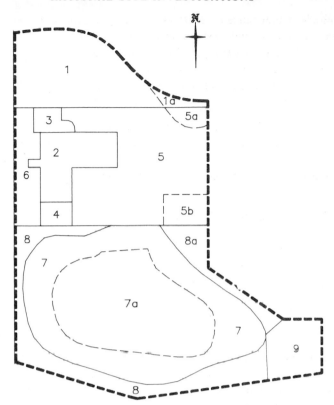

Figure 3.5 Subdivision of the example site. (1) coal storage area; (1a) filled river course; (2) main engineering works buildings; (3) boiler house; (4) vehicle maintenance area; (5) hardstanding area; (5a) filled river course; (5b) fuel storage; (6) area to west of engineering works; (7) waste tip; (7a) former quarry; (8) unfilled land around waste tip; (8a) adjacent to oil terminal; (9) ponded area.

During the reconnaissance, the following main observations should be made with reference to each area of the site:

- Surface materials, condition and appearance
- Materials exposed in excavations and side slopes
- Vegetation type and condition
- Unusual colours, fumes, odours and obvious contamination
- Presence of tanks, drums and other containers (details of any identifying marks and signs of leakage or spillages to be recorded)
- Lagoons, ponds, pits, depressions or swampy areas
- Tip materials, extent and description
- Signs of heating or combustion
- Drainage, outfalls, streams and watercourses
- Drainage interceptors and water treatment plants

- Appearance of surface waters
- Accessibility for exploratory investigations

At the same time, or more practically during a separate visit, simple on-site testing and sampling can be undertaken. This may include collection of surface soil and water samples, superficial excavation to expose near surface materials, and testing for gases and vapours using portable meters.

Having completed the site reconnaissance, the observations should be properly recorded by reference to each of the areas into which the site has been subdivided. The description of the theoretical 'model' should be adjusted and refined according to the observations made.

3.4.6 *Defining objectives*

Armed with a theoretical 'model' of the site, based on documentary records and the site reconnaissance, the next step is to define the objectives of the required exploratory work. This will be dependent on the amount of data already available and the purpose of the overall investigation. In some cases sufficient information may already have been acquired and no further investigation may then be necessary.

Primarily, the objective will be to provide specific details of the conditions in each area of the site, the level of contamination and quantitative data to permit an analysis of its implications.

3.5 Designing an investigation

3.5.1 *Information required*

The site investigation may initially be to determine whether a site is contaminated and to what level, but it will usually be required to determine what remedial works and precautions are needed. The site investigation should therefore be designed with this in mind. A result which indicates that the site is generally contaminated with a range of contaminants is of little value in specifying remedial action. The problematic parts of the site and the areas where particular contamination occurs will need to be known. In addition, those parts of the site that are not contaminated will need to be identified and this will require establishment of background conditions.

The investigation should therefore be designed to determine the extent of each of the different materials making up the site within each sub-area, and to describe their properties. The materials will be identified by:

- Physical properties
- Chemical properties

- Location
- Depth

The main physical properties to be determined are visual appearance, smell, temperature, soil classification and compaction. The investigation may also require the evaluation of various geotechnical properties at the same time. Permeability testing will provide information about the ability of the ground to allow migration of liquids and gases.

Chemical properties determined should be consistent with the objectives of the investigation and the contaminants anticipated, but should also be directed at commonly occurring contaminants and include provision for the identification of unexpected substances. It will also be necessary to test for contaminants to which the site or planned use may be particularly sensitive.

The evaluation of these properties is achieved by a combination of observations made on site during the investigation work and the results of testing and monitoring conducted, either in situ or on samples taken for laboratory analysis.

The investigation of contaminated land and the refinement of the theoretical 'model' is an evolutionary process. This often leads to the necessity of carrying out site and testing work in more than one phase. Initial tests, for instance, may show unusually high contamination at a single location, which warrants further work to delineate its extent.

3.5.2 *Sampling strategy*

The design of the investigation should ensure that observations are made and samples are obtained which are representative of the various materials making up the site. In the Department of the Environment's publication on *Problems Arising from the Redevelopment of Gas Works and Similar Sites* [12] it is stated that the sampling process should aim to:

Identify the highest levels of contamination
Describe the distribution of contamination both spatially and vertically

Preparatory work will already have provided a degree of understanding of where the different materials are located and what their properties are likely to be. Using the theoretical 'model' prepared during the preliminary investigation, all materials making up the ground within the site can therefore be subdivided into segments. Each segment will consist of materials considered to have characteristic and relatively uniform properties related to soil type, method of deposition, and exposure to contaminating processes.

For instance, the coal storage area in the example site will almost certainly comprise a coal rich surface layer overlying boiler ashes with natural alluvial deposits below forming the original sloping river valley side. These will in turn be underlain by natural clays and rock. The investigation of this area would therefore seek to quantify the physical and chemical nature of each of these materials and delineate their extent. The degree of subdivision is dependent on the level of detail necessary and the minimum size of area to which the representative parameters need to be assigned.

In addition to the investigation of the solid materials, it will also be necessary to obtain information on the location, nature and quality of groundwaters and soil atmospheres. These should initially be described with reference to the materials in which they are located.

An appropriate exploratory pattern, density and total number of samples in each of the individual segments must be specified.

In considering this section, the reader is reminded that as well as subdivision of the site by area, the ground in any area may contain a number of segments at different depths. The ashes in the coal storage area in our example site will need to be assessed as a separate segment to the underlying alluvial deposits within the same area (Figure 3.4).

The strategy to be applied to each segment may need to be different. There are two main categories:

(1) Segments within which no particular pattern or variation in conditions are anticipated, e.g. waste tips, ground into which contamination may have been deposited in a haphazard manner.
(2) Segments where contamination is related to known points of origin, e.g. ground adjacent to the doorway of a contaminating process building.

3.5.3 *Random and grid sampling*

The majority of the investigation will usually relate to contamination for which no known pattern or distribution can be predicted within individual segments.

If the intention is to interpret the data in a strictly statistical manner, a random pattern of sampling should be adopted. Unfortunately this has several disadvantages; sampling points will be unevenly distributed within the segment and constraints on the site surface will inevitably prevent completely random selection. A large number of samples are required to provide a high level of confidence and, to be truly random, samples at varying depth must be taken from separate excavations.

The most widely used method of locating sampling points is on a grid basis. This has the advantage of even distribution throughout the segment and can be easily set out on site. It is a more systematic approach to which subsequent investigations or restoration activities can easily be related. The size of the grid to be used is very much a matter of judgement. BSI DD175 recommends minimum numbers of sampling points for different site areas.

A method of calculating the probability of locating a given size of contaminated area related to the numbers of randomly located samples is reported by

Bell *et al.* [13]. Application of the binomial theorem gives the following simple relationship:

$$P = 1 - (1 - a)^n$$

where P is the probability of obtaining at least one contaminated sample, a is the proportion of the area of the site that is contaminated and n is the number of samples taken. This can also be expressed as follows:

$$n = \frac{\log(1 - P)}{\log(1 - a)}$$

Using this method, Table 3.6 gives the minimum size of contaminated area for which the sample numbers recommended in BSI DD175 would achieve 95% probability of locating. From this it can be seen that relatively large areas of contamination need to exist before this level of probability of finding them is achieved. Smaller occurrences are less likely to be located by random testing.

Table 3.6 Minimum areas of contamination located by BSI DD175 sampling frequency

Area of site (ha)	Recommended number of sampling points	Minimum contaminated area to provide one contaminated sample (at a 95% confidence level) (m^2)
0.5	15	905
1.0	25	1129
5.0	85	1732

This method of assessment is also related by Bell *et al.* [13] to grid sampling and it is not expected that probabilities will be significantly altered at higher levels of confidence.

ICRCL 18/79 [9] recommends that the grid should be no larger than the largest area of contamination that could be handled without difficulty if it were not found during the investigation, but only discovered during the development. For small sites, a grid spacing of 10–25 m is suggested and for larger sites, 25–50 m. Leaving aside the problem that there is no certainty that any subsequent development would involve activities that would enable such contamination to be discovered, the statistical method described above can be applied to this philosophy. On a 1 ha $(1\,ha = 10^4\,m^2)$ site a 10 m grid would give a probability of only 63.4% of finding a contaminated area of $100\,m^2$ (i.e. the size of the grid).

This clearly demonstrates the importance of understanding the likely distribution of contamination on the site and the careful choice of segments before determining sample locations. It is suggested that grid sampling should therefore be used to establish the overall quality of materials within the segment under consideration based on the 20% of samples with the highest concentration recorded. This philosophy, which is proposed in the Department of the Environment publication [12] also requires further investigation in areas where individual high results are recorded. The grid spacing to be applied should be based on the degree of confidence in the anticipated variability of the contamination and on the level of information necessary for the particular investigation being designed as discussed in section 3.2.

As a guide, it is worth noting that Smith and Ellis [14] reported that, in an empirical study on a gasworks site, a 25 m grid gave results that were not significantly different to grids as small as 6.25 m. Above 50 m, grid spacing results were considered unsatisfactory. This result may reflect the proportion of the study site that was contaminated. In view of the statistical argument discussed above, where contaminants affect a smaller proportion of the segment under consideration, closer grid spacings will be necessary.

3.5.4 *Targeted sampling*

In segments where the contamination source is known, the sampling strategy should be targeted around that source. Normally sampling points should be located at regular distances along lines radiating from the contaminant source. The spacing selected should be based on the anticipated distance over which the contamination may have spread. The method of translocation must also be considered in locating sample points. Drainage ditches may have carried contaminants some considerable distance from their source. Surface spreading, however, would normally be much more local to the point of deposition. A staged process may be appropriate, with a wide spacing used initially and closer spacing subsequently applied between the points where contamination falls to an acceptable level. This philosophy can also be applied to a sequence of layers below a large contaminating source when the overall quality of each layer is assessed using the technique described in section 3.5.3.

Provision should also be made in the investigation for the collection of additional samples of small isolated pockets of material which are visually suspect. In such cases, samples of the surrounding matrix of 'normal'.material, representative of the segment as a whole, should also be taken.

3.5.5 *Sampling depths*

The depth at which samples should be taken will depend on the variability of materials encountered and the level of detail required. Exploratory holes may pass through a number of individual layers of different material, which will represent different segments. These may have been predicted by the preliminary investigation or may be unexpected. In any event at least one sample should be taken in each segment encountered in the hole. Segments that extend to some depth will require multiple samples taken at different levels.

BSI DD175 recommends a minimum of three samples from each excavation, one at the surface, one at the greatest depth and one at a random point in between. The cost of taking samples is relatively cheap compared with the cost of excavation or subsequent testing, therefore it makes sense to take too many rather than insufficient samples. The testing strategy can then be determined following detailed consideration of the whole investigation.

Where it is important to establish the variability of materials with depth, samples at 0.5 m intervals should be taken, although it is very unlikely that these will all need to be tested. Testing strategies are described in section 3.7.7.

3.5.6 *Background conditions*

In addition to establishing the concentration of contaminating substances within the site the investigation should be designed to gain an indication of the general background level of these substances in the local environment. These can be expected to vary considerably from place to place and a number of sources give typical ranges of concentrations [15, 16].

A number of sampling points should be located in typical materials away from possible contaminating influences. This may require excavation outside the site and the cooperation of adjacent landowners. Alternatively it may be possible to establish areas of the site which are clearly not affected by contaminating processes. Representative samples from such areas should be taken.

3.5.7 *Water sampling*

The investigation must allow for the sampling of all waters encountered. Movement of water is the primary means by which contamination is spread from its initial source into the surrounding environment. The investigation must therefore enable the location of perched waters, aquifers and linkages between them to be established. The quality of these waters will give an indication of the mobility of the contamination and its rate of dispersal. Samples of water must therefore be taken for chemical analysis at each point that is encountered.

Similarly, the quality of surface water must be established. Samples should therefore be taken from surface pools and ponds as well as from drains, water courses and rivers. The investigation should establish the quality of waters both entering and leaving the site and any changes in quality of surface streams and rivers into which site waters drain. In the example site (Figure 3.1), samples of water will be taken from the river both upstream and downstream and from any outfalls within the site curtilage. Samples will also be taken from the surface pond south of the oil terminal.

3.5.8 *Gas monitoring*

Examination of the soil atmosphere is an essential part of the investigation of contaminated sites. Fill in the ground may contain degradable materials, which can produce flammable gases, carbon dioxide and depleted oxygen. The presence of carbon monoxide may indicate a combustion process within the ground. Spilled liquids may produce volatile vapours that may be toxic or flammable, and the presence of such vapours may indicate contaminants that have not been located by exploratory excavations.

The investigations must therefore provide for monitoring of gases during the exploratory works and the installation of monitoring wells at locations at which gases might be expected. The ash tip and filled quarry in the example site would need to be monitored for landfill gases, and the area adjacent to the oil terminal checked for volatile hydrocarbons.

A preliminary gas survey using sub-surface probes and portable equipment will give an early indication of likely problem areas and enable the planning of gas monitoring well locations prior to the main work.

A number of other techniques can be used to investigate gases and this subject is dealt with more fully in chapter 8.

3.5.9 *Testing*

The investigation will include testing of materials both in situ and in the laboratory. In situ testing will largely be restricted to the investigation of gases and measurement of temperature. However, several techniques are available, including the use of portable gas chromatography. These can be used to provide quantitative information as the investigation proceeds and enable reactive adjustments to be made to the investigation programme.

Great care should, however, be taken and major amendments avoided since there will rarely be time properly to consider all the implications. Despite this, in situ tests are extremely useful in directing the main investigative effort to the most important areas and may avoid the need for further investigation phases when, for instance, carrying out the targeted sampling described in section 3.5.3.

A laboratory testing programme, which will provide the appropriate level of accuracy and cover the range of contaminants required as discussed in sections 3.3.3. and 3.5.1, must be prepared. There is limited published guidance on methods of chemical analysis, and often different laboratories will use different methods. BS1377 [17] gives methods for tests normally undertaken as part of geotechnical investigations such as acidity and sulphates. BS6068 Part 2 [18] gives methods for testing of waters and associated materials. BS1016 [19] gives various methods for determination of calorific value and carbon content. The Health and Safety Executive provides guidance on methods of determination of hazardous substances [20] and further information is available from ADAS [21] and the HMSO 'Blue Book' series [22]. In the United States, guidance is available on a range of analytical techniques [23, 24]. There are two published methods available for the evaluation of combustibility of soils. The BRE gives details of a method that provides a qualitative assessment. A test described by Cairney *et al*. [25] gives more detailed quantitative information and enables combustibility to be related to compaction and air flow as well as material type.

The chemical determinations specified must relate to parameters that are capable of assessment against some bench mark. Guidance on acceptable levels is provided by ICRCL for a limited range of substances and these are discussed more fully in section 3.7. Most of the data available relate to total concentrations and therefore analysis for this parameter is most easily assessed against the more widely accepted guidelines; however, care must be taken to ensure that this is appropriate.

The overall cost of chemical testing can be reduced by adopting a strategy of initially determining only total concentration or even groups of substances. More detailed analysis to determine contaminants in specific forms should then be undertaken on samples representative of only those for which the simpler tests indicate a potential problem. Samples taken from the ground adjacent to the oil terminal in the example site might initially be analysed for total hydrocarbons. Only if they were detected at elevated levels would more specific determinations then be carried out.

3.5.10 *Sample size*

The size of sample taken and the method of preparation must be decided with care. The minimum amount of material required will depend on the proposed testing programme. A sample size of 1 kg is recommended by Lord [26] for most testing. However, some tests may require more. Up to 10 kg will be required for a full combustibility analysis for instance. It is therefore essential that, at the design stage, an anticipated testing programme is formulated and the appropriate sample sizes selected for each material to be sampled. This will of course need to be refined as the investigation proceeds, and during the implementation process the investigator must keep in mind what tests are likely to be needed when taking samples.

In a heterogeneous material, the size of sample taken and its preparation prior to testing will have a significant effect on the results. Any sample of soil may contain a mixture of materials, some of which will contain more of the alien contaminant than others. If the sample is tested 'as received', as is necessary when testing for volatile substances, the analyst must remove a small quantity of the overall sample for his analysis. This will give a high or low result depending on whether a more contaminated 'vein' within the sample is selected. If, however, the sample is dried, ground, mixed and quartered to obtain a sample for analysis, the result will be an average figure for the sample as a whole. Samples tested on an 'as received' basis will therefore have a wider spread of results. Similarly, if the size of the overall sample is initially small, it will be less representative of the material on site as a whole and give a wider spread of results compared with larger samples. Since, when assessing contamination levels, the primary concern is with the highest results, testing of smaller samples will tend to give higher individual concentrations, but with less frequency. Standardization of sample size is therefore urgently required and, when reporting results, the sample size must be quoted along with the preparation method and test procedure.

3.6 Exploratory techniques

There are several exploratory methods available to provide access to underground materials and to obtain samples.

3.6.1 *Hand methods*

Inspection and sampling of the surface layers can be undertaken simply and cheaply by hand. Manual excavation down to depths of about 0.5 m is straightforward and can be achieved quickly provided hard obstructions are not present. This permits good access for observation and disturbed samples are easily obtained.

Samples of soil can also be obtained using hand augers. A variety of types are available and in suitable conditions they may be driven up to 5 m in depth. However, progress is usually very slow and their use is normally restricted to fairly shallow depths. In addition, only small disturbed samples are produced.

3.6.2 *Trial pits*

Excavation of pits or trenches is perhaps the most widely used technique in the investigation of contaminated land. It is usually carried out by mechanical excavators, and pits can be extended up to 7 m in depth or even deeper if benching of the excavation is carried out. The most appropriate type of plant will depend on the required maximum depth, materials to be excavated and access constraints. A wide range of excavators is now available and details are given in the Construction Plant Annual [27]. Mini excavators can be used where access constraints prevent the use of larger plant. Wheeled back hoe excavators are capable for excavation down to 5 m, and 360° slew excavators enable trial pits to be extended down to 7 m depending on the machine. The size and power of the plant chosen will also dictate the speed at which the excavations can be advanced. Breaker attachments may be necessary if hard slabs are present.

Trial pits have the advantage of exposing the ground for inspection. They are normally about 1–1.5 m wide with the length dependent on the depth. They can easily be extended into trenches in order to expose more ground if required. Excavations deeper than 1.2 m should not be entered unless the sides are supported and the atmosphere checked for gas. Observations in pits deeper than 5 m become quite limited towards the base.

Large samples of soil are obtained and materials should be examined carefully on the surface. Wherever possible, visually different materials should be placed in separate stock piles as this facilitates examination and sampling. Details of the soils encountered can be logged as excavation of the pit is progressed. However, this should be checked after the excavation is complete, since the extent of individual strata is difficult to judge until fully exposed. There may also be variations across the width of the pit or along its length that need to be noted. A levelling staff should be placed into the pit and readings noted at the surface and at points of interest down the pit side. Trial pits should always be photographed before backfilling and the use of flash will produce much clearer results. Photographs of the main materials encountered should also be taken in the stockpiles on the surface.

When backfilling the pit, the materials should, as far as practical, be returned in the same order as excavated. This will limit the transfer of contaminants into otherwise clean ground.

Trial pits are the only method that permits examination of relatively large cross-sections of the ground. This enables a good description of the conditions and allows differentiation between continuous bands of material and small local pockets. It also allows selective sampling of different materials that occur at the same depth.

Relatively undisturbed samples can be obtained by driving sampling tubes into the side of the pit. This should be carried out before the excavation progresses below 1.2 m. At greater depths the sides of the pit must be supported and access becomes expensive and time consuming.

Trial pits are relatively cheap and up to 20 holes per day can be excavated. The main disadvantage of trial pits is the disturbance that they create. A large number of loosely backfilled holes 7 m deep across a development site will clearly create problems. Similarly, investigations within an existing development will cause extensive damage at the surface, which will need to be reinstated.

3.6.3 *Light cable percussion boreholes*

Commonly known as shell and auger, this is the traditional method of investigation for geotechnical work in soils. It essentially involves repeatedly dropping a steel tube known as a 'clay cutter' into the ground and withdrawing it to remove soil that has been wedged inside. A casing is progressively driven into the hole that has been formed in order to support the sides and enable the depth of the excavation to be determined. As well as disturbed samples from the clay cutter, undisturbed samples can be obtained by driving a tube, usually 100 mm in diameter, into the base of the hole.

Boreholes can be advanced by this method down to a depth of 50 m within soils, but will not easily penetrate hard obstructions or rock. The technique enables accurate measurements of depth to the various strata encountered and yields good quality samples. However, it does not permit examination of the ground in situ and it is not possible to differentiate between continuous strata and pockets of contaminated material larger than the borehole diameter.

Standpipes and piezometers for water monitoring and gas wells are easily installed by this method and multiple installations can be used to monitor strata at different depths in the same hole.

The equipment necessary is fairly lightweight and can be manoeuvred into difficult locations, although it cannot be operated where there is restricted head room. 'Low head room' rigs need a minimum height of 3 m. The technique is generally slower than trial pits, but is capable of greater depths. It has the advantage of creating much less disturbance of the ground or surface damage.

3.6.4 *Power auger drilling*

In the United Kingdom, flight augers are the most commonly available. The drilling tool comprises a cutting edge, above which a spiral flange is fitted. This is rotated from the surface by means of a rod or 'kelly' and draws soil into the spiral as the tool descends into the ground. In short flight augers the spiral section is of limited length and must be withdrawn after a small number of revolutions. This enables disturbed samples from known depths to be obtained and therefore allows the succession of materials encountered in the ground to be described. However, samples are highly disturbed and the method is limited to a depth of about 6 m unless a very expensive piling rig is used. It is, however, quicker than shell and auger drilling and causes less disturbance than trial pits.

Continuous flight augers allow much deeper holes to be drilled, but with this equipment the soil is pushed upwards to the surface along the continuous spiral. The depth to the various soils encountered cannot, therefore, be determined and considerable mixing usually occurs.

3.6.5 *Dynamic probes and small diameter boreholes*

This method simply involves driving a rod into the ground by percussion. It was originally developed as a means of rapid determination of soil strength by penetration testing. Most rigs therefore utilize a falling weight to drive the rod downwards. The equipment can be adapted to recover small diameter disturbed soil samples, either continuously or from specific depths, by replacing the rod with a small sampling tube. The equipment can also be used to drive gas monitoring probes directly into the ground without excavation. The use of dynamic probes to depths of 50 m are reported [28], but the rate of progress for sampling at such depths is probably unsatisfactory. An alternative technique is to drive the sample tube using hand-held hammers powered pneumatically or by hydraulics. In this case, a separate extraction tool is necessary to withdraw the probe and sample.

The equipment only produces very small samples, and the opportunity of selecting representative materials for testing is very limited. The use of tubes with discontinuous side walls (or 'windows') enables examination of samples before removal from the equipment, but because of the very small diameter necessary to facilitate driving, even very local pockets of material can appear as significant strata. The problems of testing small samples has already been discussed in section 3.5.10. This technique should not therefore be used as the primary investigation method, but is very useful to gain additional data about known strata or previously identified segments of the site.

3.6.6 *Soil gas surveys*

The presence of volatile contaminants or gas producing material can be determined by sampling the soil atmosphere within the ground. The process usually involves driving a tube to a predetermined depth and drawing out a sample of the soil atmosphere. This can then be analysed for the presence of the vapours or gases anticipated. Analysis is usually conducted on site by portable gas chromatography or photo-ionization and this permits the sampling strategy to be adjusted in reaction to results. Sampling is usually carried out on a grid or radial patterns and the measured values are then plotted on a plan. Vapour plumes can be illustrated by contouring.

Soil gas surveys can be carried out as a precursor to exploratory excavations in order to identify areas that warrant closer scrutiny. They can also be used to assist in the delineation of previously identified plumes of contamination. This technique can only be used to detect volatile contaminants and will not locate heavier substances. A survey of the oil terminal site in Figure 3.1 using this technique could fail to locate large areas of soil contaminated by heavy oils. The measurement of landfill gases is discussed in detail in chapter 8.

3.7 Implementation

3.7.1 *Specification*

To ensure that the site investigation is conducted in the manner intended and the correct information recorded, the work should be carefully specified in advance. The specification should clearly define the methods of exploration to be used, the observations to be made and the sampling strategy to be adopted. It must also define the method of sampling, size of samples required and testing strategy. Until standardized sample preparation and test procedures are produced, either nationally or internationally, these will need to be defined for each investigation. Investigators may be competent to specify these themselves or, alternatively, if an external testing agency is to be used, the analyst may be requested to propose the most appropriate procedures. In any event, a full description of sample preparation and test procedures must be prepared. Methods of sample preservation must also be specified. This will affect subsequent laboratory testing procedures and the analyst must therefore be aware of how the samples have been treated between excavation and delivery to the laboratory. This is of particular importance where the testing agency is a separate party to the sampler, since the laboratory results will be representative of the samples 'as received' and not necessarily 'as sampled'. The specification should also set out the way in which the data are to be reported, so that they can be readily understood and interpreted.

3.7.2 *Site observation and supervision*

Implementation of the site investigation work on site will involve a variety of personnel ranging from plant operators to experienced investigators. The whole operation must therefore be supervised to ensure that all the work is carried out as specified and in a logical manner.

The supervisor must be fully aware of the strategy being adopted and must be familiar with the theoretical 'model' described in section 3.3.4, and the segmentation of the site. Observations should be made of all the materials that are exposed, and compared with what was expected. As far as possible, a description should be given of the materials in each segment and any variations that occur. Unusual colour, surface textures and smells must be recorded.

3.7.3 *Reactive adjustments*

As the investigation proceeds it will usually become apparent that the distribution of materials about the site is not as was predicted. This will lead to the need to adjust the investigation to obtain the appropriate data. An example is, the clays adjacent to the quarry (Figure 3.4) were found to contain lenses of sand into which mobile contamination from the tip had migrated. These therefore had to be treated as additional segments from which additional samples were taken and observations recorded. The presence of such lenses between the tip and the housing also provided pathways for migrating gas, and monitoring wells therefore had to be installed.

Oil contamination from the oil storage terminal (Figure 3.1) was also found at a greater distance from the facility than originally anticipated. Additional trial pits were therefore necessary to delineate the extent of that sub-segment of the site. The waste tip proved to comprise a number of visually different materials, which occurred haphazardly throughout the segment. The sampling frequency therefore had to be changed to obtain representative samples of each material as they occurred throughout the tip.

3.7.4 *Recording of data*

Data obtained during the investigation must be accurately recorded, in a manner that can be subsequently understood. If this is not done, the value of the entire exercise will be seriously impaired. Records should clearly indicate the following items:

(a) *Accurate location of all exploratory holes and surface levels.* These must be sufficient to allow each location to be re-established with some accuracy.

(b) *Clear description of all material encountered.* A detailed log must describe the characteristics of the materials (appearance, colour, texture, smell) and their vertical and lateral distribution should be noted. Similar materials are easily matched on site and these similarities should be recorded. This grouping process is much more difficult if it has to be undertaken using only descriptions and laboratory samples. However, it must be recognized that visual appearance is not the only criterion and further subgrouping may occur at the interpretive stage. Conversely, variations in moisture content can radically alter the appearance of otherwise similar materials.

(c) *Record of water encountered.* The depth at which water is encountered, together with the containing strata, must be recorded. The description of the rate of flow into the excavation should also be given where possible and standing water levels should be recorded. In boreholes the recording of water strikes and levels should be as described in BS5930 [29]. The appearance and smell associated with the water should also be recorded.

(d) *Clear identification of materials sampled.* Each sample must be allotted a unique number. This should be related to the exploratory hole number from which the sample was taken. A clear description of the materials sampled should also be given, together with a note of the sample type and size. Identification by depth alone should be avoided since this can lead to confusion, particularly when the depth is at or close to a change of strata. Where a sample of an isolated pocket of material, that is unrepresentative of the surrounding soil matrix is taken, this should be noted. In this case, a sample of the soil matrix should also be taken from the same depth and clearly identified as such. The depth from which water samples are taken should be recorded, together with a note of their description and the strata with which they are associated.

(e) *Details of in situ testing.* The location of each test carried out must be logged. The precise test method used must be identified and all the results recorded, including any observations or irregularities noted.

(f) *Instrumentation for monitoring.* When instruments are installed into the excavation, their precise location with reference to the descriptive log should be given. A full description of the installation should be provided, including the method of backfilling.

3.7.5 *Sample protection*

Of equal importance to the recording of data on site is the protection of the samples that have been obtained. Deterioration due to the sampling process, the method of storage, transport and the time delay between sampling and testing may all invalidate the results. The sampling process should involve the least possible disturbance of the material. Clearly, the method of exploration will determine what is practical.

Containers used for samples should be air- and watertight and of materials that will not react with or contaminate the samples. Glass or plastic containers (polypropylene or polyethylene) are most suitable and should generally be at least 1 litre in capacity. Disturbed soil samples should be transferred to containers using stainless steel tools, which should be previously cleaned prior to each use.

Lord [26] places samples in three classes:

(1) Stable material
(2) Unstable material that can be stabilized
(3) Unstable material that cannot be readily stabilized

Since it will not be known with certainty into which class any particular sample falls, a decision will have to be taken at an early stage as to what analyses are planned so that the appropriate treatment can be adopted. All samples should be kept cool, disturbed as little as possible and tested at the earliest opportunity.

As stated in section 3.7.1, the method of preservation must be agreed with the analyst at the outset. Where testing for unstable contaminants is required, samples may need to be stabilized. This may affect other testing work, in which case separate samples should be taken. Volatile solvents are particularly difficult to stabilize and Lord recommends the following:

(1) Analysis on site or within a few hours of sampling
(2) Gas tight storage, freezing and testing within 24–48 h
(3) Sampling into a non-volatile compatible solvent

3.7.6 *Safety precautions*

Work on any site produces risks and hazards against which precautions should be taken and these are well documented elsewhere. The investigation of contaminated land introduces hazards for which special precautions are necessary. The main hazards are contact with toxic or corrosive substances, risk of fire or explosion, and dangers from unstable ground. Precautions needed, therefore, involve protective clothing and working procedures designed to prevent harmful events. The subject is fully discussed in chapter 13.

3.7.7 *Testing programmes*

The testing programme must be consistent with the overall strategy of the investigation. It should therefore be designed to identify the chemical conditions of the soil, water and atmosphere in each segment of the ground. In particular it must identify the severity and form of the contamination, significant variations within the segment and local anomalies.

The contaminants to be tested for consist of four main groups:

(1) Contaminants anticipated due to site history
(2) Commonly occurring contaminants
(3) Contaminants to which the site or planned use is likely to be particularly sensitive
(4) Unexpected contaminants

The first three will have been predetermined during the design stage, but analyses for unexpected contaminants will depend very heavily on the skill of the analyst, firstly in becoming aware of the presence of such substances and then identifying them. The testing programme, like the rest of the investigation, should follow a rational strategy in order to maximize the value of the test results and minimize the costs. Before specifying the testing programme, the record of exploratory work and in situ testing should be examined. This may indicate the need to adjust the original subdivision of the site to ensure that each segment represents materials of similar characteristics. However, assessment must not be based on visual appearance alone. A sample of apparently innocuous 'soil' tested during one investigation showed it to consist of about 40% of an explosive substance.

An initial testing schedule, which includes testing of representative samples from each segment, should be produced. Where sampling points have been distributed either randomly or on a grid pattern within each segment, sufficient samples should be tested to provided the requisite degree of confidence as discussed in section 3.5.3. Visually anomalous samples should not be included in this schedule, since they are likely to be unrepresentative of the segment as a whole. They should, however, be tested separately along with an adjacent sample of the 'normal' material for comparative purposes.

Samples from different depths within each segment will require testing and account must be taken of the irregularities that the method of transportation and deposition may have caused. Contaminants may wash down easily through coarse granular materials and concentrate at less permeable strata. An investigation of a tar works site on Tyneside, which had been constructed on top of tipped gravel ballast, showed the gravels to be relatively clean to a depth of up to 8 m. A highly contaminated layer of tar liquors approximately 1 m thick was found to be located at the base of the gravel immediately overlying a clay stratum. Similarly volatile materials may have evaporated in the upper layers, but still be present at greater depths. In disturbed or filled ground, contaminated materials may have been placed at any level. However, if the ground is undisturbed, has relatively low permeability and contamination is seen to diminish with depth, then samples should be tested sequentially downwards until 'clean' soils are proven.

Where targeted sampling has been adopted, testing should be conducted progressively outwards and downwards from the source.

The testing itself may also be structured so that simple tests such as solvent extraction methods are initially undertaken. More detailed analysis for specific contaminants may follow where appropriate. This itself may also be structured to check first for total concentration and only for specific forms where the previous tests indicate this is necessary. In order to identify the potential for leachate production, leaching (or solubility) tests may also be appropriate. A test to provide qualitative information on the composition of leachates and mobilized potentially toxic substances is given in the proposed EC directive on the landfill of waste [30].

As many samples of water as possible within the resources of the investigation should be tested. Contamination that has become dissolved in the groundwater is potentially the most mobile and usually will have the highest priority for treatment. At the very least, the investigation must identify the quality of one sample of water taken from each segment, at each and every level at which it is encountered. Testing of surface waters should include an analysis for hardness (as $CaCO_3$) in order to permit assessment against Environmental Quality Standards (see section 3.8.5).

Laboratory testing of gases and vapours will usually supplement meter readings taken in situ and should be carried out where it is necessary to know, with accuracy, the constituent gases and concentrations. The testing of landfill gases is discussed in chapter 8.

3.8 Reporting and interpretation

3.8.1 *Presentation of data*

During the course of any investigation, a considerable amount of relevant data will have been assembled. In the case of a comprehensive investigation of a large site, this may amount to thousands of individual pieces of information. It is therefore essential that the data are presented in a logical manner that can, first, be understood and second, allow easy retrieval of individual items. Reports should differentiate between information that is purely factual and information that is the result of interpretive analysis. In some cases, the factual and interpretive parts may need to be produced as separate documents, since the interpretation may be of a confidential nature. It is always helpful if a summary of the investigation is provided to give a fairly rapid overview without recourse to examination of the entire report. The overall format suggested, therefore, comprises an executive summary, factual report of data, interpretive analysis, and conclusions and recommendations.

The factual section should describe what has been carried out during the various stages of the investigation and clearly identify the sources of information provided. This is of particular importance when reporting a preliminary investigation, since the information presented will be attributable to a variety of sources of differing ages and reliability.

The reporting of works conducted on site should comprise the following detail:

- Description of work carried out
- Logs of exploratory holes, with accompanying photographs where available
- Details of samples taken
- In situ tests results

Factual details of laboratory testing should comprise a description of the preservation methods and analytical procedures. The quantitative results of the testing completed for soil, water and gases should be presented in tabular form.

The interpretive section of the report should present a description of the 'model' of the site, deduced from the investigation. This will normally be presented in the same manner as the description of the preliminary theoretical 'model' described in section 3.4.4, but refined to take account of the later stages of the investigation. The interpretation should describe the characteristics of each segment, the distribution, type and concentration of contaminants, and evaluate background conditions. Reference values for all contaminants tested should be proposed stating how they have been derived. The results of analyses undertaken should then be compared against these values (see section 3.8.5).

Finally, conclusions and recommendations that satisfy the original objectives of the investigation should be provided. These objectives should, of course, also be defined.

3.8.2 *Assessment of random and grid testing*

Detailed examination of the records of exploratory excavation and sample description should be carried out in order to adjust the subdivision of the site. Interpretation of chemical test results should be conducted by reference to the individual segments, and results should be taken as representative of the whole of each segment and not necessarily of the variations of contaminant levels within it.

It may be possible to judge that one part of a segment consistently shows different characteristics from the rest, but this deduction is only possible if there is a large number of results to compare. In a segment represented by only four samples, for instance, where two are found to be clean and two contaminated above a given threshold, it could not be confidently stated that half of the segment was free from contamination. The segment would, by definition, consist of materials with similar physical properties and exposure to contamination. Conclusions may be drawn about variations with depth only if it can sensibly be expected that contamination has advanced downwards from the surface in a progressive manner. Unless there is other supporting evidence such as visual differences, the test results for all depths should be considered together in order to classify the chemical condition of the segment as a whole.

Tests on visually anomalous materials should be excluded from this overall assessment, but should be reported separately as being present within the segment. This may be of particular importance if the anomalous material is substantially more contaminated than its containing matrix. On a former oil terminal site similar to that shown in Figure 3.1, large quantities of materials were classified as requiring removal off site for disposal. However, on excavation it was possible to selectively remove a relatively small proportion of badly contaminated 'veins' of soil, leaving the main body of material on site. Filled gravel pits in the south-east of England contained a mixture of different wastes. These were divided into groups such as clays, rubbles and ashes and it was found that only the ashes contained contaminants that were of concern. It was therefore a relatively simple matter to selectively excavate the ashes by visual identification, and only limited confirmatory testing of other materials was then necessary.

The material within any segment should therefore be described with reference to the physical and chemical characteristics of:

(1) The main soil matrix
(2) Separate, visually identifiable anomalous materials

It may also be possible to indicate general variations within the segment either laterally or vertically, provided the reservations described earlier are satisfied.

The chemical results should be presented as a range of values measured and may be interpreted as discussed in section 3.5.3, with the highest 20% of the results used to compare against reference values. The proportion of all test results in each segment that exceeds the reference value should also be expressed. If there are individually exceptionally high results, additional investigation should be recommended.

In some circumstances, the above approach will not be acceptable. If the hazard represented by the contamination is particularly acute, then the whole segment must be assumed to potentially contain the same hazard. Examples are explosives, radioactive materials and highly toxic chemicals that have a very small lethal dose.

3.8.3 *Assessment of targeted testing*

This concept relies on a knowledge of a contamination source at a particular location and the strategy is designed to establish the extent to which the contaminating substance has spread. Contaminant concentrations at various distances and depths should be examined to establish whether there is a relationship between the measured values and distance from the source. The method of translocation of contamination must be taken into account as described in section 3.5.4, but the reservations described in section 3.7.7 should also be considered.

3.8.4 *Assessment of water testing*

Water-borne contamination is potentially very mobile and its effect on the aquatic environment needs to be carefully assessed. It is therefore necessary to establish what water bodies exist within the site, where these are contaminated and where they are likely to migrate, either if left undisturbed, or as a consequence of disturbance. The effects of the contamination on river or other water courses and aquifer quality must be determined, since these will be the main criteria by which water contamination will be judged. Access to abstraction points will be of particular significance.

The results of the investigation should first be carefully assessed to establish the location and likely extent of all individual bodies of water. These will include surface waters, perched and contained groundwaters, and aquifers. If possible, their direction of flow should also be determined. The water samples that have been taken in each of these bodies of water must then be identified and the test results examined to establish overall quality and variations across the site. The quality of surface and groundwaters both entering and leaving the site should be determined, as should discharges from site drainage. Where site waters are likely to enter adjacent rivers, such as the River Swift shown in the example site, the quality both upstream and downstream should be compared.

A full assessment of the hydrogeology of any particular site is usually beyond the scope of a general contaminated land investigation, but enough information should be collected to indicate whether more detailed work is necessary.

The water quality within each segment of the site should also be assessed with reference to the contamination in the soil samples. This will provide a useful indication of the solubility of the contaminants within the soil, although it must be recognized that these will usually be in a relatively stable condition, which may be upset if the ground is subsequently disturbed.

3.8.5 *Setting interest and action levels*

Contaminant levels within the site should be considered in comparison with reference values of acceptability. These are referred to by ICRCL [2] as trigger concentrations. In order to establish appropriate values, it is necessary to consider three separate issues: (i) targets; (ii) type of hazard; and (iii) exposure routes.

There are two main target groups as discussed in sections 3.2.1 and 3.2.2:

- Users
- Environment

The main hazards to each of these groups are summarized in Table 3.7 and the main routes by which contamination can reach targets to create a hazard are given in Table 3.8.

Table 3.7 Targets and hazards

USER TARGETS

Humans and other animals	Plants	Constructional materials
Toxicity Ingestion Inhalation Skin contact Uptake in food	Phytotoxic contamination Contaminant uptake	Degradation of materials due to aggressive properties of contaminant Migration of contamination into services or drains
Injury Combustion Explosion Collapse		
Amenity Appearance Smell Taste		

ENVIRONMENTAL TARGETS

Aquatic environment	Atmospheric environment
Pollution of: Water supply Surface waters Drains and sewers Groundwater	Accumulation of gases, vapours or dust to produce: Toxic hazards Explosive or flammable hazards

Table 3.8.   Main routes by which contamination can reach targets to create a hazard

Direct contact
   Presence in surface or foundation zones
   Contact with aquifer

Translocation
   Disturbance
   Seepage of liquids, gases or vapours
   Capillary rise
   Infiltration
   Groundwater movement
   Surface drainage
   Progressive combustion

Reference values of acceptability are usually expressed at two levels:

● Trigger threshold (or interest level)
● Action level

Interest levels are those above which action needs to be considered. Action levels are those above which contamination normally requires treatment.

In setting the reference values, it is necessary to consider background levels in addition to the hazard, target and possible routes. Clearly, remediation to below the general level of contamination of the immediate environment may not be feasible. Where possible, the background levels of contaminants tested should therefore be determined.

It will rarely be practical to establish interest or action levels from first principles each time a site investigation is conducted, and a certain amount of published guidance is available. This is discussed below and data are provided in Appendix I.

Reference values for users.   In relation to users, the main sources of guidance in the United Kingdom on reference values for soils, are provided by ICRCL [2, 3] and further information will no doubt be provided in due course. These values have been developed largely with reference to specific industries and care should be taken in their application. Also, they only cover a very limited range of contaminants and, for this reason, reference to a more general classification of contamination given by Kelly [31] is often used. Dutch figures are also available [32] for a range of contaminants. However, these are based on a policy of clean-up to a level suitable for any use, a concept which was rejected by the House of Commons Select Committee [33]. Nevertheless, they do offer a quantitative means of assessing contamination levels. These sources are not consistent with one another and it is usual to adopt ICRCL values in preference to other guidance, provided they are appropriate to the situation being considered.

For contaminants not covered by these documents, or where they are not appropriate, reference must be made to other sources. For human toxicity, the criteria defined in the Special Waste Regulations for the definition of special waste, may be applied [34]. These encompass materials that have a flashpoint less than or equal to 21°C or are 'dangerous to life'. This is defined as a single dose of 5 cm³ being likely to cause death or serious damage to tissue if ingested by a child of 20 kg weight or, up to 15 min exposure being likely to cause serious damage to tissue by inhalation, skin or eye contact. Sax and Lewis [35] provide a summary of toxic properties of a wide range of substances that may be used to assess reference values. Materials likely to be aggressive to constructional materials are listed in CIRIA Report 98 [36], although no values are given.

It may be necessary to set reference values based on laboratory testing of samples. For instance, coal-bearing ashes from the example site would be subjected to the combustibility test described in section 3.5.9 [25] to establish acceptability.

Reference values for the environment.   The European Community (EC) has produced a series of directives concerning ground and surface waters. The main directives are EC76/464 EEC [37] and EC80/68 EEC [38] in which hazardous substances are grouped into two families. List 1 comprises substances which, due to properties of toxicity, persistence and bio-accumulation, should be prevented from being discharged. List 2 comprises substances considered to have a deleterious effect and discharge of which should be limited. Guidance on the implementation of these directives is given in Department of the Environment (DoE) Circulars 4/82, 7/89 and 20/90 [39–41].

There is no quantitative guidance or acceptable values for these substances in groundwater, but EC80/68 [38] draws a distinction between direct and indirect discharges. Under EC76/464 [37], discharges into the aquatic environment are to be considered with respect to Environmental Quality Objectives (EQOs) for the receiving waters. DoE circular 7/89 [40] gives both EC and UK national EQOs for a range of substances. Additional information is provided in EC directives on water quality in relation to abstraction for drinking, bathing and fresh water to support fish life [42–44], and in UK Surface Water Classification Regulations [45, 46].

The National Rivers Authority is responsible for controlling discharges and must be consulted in determining acceptable reference values for both soils and waters. In considering soils, the 'leachability' of contamination will be of primary concern in contrast to the ICRCL approach. Control criteria to be applied to the results of a test for assessing leachate potential are provided in the proposed EC directive on the landfill of waste [30]. In setting reference values, the nature and quality of aquifers likely to be affected, accessibility to abstraction points and surface waters will need to be considered.

Dutch figures [32] are available for groundwater but have no formal status in the United Kingdom. They should only be applied with reference to site specific conditions.

Discharges to sewers are subject to controlled consents and specific values will be set by the controlling authorities. The Trade Effluent Regulations [47] list prescribed substances within trade effluents and require them to be limited to background concentrations.

Guidance is available on landfill gases in Waste Management Paper 27 [48], and is discussed in chapter 8. Other gases and vapours must be assessed in relation to their individual properties and access to targets.

Application of reference values.   Reference values for each segment of the site must be considered separately. User values will be of little importance in soils at great depth, which are unlikely to be disturbed, but environmental reference values will still have to be applied. Care must be taken not to confuse the two groups. The results of proof testing of an excavated waste tip in the north-west of England were checked against ICRCL threshold values. Since the void was to be backfilled with several metres of imported clay, these were entirely inappropriate.

3.8.6 *Conclusions and recommendations*

Reporting and interpretation is arguably the most important part of any site investigation report, since it represents the culmination of all the work undertaken and will probably receive the most attention. Its content will depend on the objectives of the study, but it should in any event give a clear description of the deduced model of the site.

The conclusion should identify both contaminated and uncontaminated segments, as well as the quality and location of waters and gases. It should also identify implications of these with respect to planned uses and any possible environmental harm. Where there are areas of uncertainty or information is limited, these should be stated. Recommendations should be provided in the report with respect to future work that is necessary and the treatment options that might be available.

Execution of the site investigation on a logical and rational basis will enable the maximum value to be drawn from the information obtained. This will lead to economy in the expenditure of resources and confidence in the end result.

4 Reclamation options

M.J. BECKETT and T. CAIRNEY

4.1 Introduction

The selection of an appropriate reclamation technique depends on the legislative and soil quality standards that are in force, and on the economic framework that exists. In the United Kingdom, these major influences are currently being subjected to closer examination than at any time over the past 15–20 years. Thus, some changes in UK practices seem inevitable in the near future. As predicting the effects of differing political and economic controls is uncertain, this chapter focuses on the factors that hitherto have been important and outlines the mix of reclamation options that should prove effective in the likely future circumstances.

4.2 Evolution of the UK emphases on land contamination

In the United Kingdom, the main emphasis has been to bring disused land back into productive use as quickly and cheaply as possible. Where land contamination would preclude or limit redevelopment, remediation has to be undertaken, but within the limits of that improvement necessary to permit a particular reuse.

This attitude stems directly from Central Government's policy and resource priorities and has been emphasized by such statements as [1]:

> The base objective is to enable contaminated land to be used safely and economically, to achieve this a balance has to be struck between the risks from contamination and the need to restore the land to beneficial use

Wider environmental concerns have thus been of lesser priority than bringing land back into use. Given this guidance, a 'free-market' approach has developed in which:

- No legislation specifically related to land reclamation exists
- No governmental policies yet make remediation of contaminated land a mandatory requirement, although Integrated Pollution Control (when it comes into effect) will force the clean-up of some sites
- The limited soil quality criteria that exist are advisory in nature, and their application implies a large degree of professional judgement

- Regulation of land use decisions is left to the planning system and is applied on the basis of local priorities
- Land developers are responsible for treating contamination, although their proposals have to be acceptable to the local planning authority and to statutory consultees, such as the National Rivers Authority (NRA) and the local Waste Disposal Authority

This has led to a much larger private sector reclamation industry than exists elsewhere, that is routinely capable of recycling old industrial sites at a surprisingly high rate. Thus, very many contaminated sites have been successfully returned to use, despite the lack of a specific policy or programme on land contamination.

Obviously this situation differs markedly from conditions in the United States, Germany and The Netherlands, where more vigorous legislation and soil quality standards exist, where governmental funding for technology trials and reclamation is more abundant, and where the emphasis is on removing or reducing hazards to the wider environment.

Given its unique economic and legislative origins, the UK industry has thus come to emphasize three particular criteria that influence the choices of reclamation techniques:

(a) *Cost-effectiveness*. Contamination is usually not identified until a prospective redevelopment site is explored. Since such contamination could limit the land's use and value, remediation is undertaken to remove this obstacle. However the final end value of the reclaimed site usually imposes economic limits, which in turn encourage the use of simpler and cheaper remediation methods. Innovative or experimental techniques are seldom favoured, since land developers are disinclined to enter into research trials. Indeed the reverse is usually true in privately funded schemes, where a proportion of the agreed reclamation fee is commonly withheld until the specified end quality has been demonstrated. Any organization offering the use of an innovative technique thus has to price it competitively, and be sure that it will in fact function successfully on that particular site. This obviously is difficult to do until that new technique has been proved in the UK conditions.

(b) *Speed of reclamation*. The sooner a reclamation is completed, the quicker can redevelopment proceed and a return on the initial investment be obtained. Thus, techniques that take years to implement (such as some bioremediation approaches), or which require long-term effectiveness monitoring (as can be the case for many of the innovative stabilization methods) are unlikely to be adopted unless no other options are possible. In addition to this financial aspect, social pressure can tend to favour reclamations which are completed quickly, to reduce the duration of disruption and local inconvenience.

(c) *Flexibility*. Site specific variations are inevitable on contaminated sites, particularly where the land might have been used for a range of industrial processes over a century or more. These variations can impose limits on the use of many of the newer techniques (Table 4.1), but are of less importance to the broad spectrum engineering methods, which in most cases have to be included to remove floor slabs and buried foundations, and to improve ground bearing capacities. Thus it has been easier to avoid the need to prove whether site conditions will adversely affect one of the newer techniques, and fall back on the established and cheaper engineering solutions.

Table 4.1 Examples of site specific factors that can impair some newer reclamation techniques (after [2])

Reclamation technique	Adverse factors
(1) Bioremediation	Biotoxicity in the wastes Presence of some chlorinated organic materials can create carcinogenic vinyl chloride Will not degrade chlorinated hydrocarbons
(2) Incineration	Presence of mercuric, lead, bromide and reduced nitrogen compounds can create serious emission problems Presence of alkali metals and fluorides can damage refractory linings Presence of chlorinated hydrocarbons can create dioxins in the emissions
(3) Stabilization	Metal complexing agents (cyanides and ammonia) can make the stabilization process ineffective. Chromium(III) can be oxidized to the much more leachable chromium(VI) form
(4) Extraction/soil washing processes	Not effective on clayey soils

Obviously these reclamation criteria are not those that would have arisen had environmental concerns been the driving force for contaminated land reclamations. They are, however, entirely logical for the current UK emphases, and have resulted in a greater achievement of land recycling than has been possible elsewhere. The UK priority is to bring land back into reuse at no greater risk than that generally acceptable on a 'green field' site, and within this limited requirement the reclamation industry has been generally successful.

4.3 Challenges to current UK emphases

4.3.1 *Range of challenges*

Five distinct challenges to current UK land reclamation attitudes have recently developed:

- Scientific and technical criticisms
- The availability of newer techniques, from the United States and Europe
- Changes in policy and legislation over waste disposal and groundwater protection
- Changes in the type of land being reclaimed
- The appearance of foreign investors

4.3.2 *Scientific criticism*

Scientific criticism of the limited range of established broad spectrum reclamation techniques (section 4.4) has been voiced for some years. Comments such as "Innovative techniques are relatively poorly developed in the U.K., particularly when compared to our European partners" [3] are widely encountered, but have had very little impact, since the established reclamation techniques appear to have been successful, and certainly are far cheaper than the innovative methods. If US costs [2], for example, have to be accepted (at between $500 000 and $1 000 000 per acre), this at a stroke would destroy the UK policy of recycling old industrial sites for beneficial reuse.

4.3.3 *Appearance of innovative techniques*

A wide range of techniques that have been proved in the United States, The Netherlands and Germany has increasingly been offered within the United Kingdom in the last year. Whilst these are still too expensive to gain ready acceptance, their mere existence has emphasized that it is possible to reclaim contaminated land without freighting vast volumes of waste [4] to licensed tips, and that some reclamation problems that cannot be resolved by established techniques, can in fact be rectified, if the cost can be afforded.

Currently the use of soil washing, chemical stabilization and incineration techniques are being considered for several larger reclamations, although only vacuum extraction and bioremediation, of the newer techniques, have in fact been used. Private sector developers of smaller sites do not, however, seem to have considered the use of any of the more innovative reclamation methods.

4.3.4 *Changes in legislation and policy*

As indicated in chapter 1, the legislative and policy framework control land reclamation to a greater extent than do any other factors. Thus it could be argued that the introduction of the Environmental Protection Act 1990, which does include clauses requiring mandatory clean-ups by the new Waste Regulation Authorities, should have a significant impact. However, past legislation (e.g. the Public Health Act, 1936 c.92, and the powers given to the former Regional Water Authorities) did include similar powers to enter land, clean up environmental problems, and recover the costs from the individual or organization that had created the situation. In practice little use has been made of these powers, given the uncertainty of the litigation necessary to recover the remediation costs [5]. The reintroduction of such powers in the Environmental Protection Act may be no more than a paper exercise, unless funding and a policy for, and a commitment to, dealing with potential problems is established [6].

Other clauses of the Environmental Protection Act, however, have a more obvious immediate impact. Chapter 11 details the increased stringency of the waste disposal regulations, which already are making the use of off-site tipping more difficult and expensive. In some areas of the United Kingdom, tip charges are now at a level which could make the use of some of the newer reclamation techniques almost cost-effective. If, as seems likely, the EC draft directive on the landfilling of waste [7] becomes mandatory in the United Kingdom, tipping charges will inevitably increase, and a critical limitation on the use of many newer reclamation methods will be removed.

Equally important is the renewed emphasis by the NRA on the protection of groundwater quality. A recently issued consultation document [8] makes clear the NRA's emphasis on preventing groundwater pollution, to the point of seeking to bring legal prosecution: "In respect of any contaminated site where it can be demonstrated that a discharge into underground strata is occurring or has occurred, and threatens or causes pollution of groundwater resources". The NRA apparently intends to utilize its status as a statutory consultee in the planning process to force a proper consideration of groundwater quality protection, and this must inevitably influence the acceptable choice of a contaminated land reclamation technique, in at least those areas of the United Kingdom where groundwaters are not already so pervasively polluted that improvements are impracticable.

Thus a number of legislative and policy changes will, within the near future, overturn the past basis on which remediation methods for land contamination have been judged.

4.3.5 *Changes in the type of site being reclaimed*

A point that is often overlooked is that the 'typical' UK contaminated site more resembles an uncontrolled waste tip than the less complicated sites described in the US or European published case studies.

Decades to centuries of often very different industrial uses have resulted usually in a thick surfacing of dissimilar waste products intermingled with generations of demolition debris, and complicated by areas of non-engineered tipping of variable refuse. Such sites became available for redevelopment with the decline of manufacturing industry from the 1960s. Since they are generally located on the outskirts of urbanized areas and served with existing roads and other facilities, their redevelopment has been attractive both to developers and to local planners.

Given the variability of materials, the differences in contamination type and concentrations, and the variations in ground permeability, these sites are less than suitable for the application of the generally narrow applications spectrum that typifies the newer reclamation methods. Thus, the UK adoption of broad spectrum engineering-based solutions has a practical basis, in addition to the economic criteria discussed earlier.

In the past few years, however, a quite different range of sites has undergone remediation. Typically these are owned by the larger oil and chemical companies and are often reclaimed not for development but to remove potential environmental hazards. In many cases, contamination from only a limited range of substances may exist (e.g. the petroleum and diesel spills on oil company production and tank farm sites) and the site's sub-surface conditions may well be relatively uniform.

These sites have offered more suitable conditions to prove the effectiveness of newer techniques [9], and have allowed the first real introduction of US and European methods into the United Kingdom. Given the existence of the Environmental Protection Act 1990, such non-redevelopmental reclamations are likely to increase, and so bring greater diversity into the UK reclamation industry.

4.3.6 *Influx of foreign investors*

The final factor of change has been the influx of overseas investors into the United Kingdom, mainly from the United States and Japan. These companies have tended to bring with them higher environmental expectations, a familiarity with more severe standards, and a greater concern for environmental liability. Thus an economic force now exists, which does not find acceptable many of the established UK reclamation approaches.

4.3.7 *Summary*

Changes in policy and regulation, which make some established UK practices either more expensive or even unacceptable in some situations, have already begun. Allied to this is the presence of foreign investors who require more complete assurances that legal liabilities will not arise from land reclamations, and the availability of innovative reclamation techniques more capable of destroying and removing some contaminants than has been possible with established methods.

Thus, the predictable trend is of the increasing use of some of the newer reclamation techniques, and a consequential decline in the reliance on the established engineering-based approaches. This does not, however, mean that older techniques will be abandoned or that the newer techniques will necessarily be more successful. Site specific differences will remain the test of which technique or mix of techniques offers the most effective solution.

4.4 Engineering-based (broad spectrum) techniques

4.4.1 *Excavation and disposal*

Off-site disposal to a licensed tip has been by far the most widely used reclamation solution. The Department of the Environment in its evidence [4] to the
House of Commons Environmental Select Committee estimated that between
1×10^6 and 5×10^6 m^3 of contaminated soil was disposed of each year in this
way. These very large volumes indicate the traditionally low cost of tipping
(often as low as £10 to £15/m^3) and the technical simplicity of the process.

However, this situation is changing. Suitably licensed tips are now scarcer
and more expensive. As a result, some costs in 1991 have been in excess of
£30/m^3, and when transportation is included, together with the costs of importing clean fill to replace the excavated volumes, it is already obvious that
off-site tipping has become as expensive as some of the newer remediation
techniques, in at least a few parts of the United Kingdom. Also, the regulations
governing the disposal of wastes have become more stringent (chapter 11),
and are likely to become more so with time. Thus it is predictable that off-site
tipping must become a far less attractive option in the near future.

Tipping costs can, of course, be reduced if the site investigation has been
properly conducted and has accurately identified the horizons and materials
that carry the highest contamination levels. It is then often possible to visually
identify these more contaminated materials and to separate them when a site
is excavated. A typical example of this occurred on a disused iron works site,
mantled by a variable surface capping of demolition rubble, slags, combustion
ashes and general debris. Whilst the total volume of these wastes was in excess
of 15 000 m^3, only the combustion ashes proved to be highly contaminated.
These ashes were easily separated when the site was necessarily excavated to
improve its bearing capacity, and accounted for a total volume of less than
2000 m^3. As local and suitably licensed tips were available, transporting this
relatively small volume of contaminated wastes off site was by far the cheapest
remediation option. However, site supervision and quality control had to be
of a higher than normal standard (chapter 10).

The predictable decline in the amounts of off-site tipping has the significant
advantage that it will remove the cheapest reclamation option and so make
other newer techniques more cost-effective. It also has the wider benefit of
reducing the past policy of simply relocating environmental problems, and
limiting the possible hazards when large volumes of contaminated soils are
moved by road transport.

4.4.2 *On-site encapsulation*

As tipping costs have risen, developers have begun to accept on-site encapsulation as a more economic solution. Chapter 5 (section 5.5.3) indicates the

approach that is advocated for the construction of environmentally secure on-site encapsulations.

A recent case where encapsulation was adopted was a 2 ha Midlands site underlain by loosely compacted industrial debris, within which thin zones of biodegradable and oily waste had been tipped. To remove the landfill gas hazards, the developer had intended to excavate the site, separate out the biodegradable and oily materials, and have these tipped off-site. Detailed costings, however, showed that transportation and tip charges would account for 75% of the total reclamation cost. In view of this, on-site encapsulation, under an area of vegetated public open space, was adopted as a more economic solution.

Whilst on-site encapsulation has cost advantages in areas where tip charges have become overly expensive, it has to be noted that the National Rivers Authority is likely to require convincing proof [8] that encapsulation designs are well founded, and a waste treatment licence may soon also be necessary.

4.4.3 *Dilution of contamination*

This can be a cheap and simple process where contamination is uneven, or where only a thin capping of contaminated material overlies clean natural materials. Recently a site in central Scotland was treated in this way. The local contamination came from ashes, which averaged only 500 mm in thickness and were absent over large areas of the site. Below the ashes a clean lacustrine clay existed. The developer had to strip out the site, to remove the various concrete floors, slabs and foundations that surfaced almost its entire area, and found it convenient to also excavate the ash band together with a thickness of the underlying clean clay. These were mixed in a high speed rotary mixer and returned to the excavated areas for compaction, once site analyses had shown that the metallic contamination had been reduced to levels below the ICRCL guidelines for land to be used for domestic housing. The main environmental risk in this case was the possibility of creating air pollution, and allowing workers and nearby residents to inhale metal rich dusts. A stringent application of dust suppression methods was thus enforced. Similar results may be achieved by deep ploughing, though this is seldom possible as most sites have abundant sub-surface obstructions.

The advantages of the dilution technique are simplicity, cheapness, and the need only for widely available equipment. Thus, this approach has grown in popularity, where access to waste disposal tips is especially difficult and expensive.

Two disadvantages are, however, apparent. The need for quality assurance is especially high, and systems have to exist to identify unacceptable materials and prevent their reincorporation into the site. The second difficulty would occur if the dilution process made any contaminants more readily leachable, and so increased groundwater pollution. Given the NRA's renewed emphasis on protecting groundwater quality, this last point has to be regarded as a critical limitation.

Thus, the dilution method is likely to be best used only where non-mobile contaminants exist and where there is already a need to significantly improve a site's bearing capacity by excavation and recompaction.

4.4.4 Clean covers

Clean covers are considered in detail in chapter 5 and are a particularly cheap option (at £15–30/m^2 of site surface). Their use is limited to those sites where surface level increases can be accepted, and they are best employed after site compaction and the removal of buried tanks, vats and pipelines has been completed. Clean covers are most appropriate for the heterogeneous sites with a history of various prior uses and contamination, and should not be used to attempt to contain gaseous or oily contamination.

4.4.5 Summary

Engineering-based reclamation techniques have so far dominated the UK industry, largely because of their relative cheapness, speed of implementation and technical simplicity. Very little evidence is available to indicate that reclamations based on these methods have been other than successful. However this judgement accepts that the safe reuse of the land is the main criterion for success, and ignores some wider environmental concerns. This viewpoint is unlikely to remain acceptable (see section 4.3) and so a decline in the use of engineering-based techniques is predictable. A summary of the salient features of the various engineering techniques is given in Table 4.2.

4.5 Innovative (narrow spectrum) techniques

4.5.1 Introduction

The number of innovative reclamation methods is already large and is constantly being increased. For example, 63 innovative techniques were proposed for governmental evaluation in 1989 in The Netherlands alone [10]. It is not possible or useful to attempt to describe every possible technique, and attention is directed below only to those UK situations where a need is apparent for innovative reclamation methods.

4.5.2 Remediation of oily contamination

Corrective action at leaking underground oil storage tanks has become one of the most prominent procedures in the highly developed US waste management

Table 4.2 Salient features of various engineering techniques

	Excavation/disposal	Excavation/encapsulation	Dilution/reincorporation	Cover
Applicable to which contaminants	All	Most (except very mobile contaminants)	Relatively immobile contaminants	Contaminants in the liquid and solid phases (excluding oily contaminants)
Effectiveness?	High	High	High	High
Time necessary for reclamation	Short, if tip access available	Short, if construction programmed with the excavation	Short	Relatively short
Cost indication?	Still the cheapest solution, but costs likely to escalate	Relatively cheap	Relatively cheap	Relatively cheap
Is quality assurance critical?	Quality assurance requirements are minor	Quality assurance requirements are particularly high	Quality assurance requirements are high	Quality assurance requirements are high
Are design requirements high?	No	Especially high and detailed	No	Yes
Limitations?	Restricted only by tip availability cost, and the legalities over waste disposal	Leaves contamination in situ, long-term performance has to be demonstrated	Can increase groundwater pollution if the diluted contaminants are leachable	Leaves contamination in situ, long-term performance has to be demonstrated

industry. As a result, a wide range of remedial techniques for the treatment of ground soaked with the lighter oils, and for the recovery of free oil product from groundwater tables has been proved in field applications. Since the United Kingdom has a similar need to treat contaminants leaking from oil tanks — and also because in the wider range of contaminated land, oily contamination is a prevalent problem not easily resolved by any of the engineering-based techniques — the US experience is thus of interest.

Treatment of oil soaked ground.   Eight distinct techniques have been evaluated and ranked by Haiges and co-workers [11]. Whilst these rankings (Table 4.3) are of value, given their derivation from a large number of field trials, note has to be taken of the different soil quality and environmental standards that apply in the United States and the economic differences between the two countries. Any future UK evaluations of these same technologies may well give different rankings.

Table 4.3   Ranking of techniques for the treatment of soils contaminated with light oils (after [11])

Technique	Technical feasibility	Achievable treatment levels	Adverse impacts	Costs	Time of treatment	Overall ranking [a]
Bioremediation	3	5	1	4	7	**1**
Soil washing	6	2	4	5	2	**2**
Soil flushing	4	4	3	8	4	**3**
Land farming	5	3	2	3	5	**4**
Vacuum extraction	2	6	5	2	6	**5**
Passive venting	1	8	6	1	8	**6**
Thermal destruction	7	1	8	7	1	**7**
Stabilization	8	7	7	6	3	**8**

[a] **1** indicates best, **8** indicates worst.

Bioremediation currently is the most commonly available of these newer techniques in the United Kingdom and essentially mirrors the natural degradation of organic material to water and carbon dioxide. Use can be made either of naturally occurring microbial populations, or of specifically tailored microorganisms. In either case, nutrients and oxygen are required and the remediation can be carried out in situ (although this can cause the clogging of soil pore channel ways), or in mounds of excavated soil.

Land farming is a simpler variation of the bioremediation process. The contaminated material is mixed into the local topsoil and biodegraded. The natural degradation capacity of the topsoil can of course be increased by introducing additional microorganisms and nutrients.

Soil flushing involves abstracting local groundwater, mixing it with a surfactant and nutrients, and spraying the liquid on the contaminated soil. The infiltrating fluids wash out absorbed oily contamination, which then is bio-degraded in situ.

Soil washing is a similar process, which calls for the addition of water, solvents and/or surfactants into the excavated soil. Separation of the mixture into solid and liquid phases concentrates the oily contamination in the liquid. Typically, several stages of washing are necessary to achieve the required contaminant reduction. Coal washing, which is widely used in UK reclamations, is a simpler form of soil washing.

Vacuum extraction and passive venting rely instead on the diffusion of more volatile constituents to reduce the soil contamination. The natural diffusion rates are increased in the vacuum extraction method by imposing a vacuum at suitably located extraction wells. Where excavation of the contaminated soils is possible, natural venting by exposure to sun and wind can be very effective.

Thermal destruction is taken by Haiges to include all the various incineration methods (section 4.5.3).

Chemical stabilization includes the various solidification and stabilization processes (4.5.4) that do not destroy the oily contamination but fix it in physical and/or chemical bonds.

Table 4.3 shows that bioremediation is Haiges' preferred option, despite it being almost the slowest technique and giving lower treatment levels than most competing methods. This preference reflects US standards on environmental emissions and need not reflect any likely UK emphasis. Vacuum extraction, which has successfully been used on one UK site [9] is distinctly cheaper, but does entail air pollution risks unless complex air cleaners and filters can be included. Thermal destruction methods rate poorly in Haiges' rankings because of their costs and the difficulties in ensuring acceptable emission standards. However, this judgement may be dated, since the newer oxygen enhanced incinerators [2] are able to double treatment rates despite their reduced capital costs, although the technical complexity of the incinerator units is inevitably increased.

Removal of free oil product from water tables.   A range of pumped extraction techniques is available, together with two long established engineering techniques (Table 4.4).

As the problem of floating free product is commonly encountered on such sites as old timber treatment plants and coke works, there is a real need for effective remediation techniques in the United Kingdom. Haiges' weightings (Table 4.4) favour the use of the dual pump system, in which a deeper pump (in the groundwater) creates a cone of depression into which the free floating product migrates to be collected by a skimmer pump set at the oil/water interface. In UK conditions, trials have indicated that the simpler surface pump system, which collects both oil and water, can be effective and cheap, particularly since a range of oil separation processes can be included as required (i.e. filters, presses, thin layer biological filters, etc.).

Table 4.4 Ranking of techniques to remove floating free light oil product (after [11])

Technique	Technical feasibility	Achievable treatment levels	Adverse impacts	Costs	Time of treatment	Overall ranking [a]
Dual pump	4	2	1	4	2	**1**
Surface pump	1	6	4	2	6	**2**
Vacuum enhanced recovery	6	1	3	6	1	**3**
Scavenger filter pump	3	3	1	4	3	**4**
Single skimmer pump	1	4	6	1	4	**5**
Cyclic pump system	5	5	5	3	5	**6**
Interceptor trench	7	7	7	7	7	**7**
Sub-surface barrier	8	8	8	8	8	**8**

[a] **1** indicates best, **8** indicates worst.

An odd aspect of Haiges' ranking is the low rating given to interceptor trenches and sub-surface barriers. UK experience is that bentonite and similar cut-off works can be very effective in the short term (chapter 6) and that floating oils can be easily skimmed off once the groundwater flow is constrained. Thus doubt exists over Haiges' judgement of these techniques.

In summary, a range of proven techniques is available for the remediation of soil soaked with fuel oils and for the removal of floating free product. Some are already available commercially in the United Kingdom and the others are likely to follow if market demand increases as the need to protect groundwater quality is enforced.

4.5.3 *Removal of cyanide and heavier hydrocarbon contamination*

The cyanides and semi-solid tarry wastes that typify many gasworks sites can be rectified by a range of thermal techniques. Most currently available methods involve the excavation, sorting and crushing (to a maximum 40 mm particle size) of waste soils, which are then fed into rotary kiln thermal units. The soil is preheated, and moved to the main heater unit where combustion occurs. The cleansed soil is then moved to a mixer for cooling and moistening, and the flue gases are passed through an afterburner with additional fuel and oxygen. Finally, flue gases are cleaned through various dust collector and filter units to give the required emission standards.

Very high achievement rates in reducing cyanides and polychromatic hydrocarbons are routinely reported, and it is fair to accept that the incineration processes are relatively insensitive to material variations and contamination concentrations. However, heavy metals and most inorganic contaminants remain unaffected.

The main concerns over the use of incinerator methods are of air pollution, although most providers do claim to achieve EC air emission standards, and treatment costs (about £70/tonne). Current trials of oxygen-enhanced incinerators, however, indicate that cost reductions are likely and that reduced air emission problems will be achieved, since the use of oxygen decreases the total volume of gases involved in the combustion process. Concerns have been voiced, however, over the sensitivity of these techniques to variations in the rate and size of materials entering the burner units.

Trials are also advanced in the United States on the vitrification of wastes, including those contaminated by heavy metals. This solution has in fact been used to treat the asbestos wastes uncovered in the reclamation of the Faslane site, but the energy demands and costs (in excess of £100/tonne) would seem to preclude vitrification from any widespread UK use. The impetus behind vitrification is that it does give a permanent solution, without such environmental problems as the leaching of contaminants over a long period of time, and that the product can be shaped into blocks for uses such as the armouring of coastal breakwaters.

Trials of low temperature fusion to either destroy or encapsulate contaminants in melted asphalt are currently in progress in the United Kingdom. These offer a permanent solution for low ignition point compounds, and a potentially secure encapsulation for substances of higher melting and ignition point. The likely advantages are low energy and remediation costs, the use of standard asphalt boilers and a minimal air pollution risk.

Incineration and fusion/vitrification techniques have yet to make a significant impact on UK practice because of treatment costs and the concern over emission hazards. However, if the experimental oxygen-enhanced incinerators and the low temperature asphalt fusion approaches live up to their initial promise this situation could change.

4.5.4 *Treatment of metal contaminated sites*

Two approaches, soil washing and stabilization, are likely to be appropriate for metal contaminated sites.

The soil washing concept rests on the fact that contamination is usually concentrated in a particular soil fraction, often in the finer material. The technique thus involves passing excavated soil through various sieves and scrubbers (using water or oxidizing chemicals) to concentrate the contamination. The concentrate is then taken as a sludge to a hazardous waste tip. A basic problem in the United Kingdom would be the prevalent clay rich soils, which could give rise to an excessive sludge problem and, consequently, unacceptably high tipping charges. Variations including froth flotation have been successful and will remove low density hydrocarbons and cyanide in addition to the heavy metals. Costs, however, are far too high (at about £100/tonne) for the UK market.

Potentially more usable is the chemical stabilization approach. In this, a variety of cement, lime, thermoplastic and soluble silicate reagents have been used to fix the contamination in low permeability matrices, which hopefully will isolate the contamination permanently. All the existing processes involve excavation, sorting, mixing and injecting the reagents. Usually the product can be returned to the site and compacted to a high density (600–1800 kN/m^2 bearing capacities). The method can stabilize hydrocarbons as well as heavy metals.

Important doubts over the long-term stabilization of the contamination have been voiced from some years [12] and even some of the more newly introduced stabilization processes have failed to achieve the standards required in the US toxicity characteristic leaching procedure [13]. Thus, the permanence of a stabilization solution always has to be subject to question.

One variation of this process is currently available in the United Kingdom [14]. This uses a patented hydrophobized calcium oxide reagent, which preferentially combines with any liquid contaminants before it reacts with soil moisture. The calcium hydroxide thus formed retains the water repelling ability, unless the hydrophobic calcium hydroxide agglomerates are physically broken down, and an absorption of atmospheric carbon dioxide gradually converts the calcium hydroxide to a surface capping of calcium carbonate. Thus, this process has a degree of self-healing ability, even if excavated at some later date. The method has a record of success on German sites and whilst still rather expensive (at £50–90/tonne) is comparable in costs to bioremediation, although much faster to implement. Given its ability to deal with lighter hydrocarbons and heavy metals, it could be attractive in some situations, particularly if its long-term durability were confirmed.

4.6 Summary

Changes in the UK approach to contaminated land reclamation appear certain to increase costs, and so make the use of some newer remediation techniques more attractive. The extent to which this will occur will depend on how stringent the applications of policy and legislation become.

Some established reclamation techniques inevitably will become less widely used, though none is likely to fall into total disuse. Until UK experience with the newer treatment methods is more widely available, land developers will have to accept the achievement records attained in the United States or elsewhere in Europe. This does entail risks, since UK conditions are not always easily comparable to those elsewhere, and it will be prudent to examine all achievement claims carefully.

The most likely situation will be that a particular mix of established and innovative treatment methods will prove cost-effective. This is likely to complicate site control, and so force the need for more vigorous quality assurance.

5   Clean cover technology

T. CAIRNEY and T. SHARROCK

5.1 Introduction

In its simplest form, a clean cover need consist of no more than a thickness of a clean material, laid over a contaminated site to separate the proposed reuse from whatever contamination still exists at depth. Covers are cheaper than other reclamation options, and fit easily with the usual need to remove buried foundations, patches of more extreme contamination and improve ground bearing capacities. As their installation calls only for the equipment and experience already widely available in the construction industry, very many clean covers (e.g. 1–3]) have been installed over the past 10 years.

However, covers are no more than containment systems, whose failure can have serious consequences for the planned reuse of a piece of land. The apparent simplicity of covers often seems to have obscured this critical point, and it is quite rare to find clean cover designs fully justified, to have design lives detailed, and to be given specified efficiencies and factors of safety. Consideration of the predictable failure modes that can affect covers is also often absent.

Because of this apparently casual approach, concern has grown that clean covers need not represent the best practicable environmental option in many cases, and that some covers might indeed offer only temporary relief from contamination migration [4].

This concern is reasonable, and can be seen as a growing maturity in what is still an evolving industry. The use, so far, of only a limited portfolio of engineering reclamation solutions has been a limitation, and the introduction of the newer thermal, chemical and microbial clean-up technologies has to be welcomed. These newer methods can offer superior solutions, particularly to situations where near surface oils, fluid tars and organic pollutants exist (chapter 4).

The more extreme criticisms that are sometimes voiced, and which suggest that 'the effectiveness of all engineering solutions will inevitably decline with time' are, however, unreasonable and go far beyond any factual comment. Such comments ignore achievable engineering quality and the fact that almost no examples of cover failure are known [5].

As detailed in the following section, clean covers can be quantifiably designed to provide whatever design lives and factors of safety are required. Their cost-effectiveness and construction simplicity advantages are well worth retaining for a large proportion of contaminated sites, and it seems likely that they will increasingly be used, perhaps in conjunction with newer techniques, chosen to destroy or remove particularly hazardous contaminants.

The essential caveat in this belief is, of course, that appropriate design care will always be included, and that any suggestion of a casual approach to designing a clean cover will be avoided.

5.2 Basic design decisions

The wide variation in the details of those clean covers that have been published, has suggested that no underlying rules exist for cover design [6]. This, in fact, is not the case. Site specific variations and the range of different end-quality requirements will quite properly lead to variations in the final cover details (sections 5.4 and 5.5), but a logical design methodology has to have been followed if a particular clean cover is to provably satisfy the necessary end-quality needs.

The design methodology can be posed as simple questions:

- What does a particular cover have to do?
- How long does it have to remain effective?
- What materials can be included in the cover?
- How can the design properties of these materials be defined?
- How is the design quantified?
- Have possible failure modes been checked and potential failure pathways been closed?
- How quickly can failure occur?
- Does the client clearly understand the design basis and any possible liabilities this could present?

It is worth considering each of these in some detail.

5.2.1 *What does a cover have to do?*

Various end-use requirements can exist. In some cases, all that is wanted is a clean site surface. If this is the case, the design process is simple and calls for very little quantification. However, such covers are only appropriate for sites where contaminant mobility is extremely limited (section 5.3).

In other cases, the emphasis may be to prevent any upward migration of contamination in rising soil water, during extreme drought periods (section 5.4). Such capillary break layer covers are more complex to design and call for a detailed knowledge of the fluid transmitting properties of the cover materials, as well as those of the unsaturated layers that underlie the cover.

Preventing or minimizing the downward movement of contaminants, dissolved in infiltrating rainfall, might also be necessary in situations where groundwater pollution is to be avoided. The design of such covers requires a knowledge of the same material properties as those intended to combat upward migration of contaminants and utilizes a similar design methodology (section 5.5).

Finally, covers usually have to support some surface vegetation, and a normal requirement is to ensure, as far as possible, that plant roots do not move down into the contaminated layers (section 5.6). The soil layer into which vegetation is planted should not be seen as an element in the engineered clean cover, but as a separate layer, designed to meet specific needs.

In many cases, a single clean cover will, of course, be required to fulfil several of the above requirements.

Deciding on the functions that have to be achieved is the first and critical stage of design, and calls for a detailed consideration of the site investigation and risk analysis data (chapter 3) against the planned reuse of the land.

5.2.2 *How long does a clean cover have to remain effective?*

On first sight, choosing an appropriate design life for a clean cover appears difficult. Covers are obviously affected by climatic factors (droughts, high rainfall rates, wind erosion, etc.) whose severities will differ from year to year. For example, a cover mainly intended to combat the upward movement of contaminated soil moisture in droughts, will experience less severe droughts quite frequently, but will be tested by extreme droughts perhaps only once or twice in a century.

It could, therefore, be argued that the climatic event chosen to typify the design life should be related to the planned duration of the land's reuse. Typically [7], reclamations for domestic housing reuse have been designed for 100 years of use, and against the drought effects that will occur on a once-in-a-century return period, whilst reclamations for industrial reuse have been designed against less severe events with shorter return periods (of perhaps 50 years). The assumption behind this belief is that as a cover's design life is increased, so its thickness and costs become significantly greater.

Experience with clean cover reclamation design, however, indicates that this level of subtlety is unnecessary. The height to which a flow of contaminated soil water can be lifted in a more normal dry summer is in fact only a small amount less than that which will occur in a once-in-a-century drought in which surface soils are so desiccated that a soil suction equivalent to 1000 cm of water head is developed, and persists for 100 days. Equally, it is known that if desiccation becomes so extreme that all plants wilt and die (at soil suctions of 15 000 cm of equivalent water head), the height to which a particular flow rate of contaminated soil moisture can rise is only a little higher than that which occurred in the once-in-a-century drought (Figure 5.1). This latter level of desiccation has, of course, never occurred in recent history.

Thus, it will be appropriate to design capillary break layer covers against a once-in-a-century drought event. This gives the advantage that any change in

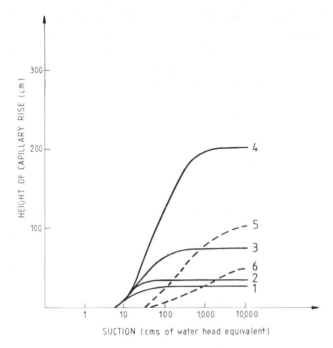

Figure 5.1 Height of capillary soil moisture rise with increasing surface suction (groundwater at 2 m depths. Materials 1–4, sands and sandy clays; materials 5 and 6, heavy marine clays. After Bloemen [12].

the planned land use, from a relatively insensitive purpose to a more sensitive reuse, will not force a re-evaluation of the cover's effectiveness.

A similar argument applies to those clean covers intended to minimize or reduce rainfall infiltration and subsequent groundwater pollution and it is reasonable to design such covers to cope with the wettest conditions likely to occur in a century.

5.2.3 *What materials can be included in a cover?*

The fact that coarse granular materials, with a minimal content of silt and clay sized particles, are less capable of permitting upward movement of contaminated soil water and more effective in diverting the downward movement of infiltration carrying washed down contaminants, is now well recognized.

Thus, the general rule is that the coarser the materials in a clean cover, the thinner (and usually the cheaper) can that cover be to perform its design requirements.

However, this is not intended to suggest that covers composed of silty or clayey materials cannot be equally effective (section 5.4). These covers will have to be thicker to achieve the same results, but that need not be an adverse feature, if locally available and cheap materials are to hand. Indeed, the above use of the words clayey and granular materials is overly simplistic, as it is the combination of the particular suction and hydraulic conductivity properties that determines whether a cover material will be more or less suitable.

Some consulting engineers seem to have preferred to avoid the costs and complexity of the clean cover design process, by adopting standard covers, specified perhaps as "washed gravel, with no particle larger than 6.3 mm diameter, laid to 300–500 mm thicknesses, and covered with a geotextile blanket" (to prevent siltation), and then "capped with a rolled clay layer" (to reduce rainfall infiltration). Such covers may well be effective (although this is not certain, since important variables, i.e. the depth to the design ground-water level, the contamination concentrations in that groundwater and the infiltration properties of the cover, have been ignored), but could be overly expensive, if their use forces the rejection of cheaper, locally available cover materials, and requires the importation of large volumes of washed gravels.

A final and obvious point is to ensure that all cover materials are themselves clean. Use is often made of materials from road schemes, adjacent reclama-tions, or from the crushing of demolished concrete foundations. Thus the possibility of importing extra contamination has to be considered. Routine sampling and analysis of all imported materials is a prudent precaution, as is the rejection of any material that is more contaminated than the ICRCL threshold trigger concentrations, devised for domestic gardens and allotments [8]. As well as the contaminants listed in this ICRCL document, it is usual to require analyses for mineral oils, PCBs and asbestos.

5.2.4 *Detailing the necessary material properties*

Once the possible sources of materials for use in a clean cover have been identified, it is necessary to obtain (except for the very simple covers dis-cussed in section 5.3) two material property curves, i.e.

(1) The hydraulic conductivity variation as a material's moisture content changes (Figure 5.2)
(2) The variation in soil suction with the state of material wetness (Figure 5.3)

not only for the cover materials themselves, but also for the unsaturated materials that will exist below the cover (Figure 5.4).

These properties are essential for covers that are intended to minimize either the upward movement of contaminants in rising soil water, or those that are to reduce any downward transference of contamination in infiltrating rainfall.

Soil suction can be seen as the driving force that encourages fluids to move, either upwards or downwards, whilst the hydraulic conductivity controls the rate of such movement. These two properties vary in different directions, as a materials saturation state changes, and are related in the appropriate versions of the Darcy flow formula, i.e.

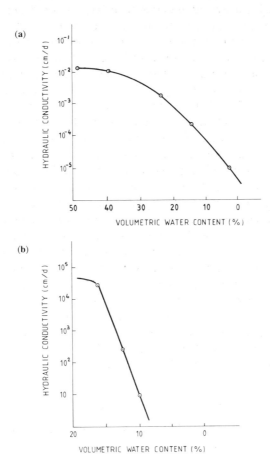

Figure 5.2 Hydraulic conductivity/moisture content relationships. (a) Stiff clay; (b) crushed concrete.

$$V = K(\psi)\left(\frac{d\psi}{dz} - 1\right) \quad \text{for upward fluid movement} \qquad (5.1)$$

$$q = K(\psi)\left(\frac{d\psi}{dz} + 1\right) \quad \text{for downward fluid movement} \qquad (5.2)$$

where V and q are the flow rates (fluxes) in cm^3/day through each cm^2 of soil, at a chosen position in or above the cover, $K(\psi)$ is the value of the material's hydraulic conductivity that relates to the moisture content at the position of interest, ψ is the value of soil suction that relates to the material's moisture content at the position of interest and z is the height above the design ground-water table of the position of interest.

Since a material's degree of saturation will change from one level to another in both the clean cover and in the underlying unsaturated materials, the conductivity and suction values will differ at each level and it is necessary to define the full range of property variation, from the completely saturated to the near totally dry states for each layer.

Sharrock [9] has defined simple laboratory tests that allow these curves to be measured at their intended compactions. The emphasis on ensuring that all tested materials are at their design compactions is an important point. Under-compaction will reduce the possible soil suction effect and increase that of the hydraulic conductivity, over-compaction will often give the reverse effects.

If direct measurement is not possible, various predictive methods (e.g. [10, 11]) do exist, which permit the hydraulic conductivity and soil suction curves to be derived from particle size analysis data, moisture saturation and saturated permeability values. Care should be taken when using these predictive methods, since none is fully accurate for all material types.

Without access to accurate and repeatable material property curves, appropriate for the planned compaction of a site and its clean cover, it is not possible precisely to design the more complex clean covers.

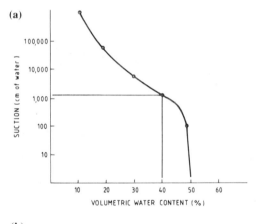

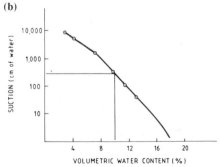

Figure 5.3 Suction/moisture content relationships. (a) Stiff clay; (b) crushed concrete.

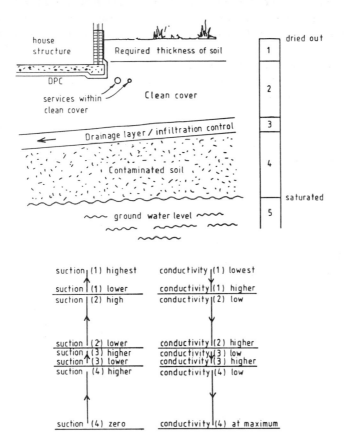

Figure 5.4 Clean cover details indicating the material properties required and their pattern of variation in a prolonged drought.

5.2.5 *Quantifying the cover design*

The assistance of an accurate mathematic model is usually necessary to quantify more complex cover designs (sections 5.4 and 5.5), particularly if a variety of possible cover arrangements is being considered.

Bloemen's model [12] is easily used for those covers intended to combat upward migration of contaminated fluids and the required computer program is fully detailed in the reference. Anders [13] evaluated the predictive accuracy of the Bloemen model against soil column studies of clean covers, and found it to be sensitive and accurate, although prone to slightly overpredict the concentration of contaminant uplift in particular imposed droughts. A graphical output from the Bloemen model is easily included and is particularly useful for a rapid evaluation of different possible arrangements of the layers in a cover (Figure 5.5).

For covers where preventing the downward movement of contamination carried by infiltrating rainfall is important, Bhuiyan's model is perhaps the most convenient [14]. Al Saeedi [15] has proved that this accurately reproduces the actual progress of a wetting front through a cover. A simpler manual computation is also possible for those cases where only the final position of the downward moving wetting front is required (section 5.5).

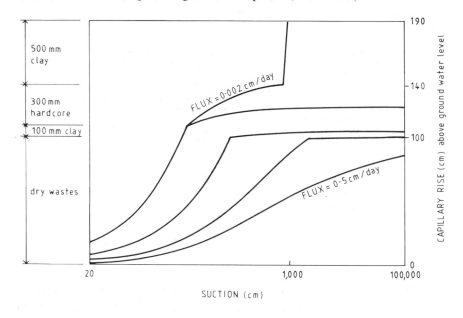

Figure 5.5   Effect of a granular break layer in reducing the possible upward flow of soil moisture.

5.2.6 *Identifying possible failure mechanisms*

Because the contamination that was of concern when a site was investigated still exists below a clean cover, considerable care has to be given to identifying and providing protection against possible failures of the cover. The more likely failure mechanisms can include:

(a) Siltation of the pore voids in granular covers
(b) Desiccation cracking of clayey covers
(c) Chemical attack on a cover's materials
(d) Settlement
(e) Erosion

(a) Siltation will clog the pores of a granular material and, where the cover is to act as a capillary break layer, will reduce its hydraulic conductivity properties and increase its suction potential. Thus, the predicted upward contaminant migration in a design drought may well be far too small (Table 5.1). Similarly, siltation of the granular material in a cover designed to minimize groundwater contamination will significantly reduce the cover's effectiveness. Avoiding siltation is not difficult, if either a designed filter layer or an appropriate geotextile blanket is included above the granular band. Filter layers are probably more robust than geotextile blankets and can be cheaper on those sites where modern crushing plants are in use, and are able to produce precisely limited granular materials.

Table 5.1 Siltation effects of a clean cover (shown in Figure 5.5)

Level of interest in the cover	Concentration of contaminant (mg/kg) added in design drought			
	Sulphate	Boron	Zinc	Free cyanide
(a) Unsilted clean cover				
Base of hardcore layer	1100	11.00	5.5	5.5
Top of hardcore layer	8.8	0.09	0.04	0.04
(b) Silted clean cover				
Base of hardcore layer	1380	13.8	6.9	6.9
Top of hardcore layer	31	0.31	0.2	0.2

Notes (1) Siltation from the overlying clay converted the hardcore to a gravelly silty sand. Presiltation state of the hardcore layer as shown in Table 5.4(b).
(2) Groundwater table in both cases 1.0 m below the base of the clean cover.

(b) Desiccation of clayey materials has been very apparent in the sequence of recent dry summers, and cracks to 600 mm depths have been not uncommon. Such cracking can be included in the designs of both the main types of clean cover, but is an additional complication that preferably should be avoided. Predicting the cracking susceptibility of clayey materials has been studied by many workers [16, 17] and it is apparent that cracking potential can be related to easily measured properties (clay type and percentage, plasticity index and bulk density). However, it is preferable to quantify the particular cracking potential by placing the clay (at its design compaction) in large diameter moulds (500 mm or greater) and exposing these to heat lamps. This will identify the precise moisture content at which the particular clayey material will just commence cracking. Use of Bloemen's model (run against the design drought's severity and length, with the design position of the groundwater table, and with the soil layers that are intended to lie above the clean cover to support plants) will then identify whether the clayey material could desiccate to the point at which cracks appear. If this occurs, the solution is to alter the type or thickness of soil above the clay cover, until it can be stated with confidence that the cover cannot dry out to its cracking point.

(c) Chemical attack on the materials used in a clean cover is a not improb-
able situation. Oil and tar vapours are known to deflocculate some
clays, and the use of dolomitic or limestone gravel over a site of very
low pH does imply a degree of acid attack on the cover. Avoiding such
risks should be a priority when considering which materials can be used
in a particular clean cover.

(d) Settlement should not occur on any well engineered reclamation if
normal compaction quality control has been exercised. However,
the risk has to be considered, since clean covers are not only relatively
thin, but also have very limited abilities to compensate for more than
trivial settlements. Avoiding this risk calls for careful attention to the
quality of the site compaction before the clean cover is installed (chapter 10).

(e) Erosion of a clean cover has so far not been reported for any UK
reclamation and would be an unusual event, given the subdued surface
topographies applied to most reclaimed UK sites. However, for sites
that include landscaped mounds (section 5.5.3), risks might be appreciable. These would call for a deliberate channelling of any storm run-off
and the inclusion of soil reinforcing fabrics in these channel ways.
Erosion where a clean cover fronts on a large water body is a much
more predictable event, and one that should be counteracted by established erosion protection methods.

In addition to these general failure mechanisms, there will be site specific
factors that could be of concern. The commonest of these seems to arise from
the location of buried services. If these are installed within the clean cover,
later maintenance and repair should not pose a hazard. Equally if the services
are within an identifiable service reservation and set in an oversized trench,
backfilled with clean cover materials, little difficulty should arise. However,
service pipes that have to be laid within the contaminated materials, below the
clean cover, can pose a very real difficulty. Later excavations to repair broken
services could disrupt the site's clean cover. Particular care should be taken
to avoid this occurrence.

Sometimes the site construction activities themselves can adversely affect
a cover. In one particular case, a cover was laid over the whole site early in
the reclamation (to prevent dust blow risks and to allow construction to go
ahead without special health and safety precautions), but a large volume of
surplus concrete foundation material had then to be crushed. Since this could
not be done off-site, the crushing plant was located on the cover and was in
operation for several months. The result of this extra loading and vibration
was that the granular clean cover became over-compacted and larger particles
were crushed. This so altered the cover's design properties that a large area
proved inadequate and had to be excavated and replaced.

Plant root migration is often seen as potentially disruptive to clean covers. Whilst this is true in the more extreme cases, the effects of plant roots in promoting upward migration of contaminated soil water are in fact easily included in the design process. All that is required is to emphasize not the soil suction at the site surface but the rather greater plant root suction (15 000 cm of equivalent water head) at the likely rooting depths (see chapter 9). For covers intended to reduce or prevent the movement of contamination washed down by infiltrating rainfall, concern is often expressed that plant roots or animal burrows could act as preferential and high capacity water channel ways. This (section 5.5) however, is an over-exaggerated concern since flow in these macropores will not usually occur until the soil around them is almost completely saturated.

5.2.7 *The time before any failure becomes apparent*

Clean covers are unlikely to display any failure until many years after their installation. Even if the extreme event against which a particular cover was designed were to occur within the first few years, the processes of moisture and contaminant movements are relatively slow and take many months to become apparent. In most cases, much less severe climatic events will occur in a cover's first few years, and no severe testing of the cover will take place.

Thus, an inadequately designed cover will tend to appear effective for some years, and the most rapid failure so far encountered took 7 years to become obvious, despite that particular cover being nothing more than 150–300 mm thicknesses of the local glacial clays, laid over acidic and cyanide rich gas works wastes, which had a near surface and highly contaminated groundwater table. No design basis for this particular cover is discernible and yet it appeared adequate for quite a long time.

The uncertainty imposed by this order of delay indicates that careful design is essential. Accelerated testing of covers is possible [18] and can be useful, but it usually will be more effective to monitor a cover for the first two or so years of life, and to record the variation of soil moisture content at chosen levels, the position of the groundwater table, and the soil moisture contamination concentration change at particular levels of interest (Figure 5.6). If an accurate mathematical model [12, 14] is available, this can be used to confirm that the measured data fit the model predictions and that the original design was in fact well founded.

The monitoring necessary is neither extensive nor expensive, requires only equipment that is commercially available, and can easily be undertaken. It is a matter of concern that very few clean covers are in fact monitored. This is in marked contrast to the routine confirmatory monitoring that is seen as necessary for most other engineering works. Without confirmatory monitoring it is difficult to see how any quality assurance can be factually claimed for a clean cover.

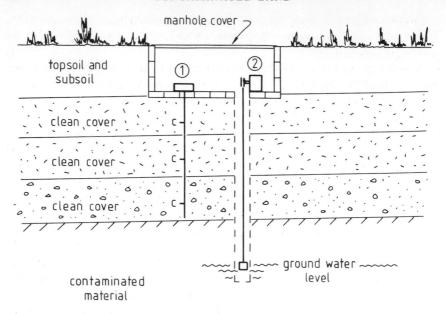

Figure 5.6   Monitoring a clean cover's performance. Soil moisture cells (c) set in vertical array and connected to an electronic readout (1). Groundwater level monitoring (2) using a punched tape recorder.

5.2.8 *Ensuring that the client appreciates the design methodology and any liabilities this may imply*

As environmental legislation becomes stricter, the liabilities on a land owner or occupier are likely to become more onerous (chapter 11). Clean covers leave contamination in place, albeit at depth and in such a condition that this should not affect the planned reuse of the land. However, if this remnant contamination affects adjoining land or water, the client might be faced with liabilities at some future date. Care should thus be taken to ensure that the client fully understands the design basis and any possible liabilities this could present.

5.3 Simple covers

The simplest of covers are those intended to separate very immobile contamination from whatever reuse is planned for the site. Such covers need consist of no more than a thickness of clean material, adequate to prevent direct contact with the buried contamination, and compacted to give the required bearing capacities and settlement characteristics.

A suitable situation for such a cover could be over a site where, below a surfacing of metallic contaminated combustion ashes, a stiff clay exists without any groundwater table or contaminated water. The hazards in such a

situation would be to human health (ingestion or inhalation of metallic contamination) and to migrating plant roots, which could feasibly abstract phytotoxic contaminants.

Covering the site to a depth great enough to ensure that no future excavations
would be likely to penetrate into the buried ashy layers, would remove the hazard
to health. The chosen cover thickness would be a matter of judgement but would
presumably be in excess of 1.5 m, plus the soil thickness needed for the planned
vegetation. Removing the plant phytotoxicity hazard would consist of no more
than providing enough soil depths, in garden and open space areas, to make plant
root migration out of the soil layer unlikely. If a particularly fertile and water
retaining soil were chosen, there would be a minimal probability of any movement
of roots into the less hospitable ashy bands. Quite obviously, such a cover would
only be suitable if contamination mobility were trivial.

For cases in which more mobile contamination does exist, but at significant
depths below a site's surface, a similar approach can be used. For example,
if highly contaminated groundwater occurs at depths greater than 5 m, then
upward migration of contaminated soil water is unlikely to be measurable,
even after extreme hot droughts. A consideration of equation (5.1) indicates
why this should be so. As the depth (Z) to the contaminated groundwater
increases, so the possible height to which capillary rise of contaminated soil
water can occur, will decline. Beyond 5 m, below the original site surface, this
is likely to be of trivial amount. In any such case, it will, however, be prudent
to check that the site is not underlain by especially sensitive materials that will
promote abnormally great heights of upward contamination. The most sensitive materials, in this context, will be sandy clays with appreciably high
effective conductivity and suction values (section 5.4).

Simple covers can include the works needed to prevent or minimize on-going
pollution of groundwater or surface water, by controlling the rainfall infiltration
that will be needed to leach out the more soluble contaminants (section 5.5).

Simple covers have been widely employed without any known problems,
but are only suitable for conditions of very limited contamination mobility, or
where the mobile contamination exists only at significantly great depths.

5.4 Covers intended to combat upward migration of contamination

5.4.1 *Designing a capillary break layer cover*

The stages in designing a capillary break layer cover are outlined in Table 5.2.
This reveals that the essential data required are:

  (a) The conductivity–moisture content and suction–moisture content
      properties of the proposed cover materials (Figures 5.2 and 5.3) and of
      those site materials that lie above the groundwater table's level, during
      the design drought conditions.

Table 5.2 Stages in the design of a clean cover

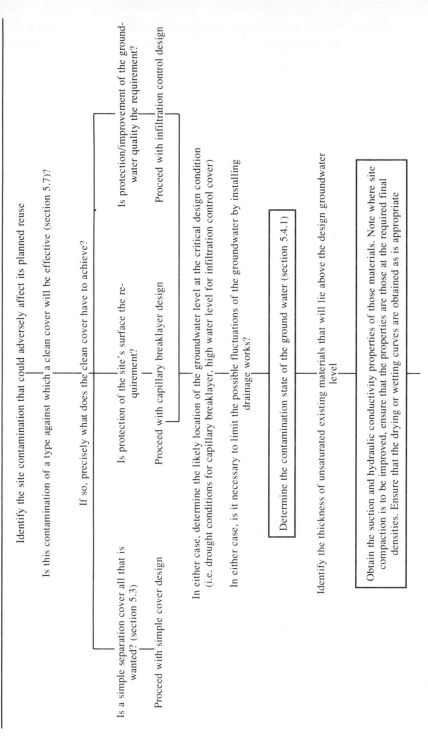

Identify the site contamination that could adversely affect its planned reuse

Is this contamination of a type against which a clean cover will be effective (section 5.7)?

If so, precisely what does the clean cover have to achieve?

Is a simple separation cover all that is wanted? (section 5.3)

Proceed with simple cover design

Is protection of the site's surface the requirement?

Proceed with capillary breaklayer design

Is protection/improvement of the groundwater quality the requirement?

Proceed with infiltration control design

In either case, determine the likely location of the groundwater level at the critical design condition (i.e. drought conditions for capillary breaklayer, high water level for infiltration control cover)

In either case, is it necessary to limit the possible fluctuations of the groundwater by installing drainage works?

Determine the contamination state of the ground water (section 5.4.1)

Identify the thickness of unsaturated existing materials that will lie above the design groundwater level

Obtain the suction and hydraulic conductivity properties of those materials. Note where site compaction is to be improved, ensure that the properties are those at the required final densities. Ensure that the drying or wetting curves are obtained as is appropriate

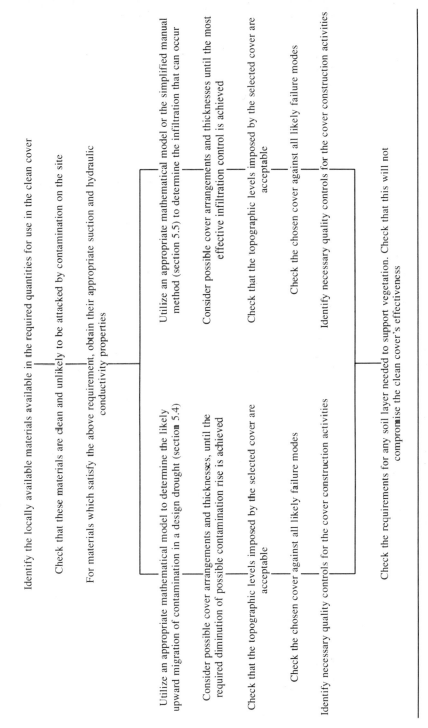

Identify the locally available materials available in the required quantities for use in the clean cover

Check that these materials are clean and unlikely to be attacked by contamination on the site

For materials which satisfy the above requirement, obtain their appropriate suction and hydraulic conductivity properties

Utilize an appropriate mathematical model to determine the likely upward migration of contamination in a design drought (section 5.4)

Consider possible cover arrangements and thicknesses, until the required diminution of possible contamination rise is achieved

Check that the topographic levels imposed by the selected cover are acceptable

Check the chosen cover against all likely failure modes

Identify necessary quality controls for the cover construction activities

Utilize an appropriate mathematical model or the simplified manual method (section 5.5) to determine the infiltration that can occur

Consider possible cover arrangements and thicknesses until the most effective infiltration control is achieved

Check that the topographic levels imposed by the selected cover are acceptable

Check the chosen cover against all likely failure modes

Identify necessary quality controls for the cover construction activities

Check the requirements for any soil layer needed to support vegetation. Check that this will not compromise the clean cover's effectiveness

(b) The chosen drought severity (section 5.2.2).

(c) The depth to the groundwater table in the design drought

(d) The worst contamination concentrations in the local groundwater

Selecting the design level for the groundwater table is usually not difficult. In some cases, the control of groundwater levels by adjacent rivers or by the sea will indicate what can be anticipated in the design conditions. In other cases, enough groundwater monitoring in the area might allow the design level to be predicted. In either case it should be noted that the shallower the depth to the design groundwater table, the greater will be any contaminated upward migration, thus a conservative design choice of as high a groundwater level as is reasonable should be adopted. Care should also be taken to determine if tidal effects on the groundwater are likely, and if so to use a groundwater level raised by the influence of high tides.

Selecting the worst likely contamination concentration of the local groundwater is a less precise matter. Groundwater quality tends to be less well defined than are the concentrations of the solid contaminants and, as chapter 3 indicates, the confidence possible from site investigation data is often much less than is sometimes assumed. Thus, a conservative approach is usually essential, which entails examining every groundwater analysis, determining if these are likely to typify the site, or if it is likely that more contaminated waters might underlie areas that have not been explored. Usually such an examination suggests that higher water contamination levels could occur in unexplored areas of the site, and this in turn suggests that a particularly contaminated groundwater quality, which includes at least the highest measured concentration of each contaminant, should be used (Table 5.3). Once the four necessary items of data are available, as many clean cover variations as are required, can be evaluated.

5.4.2 *Example*

An example of the design of such a cover is that installed over the site of a former metal refinery. Site exploration had revealed a near surface (0.5–0.8 m depth) contaminated groundwater, which was affected by the water depth in the adjacent main river. From river low flow records, it proved possible to predict that the groundwater table would exist at depths of 1.0 m or more, during an extreme design drought. Thus, the conservative choice of a 1.0 m depth to the design groundwater level was made.

Groundwater contamination concentrations proved to be very variable, and obviously related to the contamination from specific production units that had existed on the site. Since the density of the site investigation works had been inadequate to give a 95% confidence that the likely worst contaminated water had in fact been encountered, the groundwater quality detailed in Table 5.3 was utilized.

Table 5.3 Design groundwater contamination quality

	Selected design concentration (mg/l)	Highest known concentration (mg/l) in any sampled groundwater
pH	6.9	6.9
Sulphate (as SO_3)	2500.00	632.00
Phenol	100.00	72.20
Copper	161.00	116.20
Lead	116.30	84.00
Chromium	99.70	72.00
Total PAH	27.90	20.12
Cadmium	6.00	4.30
Arsenic	4.70	3.40

The site itself was mantled by up to 5.5 m thicknesses of decomposed ashes and mine tailings, which now behave as silty, sandy clays, with meaningfully high suction and hydraulic conductivity values. Thus, the potential for upward movement of contaminated soil water is high if the site were exposed to surface desiccation.

Sources of clean cover materials in the necessary quantities were limited, and finally were reduced to a choice of two:

(a) A stiff local clay (Figures 5.2(a) and 5.3(a)) whose properties are summarized in Table 5.4(a)

(b) Crushed concrete and brick derived from the demolition of the refinery buildings (Figures 5.2(b) and 5.3(b) and Table 5.4(b))

Table 5.4 Properties of the materials considered for use in the clean cover

(a) **Clay material**						
Liquid limit	75					
Plastic limit	36					
Saturated permeability	0.018 cm/day					
Particle size distribution	–					
Sieve size (mm)	600	200	60	20	6	2
% passing	100	99	98	84	74	64

As laid moisture content   40% (volumetric)
Hydraulic conductivity and suction properties shown in Figures 5.2(a) and 5.3(a).
Dry density (as laid)   1330 kg/mg^3

(b) **Crushed concrete material**							
Saturated permeability	66 000 cm/day						
Particle size distribution	–						
Sieve size (mm)	37.5	20	10	6.3	2.0	600 µm	63 µm
% passing	100	69.5	33.4	17.4	3.8	1.8	0.3

As laid moisture content   10% volumetric
Hydraulic conductivity and suction properties shown in Figures 5.2(b) and 5.3(b)
Dry density (as laid)   2170 kg/m^3

The prime requirement for this cover was to ensure that the site's new higher surface would not be adversely affected by upward migration of fluid contaminants.

If the site were not covered, its surfacing of silty sandy clay materials would (in a 100-day-long design drought whose severity were such as to desiccate the surface to produce a suction equivalent to 1000 cm of water head) permit very significant upward movement of contaminated soil water. For the phenolic contamination, whose concentration is 100 mg/l in the groundwater, this would lead to the addition of an extra 42 mg/kg of phenol to the site's surface materials.

The planned reuse of the site called for the worst contamination concentrations to be no more than 1% of the levels in the design groundwater, and the level at which this reduced contamination could be allowed was specified as the top of clean cover (i.e. below the soil layers and inert fills brought in to provide the required final topography).

Using the crushed concrete material, a cover of 25 cm thickness produced the required reduction. With the stiff clay cover, a rather greater thickness of 100 cm was necessary to give the same effect.

In each case, the results were obtained by identifying the various fluxes that would rise to different heights above the design groundwater level (see Fig. 5.5). The fluxes were graphed against the heights to which they rose and the various dv/dh ratios established. Equation (5.3) gives the amount of contaminant that can rise to whatever level in or above the clean cover is seen as important.

$$\text{Contaminant addition} = \frac{dv}{dh} \times d \times \frac{\rho w}{\rho s} \times c \qquad (5.3)$$

(mg/kg)

where d is the design drought duration (days), ρw is the groundwater density (kg/m^3), ρs is the compacted density of the cover (kg/m^3) and c is the concentration (mg/l) of a particular contaminant in the design groundwater. For any other contaminant, it is only necessary to scale the above listed phenolic contaminant additions in the groundwater contamination ratios, as is obvious from equation (5.3).

The selection of the top of the clean cover as the level of interest for contaminant addition came entirely from the planned reuse of the site and an identification of the most sensitive targets for contaminant attack and hazard. If any other level of interest or any different allowable contamination addition had been chosen, the same process would have been followed.

This example confirms the view expressed earlier that non-granular covers can be as effective as granular covers, though as rather thicker layers. Granular covers do, however, have the real advantage that minor increases in thickness give sizeable improvements in cover efficiency. If, for example, an allowable contaminant concentration of half that chosen above had been selected, the granular cover thickness would have had to be increased by some 7%. For the cohesive material cover, however, a thickness increase of 47% would have been necessary to give the same effect. Thus, if the crushed concrete cover had been laid as a slightly thicker (27 cm instead of the planned 25 cm) band than required, the cover's effectiveness would have been 100% increased.

5.5 Covers designed to minimize groundwater pollution

5.5.1 *Introduction*

The emphasis that a clean cover's primary function is to prevent the new and higher surface of a site becoming contaminated is increasingly seen as too limited. Although most contaminated sites are underlain by polluted groundwater (due to past spillages and to the ongoing leaching out of more soluble contaminants by infiltrating rainfall), reducing this pollution locally can be important. This is particularly so where the groundwater flows to surface water bodies or to groundwater abstraction points. Thus, covers intended to control rainfall infiltration are becoming more widely used.

The greatest experience with infiltration control is in the design and construction of clay caps for landfill sites. The priority in these is to reduce infiltration and so minimize the generation of polluting leachates. The UK guidance on this is contained in *Waste Management Paper, No. 26* [19] which recommends:

- A cover of topsoil/subsoil thick enough to protect the clay capping below from desiccation and cracking. This cover must also be of a thickness suitable to support whatever vegetation is anticipated.
- A rolled clay cap of about 1 m in thickness and laid to a permeability of 10^{-9} m/s.

It could be argued that this guidance is appropriate for a clean cover intended to minimize groundwater pollution, but this view ignores the ineffectiveness of many landfill cappings. Knox [20] has summarized the infiltration results available from a number of monitored landfill caps (Table 5.5), and has shown that high infiltration rates can occur in some cases. It thus is more appropriate to examine the causes and mechanisms of rainfall infiltration, to identify where solutions are possible.

5.5.2 *Rainfall infiltration and cover design*

Infiltration is a complex process affected by a number of variables, including:

- Rainfall duration and intensity
- Absorptive capacity of the surface soils
- The prior state of the surface soil's wetness
- The permeability of the surface soil

Table 5.5 Infiltration data from experimental studies and completed landfills (modified from [20])

Site	Capping	Infiltration (mm/year)
Undrained sites		
1(a)	900 mm topsoil/subsoil	>84
2	750 mm topsoil/subsoil	11– > 66
	500 mm clay	
	200–500 mm weathered shale	
3	1 m topsoil	200
Drained sites		
1(b)	900 mm topsoil/subsoil	< 20
	600 mm Gault Clay	
4(a)	Soil	5–7 (on land slopes
	300 mm drainage layer	of 4%)
	500 mm clay	
5	Topsoil 700 mm	56–89
	Drainage layer 300 mm	
	Clay 500 mm	
Sites with steeper land forms		
4(b)	Soil	1–2 (on 20% land
	300 mm drainage layer	slopes)
	500 mm clay	
Sites without a surface soil layer		
6	300–600 mm boulder clay	20
7	> 3 m silty clay	90–140
8	1 m Keuper Marl	40– > 86
9	1 m Oxford Clay	49
10	500 mm weathered shale	30–120

- The topographic slopes
- Ambient temperature and season

It is extremely difficult to quantify over other than a relatively long time span. If an annual quantification is acceptable the task becomes more achievable, and it is obvious that two quite separate forces encourage rain to enter a soil and then move down the profile.

The first of those is the surface soil's suction, which (Figure 5.3) is highest when the soil is dried out and pulls in moisture. Obviously, as the infiltration continues, the surface material becomes wetter, its suction falls, and so the widely known decline in infiltration rate with the duration of rainfall occurs. As the suction reduces, the second force (the gravity potential) comes to dominate the process, and within a relatively short time the progress of the wetting front down a soil profile becomes controlled by that soil's saturated hydraulic conductivity value. This allows the speed of advance of the wetting front (Figure 5.7) to be evaluated.

The rate of rainfall entry into a soil will be reduced if the surface layers have as low an absorptive capacity as possible. Thus, the use of porous, granular, or high organic content soils above a clay cap will inevitably increase the ultimate amount of infiltration. This indicates that the usual practice of laying topsoil and subsoil over a clay cap is bound to increase the infiltration (Table 5.5), but omitting these soil layers would expose the clay cap to erosion and desiccation, and would also limit the vegetation cover that is usually required. Thus, an acceptable compromise has to be made.

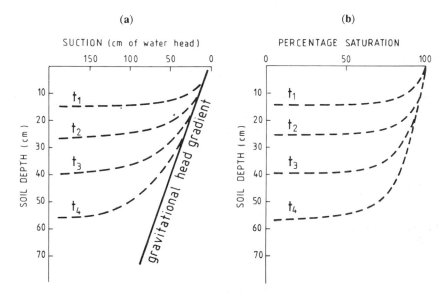

Figure 5.7 Progress of a wetting front at times t_1, t_2, t_3 and t_4 in a uniform soil. (a) Suction/depth relationships; (b) water content profiles. After Hillel [23].

Increasing the topographic slope of the cover will, of course, minimize the rate of rainfall entry (Table 5.5, site 4(b)) by encouraging rainfall to move off as surface run-off.

The initial progress of infiltrating rainfall through a clay cap is thus controlled by its surface condition and type, and if a clean cover of the *Waste Management Paper No. 26* type was installed, all that would be achieved would be the delay until the wetting front had moved down into the materials below the cap. Thus, adequate infiltration and leachate formation control would not occur:

If, however, a coarser layer with large, open, air-filled pores is laid below the clay cap (Figure 5.8), infiltrating moisture moves down in the wetting front until the interface of the finely porous clay and the underlying coarsely porous granular layer is reached. At this stage, downward progress of the wetting front is halted for a time, since moisture cannot move out of a material that still has a high suction into one with a lower suction. If the materials detailed in Figure 5.3 and Table 5.4 had been used, the crushed concrete laid at a 10% moisture content would have a suction of 500 cm, and infiltration down from the stiff clay would not take place until the clay's suction had fallen to at least this value.

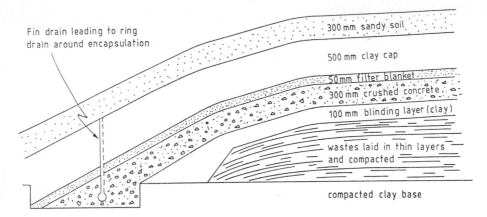

Figure 5.8   Encapsulation design.

Halting the wetting front would not affect infiltration into the site's surface, and so the overlying clay would become wetter and its suction would decline.

At the 500 cm suction level, the stiff clay's moisture content would be close to its saturated level. In practice, an even greater moisture content in the clay would be necessary to move moisture into the underlying crushed concrete. This is because the crushed concrete's pore voids are still air-filled and this air has to be forced out before water entry can take place.

Thus (with the materials shown in Figure 5.3), the stiff clay would be almost totally water saturated and the crushed concrete would become entirely water saturated when moisture finally crosses the material interface.

It could be said that the above argument is flawed, since the clay band (layer 5) below the coarse granular layer (Table 5.6) will be relatively dry, and thus have a very significant suction ability. This, it could be argued, would pull moisture down from the saturated granular layer, and so allow the wetting front to continue its downward progress to leach any soluble materials from below the cover.

This, however, ignores the facts that:

(a) The speed of outflow in a saturated granular material is extremely rapid, whilst the progress of a wetting front in a clay is very slow and equivalent to its saturated hydraulic conductivity (i.e. in this case to a mere 0.18 mm per day)

(b) The suction ability of the basal clay layer will decline as it manages to abstract some moisture from the overlying coarse granular band

The net result should be a near total interruption of the downward movement of the rainfall infiltration.

Table 5.6 Functions of the layers in the encapsulation capping

Layer	Functions
No. 1. Top- and subsoil (300 mm)	To support grass cover; to minimize infiltration
No. 2. Clay (500 mm)	To act as soil moisture storage reservoir of high suction capacity material to be compacted to 10% air void ratio and maximum practicable compaction (Table 5.4(a))
No. 3. Filter blanket (50 mm)	To prevent siltation of the crushed concrete layer
No. 4. Crushed concrete (300 mm)	To underdrain the overlying clay by virtue of low saturated moisture content and high saturated permeability Material compacted to maximum practicable compaction (Table 5.4(b))
No. 5 Blinding layer (100 mm)	To cap the contaminated material and prevent airborne contamination; to underseal the crushed concrete layer Material to be compacted to 10% air void ratio and maximum practicable compaction

However, this will only be achieved if suitable drainage outflow facilities are provided for the coarse granular band, and that usually will imply that the cover as a whole will have been laid to at least a 10% topographic slope.

As the coarse granular band is the critical layer in achieving the required control over infiltration, it is valid to consider where this should be placed within the cover profile. Arguments are possible that the drainage layer should be laid below the surface soils and above the clay cap (Table 5.5, site 4(a)), but it seems preferable to locate the drainage layer below the bulk of the clay capping, since this ensures that the moisture storage capacity of the clay is utilized (Table 5.7), and that a maximum inflow to the coarse granular band takes place.

Table 5.7 Moisture storage capacity of the capping layers

Layer	Initial volumetric moisture content (%)	Saturated volumetric moisture content (%)	Equivalent rainfall infiltration to achieve saturated moisture content (mm)	
1 Sandy topsoil (300 mm)	30	50	60	
2 Clay (500 mm)	40	50	50	
3 Filter blanket (50 mm)	10	32	11	187
4 Crushed concrete (300 mm)	10	32	66	
Outflow to ring drain comes into operation				
5 Blinding layer (100 mm)	40	50	10	

5.5.3 *Encapsulation covers*

Covers of the type outlined in Figure 5.8 are particularly appropriate for those cases where a developer requires a provably clean site, and so has the contaminated surfacing scraped off and deposited in a capped mound (chapter 10, section 10.3.1)

In such cases, it is relatively easy to define the functions required of each layer in the cover (Table 5.6) and to demonstrate the infiltration storage that will occur (Table 5.7). With relatively steep topographic slopes to encourage run-off and with care taken to select a topsoil cover that will not encourage excessive water entry, infiltration is likely to be no more than a maximum of 5% of the winter rainfall value (infiltration from summer rainfall will normally be minimal given the water consumption by plant cover and the higher evaporation rates), and the clay layer (layer 2) will provide several years' water storage, before the outflow drainage from the granular band commences.

A thorough compaction of the contaminated wastes that are to be encapsulated, will ensure that the surface area available for leaching will be minimal.

Whilst little or no infiltration is likely to move down to the encapsulated wastes, these encapsulated contaminated materials were of concern when they occurred on the site's surface. When concentrated in an encapsulation, they have to be regarded as potentially more hazardous, and particular care has to be taken to minimize the effects even of the negligible rainfall infiltration that might reach them. If tarry deposits are present on the site and are air-dried and oxidized prior to encapsulation, it is possible to compact such materials to highways quality, and so reduce any leaching effects [21] to trivial proportions (Table 5.8).

Table 5.8 Leachate from oxidized tarry wastes (W. Midlands site)

Parameter	When wastes finely ground (mg/l)	When wastes compacted to highways standards (as would be the case in an encapsulation cell) (mg/l)
Naphthalene	84–24	< 1
Acenaphthylene	12–10	< 1
Acenaphthylene	7–2	7–2
Fluorene	19–7	2– < 1
Anthracene	15–7	< 1
Phenanthrene	20–5	3– < 1
Fluoranthene	18–5	3– < 1
Pyrene	27–6	4– < 1
Benzo anthracene	43–7	< 1
Chrysene	16–7	< 1
Benzo [b] fluoranthene	7	3– < 1
Benzo [k] fluoranthene	< 10	
Benzo [a] pyrene	< 10	< 1
Indeno [1, 2, 3 – cd] pyrene	< 30	< 1
Dibenzo [a, h] anthracene	< 30	< 1
Benzo [g, h, i] perylene	< 30	< 1
Toluene extract	46–96	6–1
Phenol	420–610	1.9– < 1
Sulphates	80–312	30–95

Encapsulation covers thus necessarily include the works to minimize rainfall infiltration, and the treatment to reduce contaminant leachability. They have to be evaluated against potential failure mechanisms (section 5.2.6), and their efficiencies as capillary break layers should be confirmed, to guard against any upward movement of contaminations, in the unlikely event that contaminated leachate were to collect in the encapsulated wastes.

It has been argued that covers of this type can be compromised by the presence of high water capacity channel ways (e.g. rabbit burrows, decayed tree roots, etc.). Various researchers (e.g. [22]) have shown, however, that water will not enter such air-blocked macropores until the soil around these is almost totally saturated. Macropore flow is only significant in circumstances where rainfall intensity exceeds the surface material's infiltration capacity and surface run-off into these voids can occur. This is usually easily prevented by minimizing surface ponding and channelling run-off into defined channel ways.

When designing such covers, it is necessary to use material curves that accurately represent the changes in suction and hydraulic conductivity properties as the material's moisture contents increase, since this will be the actual condition that will occur. The analogous material properties needed for a capillary break layer design (section 5.4) should, in contrast, be obtained from tests carried out as the materials are progressively dried out. For many materials, the wetting and drying curves will differ to a significant extent.

5.6 Soil cappings

In most cases, it is necessary to allow for a vegetation cover above at least parts of a reclaimed site. This obviously necessitates the provision of an appropriate type and thickness of soil capping. As stressed above (section 5.2.1), such cappings should be regarded not as part of the clean cover, but as separate works, designed to meet specific and defined horticultural requirements. Chapter 9 gives appropriate guidance on rooting depths for various plant species and on the soil characteristics that are required. All that need be added is that care should be taken to ensure that the soil cappings do not compromise the underlying clean cover, by allowing either siltation or desiccation. The migration of plant roots into an engineered cover is only likely if the plants have not been provided with an appropriate type and thickness of soil for their root development.

5.7 Appropriate and non-appropriate applications of clean covers

Clean cover design philosophy is based on the proven principles of soil moisture movements. Thus, the use of clean covers is most appropriate when the contamination mobility is mainly in a liquid phase. When gaseous contamination is a particular hazard, clean covers can be used, although only on sites where the gas hazard is trivial, and slotted gas collection pipes can be set in a coarse granular cover, and then vented to the atmosphere. On sites with more severe gas hazards (i.e. where measurable outflow rates occur, or where high gas pressures exist), clean covers will not be a successful solution.

Equally, it will be inappropriate to expect any clean cover to deal effectively with organic vapours (from oil soaked ground or from free product floating on the groundwater) or with layers of semi-liquid tars, which can move up through the cover as the load on the site's surface is increased. Technologies are now commercially available to resolve such contamination problems (chapter 4) and these should be applied before any clean cover is installed to cope with any remnant and non-organic contamination.

Clean cover technology is an example of the wide spectrum reclamation technologies, and so is particularly useful on those diversely contaminated UK sites that have had several prior industrial uses. If a site, however, is affected only by a contaminant from a point pollution source, it will be more appropriate to select one of the newer technologies that can destroy or immobilize that particular contaminant.

6 In-ground barriers

S.A. JEFFERIS

6.1 Introduction

Adjacent to any centre of population anywhere in the world there is likely to be a landfill site. Many of these sites will have been used because they were in a convenient position or simply because they were available (for example holes left by gravel, chalk, clay extraction, etc.). Inevitably some of these sites will be in areas where the hydrogeology is such that they pose a threat to groundwater resources and thus remedial action is necessary. Furthermore, developments in landfilling practice, such as daily cover layers and greater compaction of the waste, have tended to reduce the vertical permeability within the landfills. This has promoted horizontal migration of landfill gas, which may need to be controlled. Ironically many carefully operated landfill sites today pose a greater hazard from landfill gas migration than those where the material was deposited more loosely and the gas could escape vertically. This serves to demonstrate that there has always been a crucial lack of information on the behaviour of contaminated sites. Currently, new specifications for landfills are being developed and yet very few landfills have been exhumed to see how man-made barrier materials or even native clays perform in situ.

Landfill sites are now being selected much more carefully and liner systems installed if they do not exist naturally. Thus, it is to be hoped that the number of problem landfills will reduce with time although this does presuppose that the current design procedures will pass the test of time. Whether or not the number of problem sites is actually reducing, the absolute number of problem sites is still large and this has stimulated a substantial amount of work on remedial treatment.

Fundamentally, existing technology allows only two treatments for problem landfills: (a) excavation and re-burial in safe(r) storage; or (b) installation of containment systems. In general, excavation and re-burial will be extremely costly and generate substantial secondary pollution for all but the smallest landfills. Thus techniques are necessary to contain pollutants in situ. In principle, three types of barrier may be needed at any contaminated site: (a) vertical barriers to form a cut-off wall around the site; (b) horizontal barriers to provide a low permeability base (although for small sites there may be potential for inclined interpenetrating walls formed by a mix-in-place process [1];

(c) cover layers to control the infiltration of rainwater. Cover layers are addressed in chapter 5. In practice, it is often found that there is a natural base layer of low permeability material beneath the contamination and thus all that is needed is a vertical wall that connects with this layer.

If no natural low permeability layer is present, a base may be installed by grout injection or by jet grouting and clearly developments in this area may considerably extend the use of barriers systems generally.

It should be noted that although a base layer is normally a prerequisite before a vertical barrier can be effective, there are some situations where it is not required.

Figure 6.1(a–d) shows four barrier layouts that have been used in the United Kingdom. Figure 6.1(a) shows a standard barrier taken down to an aquiclude. This type of barrier will be suitable for liquid control, although for contaminated groundwater it will be necessary to consider the durability of the wall material. Figure 6.1(b) shows a standard wall with a vertical membrane. This type of wall may be suitable for gas and the more aggressive groundwaters (although the durability of the membrane will now have to be considered). Figure 6.1(c) shows a partially penetrating wall for gas control. The design groundwater level will need to be carefully assessed (it may be seasonal) as will the maximum gas pressure. The wall need not extend below the membrane unless leachate is also to be contained (the membrane is necessary for gas control, see section 6.3). When using a partially penetrating membrane, careful attention must be given to the integrity of the wall at the toe of the membrane. Figure 6.1(d) shows a partially penetrating wall without a membrane. This

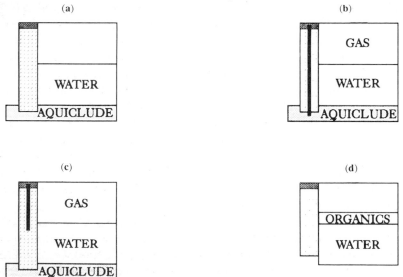

Figure 6.1 Types of cut-off wall. (a) Standard cut-off; (b) membrane barrier; (c) partially penetrating gas barrier with leachate barrier; (d) partially penetrating light organics barrier.

might be used for the control of light non-aqueous liquids in a region where there is no aquiclude within a reasonable depth; for example for oil spill control at a tank farm founded on permeable soils.

Barriers may also be required as part of the clean-up strategy for contaminated industrial sites or as preventive measures at new sites. At the present time in the United Kingdom, most barriers are used at landfills and particularly for the control of gas migration. However, it is to be expected that they will find increasing use at contaminated sites in general as controls on groundwater pollution are progressively tightened.

6.2 The requirements for a cut-off

The contamination at landfills and other sites need not be limited to aqueous solutions. A barrier may have to retain not only aggressive chemical solutions but also light and dense (relative to water) non-aqueous phase liquids (NAPLS) and gases. For all of these the cut-off should represent a sufficient barrier that the rate of escape of pollutants (or transit time through the barrier for degradable materials) is such that it will not adversely affect the surroundings. Thus the barrier performance ought to be related to the local geology, hydrogeology, the nature of the contained materials and the land use adjacent to the site.

For the control of aqueous liquids there is some international consensus that clay liners for landfills should have a permeability equivalent to 1 m of 10^{-9} m/s material or less (i.e. a permittivity of 10^{-9} s^{-1}; permittivity is the permeability divided by the thickness of the barrier and thus an indicator of the overall hydraulic resistance). Although 1 m of 10^{-9} m/s material seems to be generally accepted, there can be no confidence that this will remain the standard. For example the proposed EC Directive on the Landfill of Waste [2] requires that the base and sides of a landfill should have a permeability equivalent to not less than 3 m of 10^{-9} m/s material, i.e. a permittivity of 3.3×10^{-10} s^{-1}. However, for slurry trench cut-offs, which are often thinner than 1 m (typically the thickness may be 0.6–0.80 m depending on depth, deeper walls being thicker in order to be able to accommodate a thicker excavator arm), the specified permeability is not usually reduced to compensate for the reduced thickness and thus permittivities rather over 10^{-9} s^{-1} seem to be permitted for such vertical barriers.

In fact there can be little case for requiring permeabilities less than 10^{-9} m/s, not because this represents the level at which all escape is insignificant but rather because it is about the level at which permeation ceases to be the most important vector for escape. As the permeability drops below 10^{-9} m/s, other transport mechanisms become progressively more important (see section 6.4).

It will be interesting to see how the permeability/thickness debate develops. It is to be hoped that the actual engineering of landfills will be addressed rather than the simplistic parameters of permeability and thickness. In addition, the overall environmental balance needs to be considered. The lining of a landfill with 3 m of clay may leave a substantial hole somewhere else, a hole which itself may represent substantial environmental damage.

6.3 Gas permeability

For the control of gases, the *Waste Management Paper 27* [3] says "reworked natural clay or calcium bentonite linings are probably the most suitable commonly available materials for gas barriers. They should be laid and compacted to achieve a maximum water permeability of 10^{-9} m/s". Thus, gas barriers are to be designed to achieve a specified water permeability although it is noted that "the effectiveness of liners in preventing gas migration is not yet fully understood and their permeabilities to gas have yet to be determined. It would seem likely that they would be more permeable to gas by several orders of magnitude compared to leachate". Indeed they will be! With the current state of knowledge, to design gas control measures on water permeability is unsatisfactory. For clay, bentonite–cement and most mineral-based materials, the gas permeability will be very sensitive to the moisture content and there can be no correlation between the gas and water permeabilities.

The proposed EC Directive on the Landfill of Waste [2] gives no guidance on the formation of gas barriers and merely requires that "appropriate measures are taken to control the accumulation and migration of landfill gas" and for biologically active landfill sites receiving or having received more than 10 000 metric tons per year of waste that "landfill gas shall be collected and properly treated and preferably used".

The gas permeability of mineral materials is complex. If the material is fully saturated, there can be no bulk gas flow and methane, etc. can move only in solution by diffusion through or advection with the water. (Gases could be produced downstream of a barrier by degradation of a leachate that has passed through the barrier. The liquid permeability must be designed to be sufficiently low to ensure that the gas generation rate is minimal.)

However, if the material dries out it may shrink, crack and so become highly gas permeable. In the partially saturated state, unless the gas forms a continuous phase within the material, then the flow may be effectively blocked unless the pressure is sufficient to force the water from the pore space and establish a flow path. The necessary pressure may be large and thus the partially saturated material may still behave as an effective barrier. However, a small further loss of water (from the blocked partially saturated state) may open up flow paths of quite significant permeability. For example a 1 m thickness of good quality structural concrete when air-dried might show a gas flow rate of 50 000 ml/m^2 per day per atm of pressure differential (compared with perhaps 50 ml for a membrane). Clearly, the permitted gas flow rate through a membrane must be related to the venting arrangements, etc.; i.e it must be designed to match the site conditions. Appropriate design procedures need to be developed.

In many situations there is probably sufficient moisture in the ground to ensure that most fine grained materials such as clay or a bentonite–cement are of low permeability to gas. However, droughts do occur and these have led to the shrinkage of clays to significant depths. The author therefore prefers to regard the in situ performance of mineral materials such as clay and bentonite–cements as unproven for gas control. Unless or until it is shown to be satisfactory, a membrane should always be included when designing a gas barrier unless the full depth that may be subject to drying is adequately protected by vent trenches (although these will exacerbate drying and thus the problems become circular). For exposed barriers such as clay capping layers, some gas permeability must be expected in dry seasons.

If a high density polyethylene (HDPE) membrane is used, the gas permeability might be of the order of 50 ml/m^2 per day per atm of differential pressure for methane, or 200 ml/m^2 per day per atm for carbon dioxide [4]. These are very low figures and indeed it is not strictly correct to consider them as permeabilities. Furthermore, it does not follow that, simply because they can be achieved with an intact membrane, such low values are actually required. In practice, there may be some additional leakage at joints between membrane sections and possibly at the toe of the wall where the membrane links with the formation. Clearly, when designing membrane systems, careful attention must be given to the joints and mechanical protection of the membrane. There is minimal design data on what represents an acceptable gas flux through a barrier.

6.4 Potentials tending to cause flow

The following are some of the potentials that ought to be considered when assessing the magnitude and direction of flow through a barrier. A fuller treatment is given by Mitchell [5].

(a) *Hydrostatic pressure* due to difference in liquid or gas pressures. The flow may be into or out from the contaminated area depending on the relative pressures. Furthermore different phases may move in different directions. For example, if the leachate level in a landfill is below that of the surrounding groundwater there will be a tendency for water flow into the landfill, whereas gas generated within the landfill will tend to move outwards.

(b) *Chemical potentials* tending to cause diffusion. In general, the chemical flux will tend to be outward as concentrations of chemical species should be higher inside the contaminated area than outside.

(c) *Osmosis*. In general, this will promote inward flow of water as osmotic potentials within a contaminated area generally will be higher than outside.

(d) *Electrical potentials*. These may be generated by the difference in chemical conditions between the contamination and the surrounding soil. For example, the environment within a landfill may be significantly more reducing than outside. Electrical potentials may induce electro-osmosis (movement of the water), electrolysis (movement of ions) or electrophoresis (movement of colloidal material). The effect of any electrical potential will depend on the charge of the migrating species (e.g. in the reducing environment of a landfill, cation migration rates may be reduced and anions accelerated, although the local charge balance must be maintained).

(e) *Temperature differentials* may drive flow. Such differentials may result from the generation of heat by microbiological activity within a landfill or the microbial oxidation of methane escaping from a landfill.

(f) *Gas flow* may also be influenced by changes in atmospheric pressure and, more particularly, the occlusion of flow paths as a result of rainfall or other changes in soil moisture content.

Thus, there are many different potentials and, as shown above, they may act in contrary directions and influence the water and ions differently. Furthermore, in a multibarrier system, each component of the barrier will have a different response to the potentials. Fluid could accumulate between layers or there could be a net removal. For example, for a waste underlain by a membrane-compacted clay-membrane sandwich, the clay layer could be desiccated by osmotic flow.

At the present time, permeability seems to regarded as the fundamental parameter for barrier design. Clearly this is simplistic and the other driving forces must be considered. In addition, the chemical and electrical as well as the physical boundary conditions need to be established.

6.5 Types of vertical barrier

Vertical barriers are now regularly used for the containment of pollutants. In contrast, very few horizontal barriers have been installed except after bulk excavation. Vertical barriers may be broadly classified into three groups: (i) driven barriers, (ii) injected barriers; and (iii) cut and fill barriers.

6.5.1 *Driven barriers*

These barriers may be formed by driving steel or concrete elements into the ground. The fundamental requirements are that the elements are durable and can be joined to form a suitably impermeable wall. Driven systems may have considerable potential in situations where barriers are required to both control pollution migration and provide some mechanical support. Furthermore, they have the advantage that there is no need for excavation or the disposal of possibly contaminated arisings. With the rapidly developing demand for barrier systems the essential simplicity of driven systems suggests that new developments are to be expected in the coming years.

6.5.2 *Injected barriers*

Grout injection has been regularly used to form seepage barriers under earth dams, etc. To obtain a low permeability barrier, it is necessary to use several stages of grouting at reducing centres and therefore costs are relatively high. A permeability of order 10^{-7} m/s is often regarded as reasonable for grouting operations but the actual value will be very dependent on the soil type, heterogeneity, etc. By its nature, the injection process may lead to a grout curtain that contains a number of ungrouted flow paths. These may allow relatively rapid erosion of the adjacent grout material, especially if there are aggressive chemicals present in water. Grouting is therefore not a first choice for pollution control barriers and it is unlikely to find major application except in circumstances where other procedures are impracticable. However, it may find considerable application for repairs to barrier systems and also for the local encapsulation of toxic wastes.

6.5.3 *Cut-and-fill type barriers*

For convenience, all processes where soil is excavated and replaced by a cut-off material may be termed cut-and-fill procedures.

Secant piling. Secant piling is often used in civil engineering to provide structural cut-off walls. The development of systems using soft primary piles and structural secondaries has allowed significant cost savings in situations where a full wall of structural piles is not required [6]. For pollution control, there must be concern about the joints between piles. In general, secant pile walls are unlikely to be competitive as barriers, unless there is a need for structural strength as well as impermeability.

Secant pile type walls can also be formed by mix-in-place techniques, for example using counter-rotating augers. Developments in this area are to be expected but the process may be restricted to a rather narrow range of soil types.

Jet grouting.   Jet grouting may be used to form secant piles or thin walls. The process may involve cutting with a high pressure air-shrouded jet followed by filling with cement grout, or direct cutting with a cementitious fluid, in which case the process becomes effectively a mix-in-place procedure. Jet grouting may find particular application in situations where there is limited surface access such as under a structure.

The vibrated beam wall.   For this process an 'H' pile is vibrated into the ground and then extracted. During extraction, a cement-based grout is injected at the toe of the pile. The technique produces a relatively thin wall, which is not ideal for pollution control. However, a more secure system may be achieved if a double wall is installed with cross walls to form a system of independent cells. If pumped wells are installed in these cells and the internal groundwater level maintained below that of both the contaminated area and the surrounding soil, then there should be minimal potential for pollution migration and the well flows can be used to monitor the performance of the system.

Shallow cut-off walls.   Shallow cut-offs may be formed by excavating a narrow slot with some form of trench cutter and inserting a geomembrane. Clearly, the excavation must be of sufficient depth to reach an aquiclude and the membrane must be sealed to it. An alternative way of forming a shallow cut-off is to excavate a wide trench, and backfill with compacted clay. The trench will need to have battered sides to avoid any risk of collapse during the compaction works. In general most cut-off walls must be taken down to below the groundwater and this may lead to instability of the trench and/or difficulties with compaction of the backfill. Whilst it may be possible to control the groundwater by pumping, it may be more economic to pursue a slurry trench solution.

6.6 The slurry trench process

Slurry trench cut-offs are currently the most widely used form of remedial barrier system and therefore the procedure will be discussed in detail. It would seem that the first slurry trench wall was built at Terminal Island, California in 1948, although Veder had conceived the idea of a contiguous bored pile wall formed under slurry in 1938. However, trials were not carried out until the 1940s and structural walls not achieved until the 1950s [7]. Many variants of the process are now available and it is one of the most versatile techniques for the formation of vertical barriers.

   If a trench is excavated in the ground to a depth of more than about 2 m it is likely to collapse. However, if filled with an appropriate fluid the trench can be kept open and excavated to almost any depth without collapse. For this purpose the fluid must have two fundamental characteristics: (i) it must exert sufficient hydrostatic pressure to maintain trench stability; and (ii) it must not drain away into the ground to an unacceptable extent. These requirements are typically met with a bentonite clay slurry, a bentonite–cement slurry or a polymer slurry.

Once a trench has been formed, the impermeable element may be installed. This may be concrete (although the costs tend to limit the use of concrete to structural walls), a clay–cement–aggregate plastic concrete, a soil–bentonite mix, a geomembrane or even specially shaped glass panels [8]. A further variant is to excavate the trench under a bentonite–cement slurry, which is designed to remain fluid during the excavation phase and to set when left in trench to form a material with permeability and strength properties similar to those of a stiff clay. These slurries are known as self-hardening slurries and are the most widely used form of cut-off in the United Kingdom. In the United States, soil–bentonite cut-offs are often used for groundwater control but it appears that membranes are preferred for pollution control. In Germany, a chemically modified high density soil system has been developed and offers an interesting additional wall type.

6.7 Types of slurry trench cut-off

There is now a significant number of different types of slurry trench cut-off and, as demands for pollution control increase, further developments must be expected. Currently the principal types of cut-off wall are as follows.

6.7.1 *Soil–active clay cut-offs*

These are formed by excavation under an active clay slurry (almost invariably bentonite) and backfilling the trench with a blended soil–active clay mix, which forms the low permeability element. Design procedures for soil–bentonite backfills are given by D'Appolonia [9].

The backfill in a soil–active clay wall must be at a relatively high moisture content so that it will self-compact when placed in the trench. Unfortunately, high water content clay systems will always be vulnerable to chemical attack, as any change in the chemical environment may lead to substantial changes in the interactions between clay particles. This may cause significant shrinkage and so have a dramatic effect on the permeability. For good chemical resistance, dense non-swelling clay systems are required but these cannot be installed by the conventional slurry trench process.

6.7.2 *Clay–cement cut-offs*

Clay–cement cut-offs are now regularly employed for groundwater and pollution control. With UK materials, typical mix proportions are illustrated in Table 6.1. The use of bentonite contents above 40 kg appears to give a marked improvement in permeability, although with a slight increase in strength (this suggests that there is an interaction between the cement and the bentonite). The 40 kg figure is not absolute and will vary with the source of the bentonite and the nature of the cement. Above 55 kg, the slurry may become so thick as to be unmixable and unusable in the trench. Increasing the cement content significantly increases the strength but also reduces permeability (it may have rather little effect on the fluid viscosity). The best design compromise appears to be to use the maximum possible bentonite content and the minimum cement content. The cement may be a mixture of ordinary Portland cement and replacement materials such as ground granulated blast-furnace slag or pulverized fuel ash.

Table 6.1 Clay–cement cut-offs

Material	Quantity (kg)
Bentonite	40–55
Cement	90–200
Water	1000

Clay–cement slurries are usually used as self-hardening slurries and simply left in the trench to set at the end of excavation. If the excavation rate is likely to be particularly slow (for example for a very deep wall), it may be necessary to excavate under a bentonite slurry and replace this with a bentonite–cement slurry or a backfill mix. Normally it will be necessary to install such walls as a series of panels, panels 1, 3, 5 being excavated and filled in sequence, and the secondary panels 2, 4, 6 installed once the primaries have set.

6.7.3 Clay–cement–aggregate cut-offs

For deep walls, or if for some reason a more rigid wall is required, the slurry may be replaced with a clay–cement–aggregate plastic concrete prepared by blending the displaced slurry with a suitably graded aggregate (and cement if there was no cement in the original excavation slurry) to form a backfill mix.

For clay–cement–aggregate mixes, the bentonite and cement concentrations of the slurry phase (excluding the aggregate) generally will be towards the lower end of the ranges listed in Table 6.1. Despite this, the mixes tend to be much stronger and stiffer than simple bentonite–cement systems and the strain at failure may be low especially if the aggregate particles are in grain-to-grain contact (this may be difficult to avoid unless the slurry is specially thickened so that the aggregate particles can be held in suspension prior to set). For pollution control, the reduced content of clay and cement and the potential for cracking at the slurry–aggregate interfaces tends to militate against their use except for systems using specially designed aggregate gradings.

The addition of aggregate will reduce the volume of clay–cement slurry required to form the wall and, as already noted, the slurry may be of lower clay and cement content than that for a self-hardening slurry wall. Despite these savings, backfill walls are likely to be more expensive than self-hardening walls because of the extra materials handling.

6.7.4 *Cut-offs with membranes*

After excavation, and before the slurry sets, a membrane may be lowered into the trench. The use of such membranes may be appropriate where high levels of pollution exist or very aggressive chemicals are present, as membranes can be obtained to resist a wide spectrum of chemicals. For gas migration control, membranes may be essential. Most slurry-based cut-off materials have a relatively high water content. If this water is lost they may become gas permeable and thus the potential for such materials for gas control must be regarded as not proven.

The major problems with membranes relate to the sealing of the membrane to the base of the excavation and the joints between membrane panels. The base seal has been attempted by mounting the membrane on an installation frame and hammering the whole assembly a predetermined distance into the base layer. This operation has to be undertaken with some care to avoid damaging the membrane and the inter-panel joints. An alternative is to over-excavate the trench and use a longer membrane in order to extend any flow path around the toe of the membrane.

A number of different joint systems are available to form the inter-panel joints but all require considerable care in use. A membrane panel may be over 10 m high by 5 m wide and mounted on a heavy metal frame. The panel joints therefore need to be robust and simple to use. Details of a number of jointing systems are given by Krause [10]. Typically a bentonite–cement trench will be of the order of 600 mm wide. The insertion of a membrane will substantially disturb the behaviour of the slurry in the trench and the membrane may act as a sliding plane and encourage cracking. Furthermore, the membrane may take up a meandering path both horizontally and vertically within the trench. This can be advantageous as it will provide some flexibility if the membrane is stretched as a result of ground movements. However, it will mean that the bentonite–cement material may be divided into isolated units that are not continuous along either side or throughout the depth of the trench. If a membrane is used, therefore, it should be treated as the primary impermeable element, and the bentonite–cement regarded as fundamentally providing only support and mechanical protection.

6.7.5 *High density walls*

These are relatively new cut-off materials developed specifically for polluted ground. The hardened material is hydrophobic and the water permeability may be very substantially lower than with conventional soil–bentonite or bentonite–cement walls. Indeed, it is held that the permeability is so low that diffusion is the most important transport process. The chemical resistance to aqueous pollutants may be better than for conventional slurry walls. Material costs are higher and a special excavation plant may be necessary because of the high density of the slurry that follows from the high solids content. Details of the system are given by Hass and Hitze [11].

6.7.6 *Drainage walls*

Pollution migration (liquid or gaseous) may be controlled by drainage ditches. These ditches must extend to a sufficient depth to intercept all the pollutants. Deep drainage walls may be formed by excavating a trench under a degradable polymer slurry. On completion of excavation, the trench is backfilled with a gravel drainage material and the slurry broken down with an oxidizing agent or left to degrade naturally. In principle, the drainage wall system could be useful for the formation of gas venting trenches. However, such trenches need not extend below the water table and, unless this is exceptionally deep, direct excavation will be cheaper than a slurry trench procedure. There is potential for driven drainage wall systems especially as there would then be no need to excavate and dispose of possibly contaminated soil.

6.8 Slurry preparation

Good mixing and accurate batching of the mix components are fundamental to the performance of all slurries Figure 6.2 shows a schematic diagram of a typical plant layout for bentonite–cement slurry preparation. The first step is to prepare a slurry of bentonite and water. This slurry is then pumped to a storage tank where it should be left to hydrate for at least 4 h and preferably 24 h. If the bentonite is not properly hydrated prior to the addition of the cement, the resulting bentonite–cement slurry may bleed excessively.

Before any hydrated bentonite is drawn off for use, the contents of the storage tank(s) should be homogenized by recirculation. This is necessary because some of the bentonite solids may settle before the slurry has developed sufficient gel to prevent sedimentation. Without recirculation, the first slurry drawn off could be excessively concentrated and the last almost entirely water.

From the hydration tank the slurry is pumped to a second mixer where cement (and, if appropriate, cement replacement materials such as slag or pulverized fuel ash) are added. When cement and clay slurry are mixed there is a mutual flocculation, which leads to a flash stiffening that normally lasts only a few moments provided the slurry is kept moving. The mixing plant must be capable of handling this flash stiffening without stalling or loss of circulation.

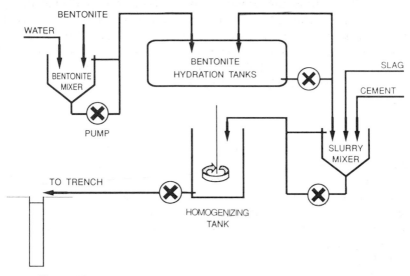

Figure 6.2   Typical plant layout for bentonite–cement slurry preparation.

From this mixer, the bentonite–cement slurry is pumped to a homogenizing/holding tank, which should have a capacity of the order of three batches from the bentonite–cement mixer. The slurry is then pumped to the trench. The pumping distance often will be in excess of 1 km. Centrifugal pumps are often sufficient but occasionally it may be necessary to use positive displacement pumps.

The type of mixing plant used can have significant effects on the fluid and set properties of a slurry. Thus, whenever possible, trial mixes should be prepared in the actual site mixers as it is difficult to simulate site mixers in the laboratory.

6.9   Requirements for the excavation slurry

The fundamental requirements for the excavation slurry are that:

(a) It exerts sufficient hydrostatic pressure to stabilize the excavation
(b) It controls the loss of slurry to the ground by penetration and filtration
(c) It should not bleed, i.e. the solids should not settle to leave free water at the surface
(d) If a self-hardening mix is used, the setting time should be compatible with the excavation procedure, and typically at least 8 h and the slurry should not entrain excessive amounts of spoil as a result of stiffening during the excavation period
(e) If a backfill mix is to be used, the slurry is sufficiently fluid to be fully displaced by the backfill

6.9.1 *Excavation stabilization*

For bentonite slurries, satisfactory stabilization of the excavation is normally achieved if the slurry level is about 1.5 m above the groundwater level although with some soils and slurries a difference of less than this may be sufficient. Occasionally, and particularly in fine non-cohesive soils, a greater difference in level may be necessary. When clay–cement excavation slurries are used, the density of the slurry will be greater than that of a simple clay or polymer slurry and this will increase the stabilizing pressure. However, this extra pressure may be offset by the increased hydrostatic pressure of the groundwater due to fluid lost by filtration from the slurry.

6.9.2 *Slurry loss*

Fluid may be lost by filtration, with the deposition of solids to form a filter cake on the soil surfaces, or by bulk penetration of slurry into voids in the soil. Penetration may be reduced by increasing the gel strength of the slurry. The penetration, L, of the slurry into the soil may be estimated from the formula

$$L = 0.15DP/t$$

where P is the hydrostatic pressure difference between the slurry and the groundwater, t is the gel strength and D is the effective diameter of the soil particles (which may be approximated by the D_{10} of the soil). Thus, for a slurry of gel strength 10 N/m^2 in a soil of effective particle diameter 10 mm, penetration might be of order 1.5 m/m of head difference between slurry and groundwater levels. If penetration is a problem it may be necessary to increase the clay concentration.

Filtration loss may be controlled by the addition of a fluid loss control agent. For pure bentonite or polymer slurries, fluid loss will seldom be a problem. With clay–cement slurries, the filter loss will be much higher than for pure bentonite slurries (very often by an order of magnitude or more). Occasionally this has led to problems of excessive reduction in slurry level in the trench prior to set. Despite this, fluid loss control agents have been used rather rarely in civil engineering (they are regularly used in oil well drilling muds). The reasons for this are probably partly economic and partly because most specifications prohibit the use of admixtures unless there is a proven track record of their use.

6.9.3 *Bleeding and settlement*

Bentonite–cement slurries may show some bleed, i.e. settlement of the solids to leave clear water at the surface. Bleed water represents a loss of useful cut-off volume. Severe bleeding suggests an improperly designed/prepared mix and, as a first step, the bentonite concentration, mixing procedure and, if necessary, source should be checked. Severe bleeding may lead to instability of the trench as the bleed water drains into the ground, thus producing a rapid drop in the fluid level in the trench and hence in the stabilizing pressure.

With backfill mixes containing aggregate there may be a tendency for the solids to settle prior to set of the slurry. This will leave a layer of thin slurry at the top of the trench. The formation of such a layer implies some instability in the mix but it may be difficult to prevent all settlement. The thin slurry layer should be removed during the capping operation. If excessive settlement occurs, this must be countered by modification of the mix design either by thickening the interstitial slurry so that the aggregate can be held in suspension, or by increasing the aggregate content to ensure grain-to-grain contact. Both of these measures may adversely affect the workability of the mix.

6.9.4 *Setting time*

The required setting time will depend on the method of excavation and the depth of the wall. In practice, set is rather difficult to define for these soft materials. The practical requirements are that excess slurry should drain freely from the grab during the excavation phase and that the integrity of the wall should not be compromised by the inevitable disturbance that occurs at day joints. Set time may be increased by the use of admixtures such as lignosulphonates. However, admixtures must be used with great care as most retarders increase bleed, and overdosing may kill all set.

6.9.5 *Displacement*

If used, a backfill mix must fully displace the excavation slurry. This requirement should not present any problem if non-setting bentonite or polymer excavation slurries are used. With clay–cement excavation slurries, stiffening will occur due to setting and thus backfilling should follow excavation as soon as practicable and certainly within the day of excavation.

6.10 Compatibility of materials

6.10.1 *Clays and water*

The predominant clay used in clay–cement slurries is bentonite. Indeed, unless bentonite (or more rarely attapulgite) is used it is very difficult to get satisfactory bleed control at a reasonable solids concentration.

Bentonite is a natural material, and supplies from different sources can show substantial differences in gel strength, viscosity, bleed control, etc. In general, it is necessary to carry out trial mixes with the actual bentonite and cement to be used in any application to ensure compatibility, low bleed and satisfactory permeability, etc. of the hardened material.

In the United Kingdom, it is normal practice to use a sodium exchanged calcium bentonite as the base for the excavation slurry. Such bentonite is quarried as a calcium bentonite and processed with sodium carbonate to produce a sodium bentonite. Generally it is not necessary to use a natural sodium bentonite such as Wyoming bentonite. Because of the sodium carbonate treatment, the pH of ion exchanged bentonite slurries tends to be higher than that of natural sodium bentonites, which may be near neutral. In the United Kingdom, most ion exchanged bentonites have a pH of about 10.5, although elsewhere pH values of 9.5 are not unusual.

In designing the slurry it is essential to check the behaviour of the bentonite and its compatibility with the proposed mix water and cement material. An outline procedure is as follows:

(a) Prepare a slurry of the bentonite in distilled water using concentrations in the range 3–7 g bentonite/100 g water. Ideally the laboratory mixing time and mixer shear rate should be matched to that of the site mixer. High shear mixers should not be used for laboratory trials unless they will be used on site. The mixed slurries should be poured into 1000 ml measuring cylinders and left undisturbed for 24 h. After this time the slurries should be inspected for bleed (separation of free water and settlement of solids). If there is any bleed, the slurry should be remixed and left quiescent for a further 24 h. If there is significant bleed at this stage, it is unlikely that the bentonite will be suitable for use in a slurry.

(b) If the bentonite has passed the test with distilled water, prepare a slurry of the bentonite using the mix water from the site. Again inspect the slurry for bleed at 24 h and, if necessary, at 48 h (after remixing). If there is bleed, it is likely that the site water contains dissolved ions at excessive concentrations (dissolved ions, particularly magnesium can inhibit the swelling of the clay). It is possible that unsuitable water can be treated or dispersants used in the slurry preparation. However, often it is cheaper to find an alternative supply of water.

(c) Prepare slurries of 24 h hydrated bentonite and the cementitious material. When the bentonite and cement are first mixed, there is a rapid and substantial thickening of the slurry. The slurry will thin after a few moments of further mixing provided that the mixer is of sufficient power to maintain circulation within the mix vessel. (If full-scale mixers are used for the trials, it may be prudent to add the cement slowly as the thickening and subsequent thinning will occur after only a very little cement has been added; if the mixer stalls with a full batch of cement it may cause considerable problems.) The mix should be checked for bleed over a period of 24 h again using a 1000 ml measuring cylinder. Mixes that show more than 20 ml of bleed water at any time may be unsuitable. Generally the amount of bleed can be reduced by increasing the solids content and particularly the bentonite content. However, increasing the solids content may produce a very viscous slurry and an undesirably strong set material. If a low bleed cannot be obtained at a reasonable solids content the bentonite is probably unsuitable.

Bleed has been selected as the indicator of mix compatibility in the above procedures as it is the simplest and often the most informative parameter for a fluid slurry. Filter loss, viscosity, gel strength tests, etc. also ought to be carried out at the design stage to check that the fluid is suitable for use on site. However, considerable experience is required to interpret the results of these tests. Bleed can give a more unequivocal indication of materials compatibility. It should be noted that the above is a general procedure. Step (a) should not be necessary unless there is particular concern about the quality of the bentonite. Step (b) should not be necessary if UK mains water is to be used to prepare the mix.

Clearly, all the raw materials should also meet appropriate local standards. Thus, the bentonite is often required to meet the Oil Companies Materials Association Specification [12]. However, this is not always a sufficient specification for cut-off slurries. Similarly the cement and slag (or pulverized fuel ash if used) should be to appropriate standards to ensure consistency of the product.

6.10.2 Cements and cement replacement materials

The nature of the cement can have a profound influence on the properties of the slurry. In particular it has been found that slag cements give much more impermeable materials. In the United Kingdom, slag cements are not generally available and so it is normal practice to use a blend of ordinary Portland cement and ground granulated blast-furnace slag. The optimum percentage of slag in the mix is usually in the range 60–80% but may vary with the source of the slag. Below 60% there is little improvement in permeability and above 80% the material may be excessively strong and brittle. The use of slag enables permeabilities of the order of 10^{-9} m/s to be achieved at total cementitious contents (cement plus slag) of the order of 90–150 kg/m^3 of slurry. Without slag, such permeabilities may not be possible even at cement contents of 250 kg/m^3.

Pulverized fuel ash (PFA), if used as a cement replacement, has rather little effect on the properties of bentonite–cement slurries except to improve sulphate resistance. At replacement at levels over about 30%, the resulting slurry may be rather weak. However, it may be added, not as a replacement, but in addition to the cement to increase the fines content. High solids content mixes will show better resistance to drying than their low solids counterparts.

However, long-term strengths may be rather high and the mixes thus show poor plasticity (see sections 6.13.2, 6.14.2). Proprietary hydration aids may be used to improve the performance of PFA.

A particular feature of PFA is that it will contain some unburnt carbon. The adsorption of organics from groundwater is well correlated with the organic carbon content of soils. Whilst coal may be a rather refractory form of carbon, it is to be expected that PFA will offer better organic absorption and retardation than pure clay, cement or slag.

6.10.3 *Mix water*

As already noted, in the United Kingdom, water of drinking quality is usually satisfactory for the preparation of slurries. If there is any concern about the quality of the water, the trial mixes should be prepared. Chemical analysis can provide some guidelines on the acceptability of a water. However, it is difficult to predict how individual clays and clay–cement systems will perform. Of the common ions found in water, problems may be encountered for bentonite slurries at over 50 mg/l of magnesium, or 250 mg/l calcium, or 500 mg/l of sodium or potassium. These figures should not be regarded as safe limits and if concentrations approach these values, hydration trials should be carried out.

Attapulgite has been used in situations where fresh water could not be used [13]. However, the design procedures for attapulgite–cement–salt water slurries are quite different from those for conventional bentonite–cement slurries and the salts may have unexpected effects on the cement hydration [14].

6.11 Sampling slurries

Samples for testing the fluid or hardened properties of cut-off slurries must be taken from the trench or the mixers during the cut-off wall construction. Coring the wall after the slurry has hardened may damage the wall, and the resulting samples are generally so cracked/remoulded as to be unrepresentative. A number of sampling systems are available and the only requirements for site operation are simplicity, reliability and the ability to perform at depth in a trench (the hydrostatic pressure can jam some valve mechanisms). The sampler should be of reasonable volume as significant quantities of slurry are needed for tests (the Marsh cone requires at least 1500 ml of slurry, while samples for testing hardened properties may require considerably more). The properties of the slurry may vary slightly with depth in the trench due to the settlement of solids. It is therefore standard practice to test samples from at least two depths and, for deep walls, from three depths (the top 1 m, the middle and the bottom 1 m).

6.11.1 *Sample containers, storage and handling*

It is most important that all containers are compatible with the slurry. For example, aluminium components must not be used with alkaline systems such as converted sodium bentonites or cement as they will react to liberate hydrogen gas. Plastic tubes are usually satisfactory. Sample tubes should be wiped with mould release oil prior to casting to ease de-moulding. When preparing samples for testing in the laboratory, it may be necessary to trim a significant amount of material from the ends of the samples. Thus, sample tubes should be at least twice as long as the required length of the test specimen.

When preparing clay–cement samples, care should be taken to avoid trapping air. After filling, the tubes should be capped and stored upright until set. Thereafter they should be stored under water in a curing tank until required for test. Bentonite–cement slurries are sensitive to drying and, even in a nominally sealed (waxed and capped) tube, some drying will occur. Storage under water is the only sure procedure.

Great care should be taken when transporting samples from the site to the test laboratory. They must not be dropped or subjected to other impact. Ideally the tubes should be packed in wet sand during transit, for protection against drying and mechanical damage. Samples should not be de-moulded until required for test.

Samples of cut-off wall slurries will be relatively fragile and should be treated as sensitive clays. Before samples are extruded, the ends of the tubes should be checked for burrs etc. to ensure that the samples can be extruded cleanly. Specimens must not be sub-sampled, for example, to produce three 38 mm diameter samples from a 100 mm sample as this will lead to unacceptable damage.

6.12 Testing slurries

Quality control on site should be to confirm that the slurry has been batched correctly and that its properties are as designed. Experience shows that control of batching and mixing is fundamental. Checking fluid properties such as density and viscosity will only identify rather gross errors. Thus, the monitoring of these properties is not sufficient as a quality control procedure.

6.12.1 *Density*

For slurries, the most usual instrument for density measurement is the mud balance. This is an instrument similar to a steelyard except that the scale pan is replaced by a cup. The instrument thus consists of a cup rigidly fixed to a scale arm with a sliding counterweight or rider. In use, the whole unit is mounted on a fulcrum and the rider adjusted until the instrument is balanced. Specific gravity can then be read from an engraved scale. The balance was developed for the oil industry and the range of the instrument is rather wider than is strictly necessary for civil engineering work (typically 0.72–2.88), whereas construction slurries will usually have specific gravities in the range 1.0 to perhaps 1.4. The smallest scale division is 0.01; with care the instrument can be read to 0.005, although the repeatability between readings is seldom better than 0.01. A resolution of ±0.005 corresponds to a resolution of order 15 kg/m^3 for bentonite or cement in a slurry. Thus, the instrument is not suitable for site control of mixes as greater precision is needed (for example it would detect only rather gross errors in batching a typical bentonite–cement slurry with 45 kg/m^3 bentonite, 90 kg/m^3 slag and 45 kg/m^3 cement).

6.12.2 *Rheological measurements*

The most common instrument used for measuring actual viscosities (rather than ranking slurries for example by flow time) is the Fann viscometer (sometimes referred to as a rheometer). This is a co-axial cylinders viscometer specially designed for testing slurries. Two versions of the instrument are generally available: (i) an electrically driven instrument; and (ii) a hand cranked instrument. All versions of the instrument can be operated at 600 and 300 rpm and have a handwheel so that the bob can be slowly rotated for gel strength measurements. Some also have additional speeds of 200, 100, 6 and 3 rpm. For all versions of the instrument there is a central bob connected to a torque measuring system and outer rotating sleeve. The gap between bob and sleeve is only 0.59 mm and thus it is necessary to screen all spoil-contaminated slurries before testing.

The instrument is not entirely satisfactory for testing cement-based systems. In particular, the gel strength readings tend to be very operator dependent and are generally of such poor repeatability as to make their measurement irrelevant.

The viscosity readings tend to be more repeatable but are very dependent on the age of the sample (due to the hydration of the cement). Thus, it is very difficult to demonstrate repeatability of behaviour for a cementitious system with the Fann viscometer. However, the instrument can be useful in identifying gross variations in mix proportions, etc. but for this a much cheaper test, such as the Marsh funnel flow time could be equally effective.

The Fann viscometer is an expensive instrument and must be used with care by a trained operator if reliable results are to be obtained. It is best suited to use in the laboratory for investigation of mix designs, etc. where the detailed rheological information that can be obtained from it may be of great value.

6.12.3 *The Marsh funnel*

For general compliance testing on site, the Marsh funnel is more convenient, although the results cannot be converted to actual rheological parameters such as viscosity or gel strength.

It should be noted that the Marsh funnel is just one of a wide variety of different flow cones in common use. It is therefore important to specify the type of cone (dimensions, volume of slurry used, volume discharged and time for flow of water) when reporting results. The Marsh funnel should be filled with 1500 ml of slurry and the discharge may be 1000 ml or 946 ml (1 US quart). When reporting results, the flow quantity as well as the flow time should be reported.

For clean water at 21°C (70°F) the times should be as follows:

- For 946 ml, 25.5–26.5 s
- For 1000 ml, 27.5–28.5 s

The Marsh funnel is suitable for testing most bentonite and bentonite–cement slurries. However, for some thick slurries, the flow time may be very extended or the flow may stop before the required volume has been discharged. It does not follow that these slurries are necessarily unsuitable for use in a trench, merely that a different cone must be used to test them.

6.12.4 *Other parameters*

Other parameters that may be specified for an excavation slurry include pH, filter loss and bleeding.

For UK or Mediterranean bentonites, prior to the addition of cement, the pH should be of the order of 9.0–10.5 and results outside this range should be investigated. The pH should be consistent between batches and again any variation should be investigated. The pH for slurries containing cement is almost always over 12.0 and gives little useful information about the slurry.

Filter loss is always high for cement-based systems and again gives rather little useful information unless the application requires the use of a fluid loss control agent.

Bleeding represents a loss of useful volume of the cut-off material and thus is important. Typically, bleeding will be measured with a 1000 ml measuring cylinder, and a limit of 2% loss of volume in 24 h is often specified. Some specifications require that the measurement of bleed is carried out in a volumetric flask with a narrow neck. The intention is to concentrate the bleed from a large volume of slurry into a narrow column and so improve the accuracy of reading. Unfortunately, this procedure does not work as the bleed water will not migrate up the neck of the flask but remains on the upper surface of the bulb. Thus, the procedure gives a false low reading.

6.13 Testing hardened properties

Tests on hardened clay–cement materials will normally be carried out in a soils laboratory or, more rarely, a concrete laboratory. However, the materials are rather different from soils or concrete. It is important that laboratory staff are familiar with them. Very often, apparently unsatisfactory results have been traced to unsuitable sample preparation/testing procedures that have been borrowed from other disciplines without sufficient consideration of the nature of the material.

For cut-off wall slurries tests that are often specified include: (i) unconfined compressive strength; (ii) confined drained stress–strain behaviour; and (iii) permeability.

6.13.1 *Unconfined compression tests*

For bentonite–cement slurries, the samples will normally be in the form of cylinders. The sample should be trimmed so that the ends are smooth, flat and parallel and then mounted in a test frame such as a triaxial load frame. The sample should not be enclosed in a membrane but should be tested as soon as possible after de-moulding to avoid drying. The loading rate is often not especially important. About 0.5% strain/min may be convenient as it gives a reasonably short test time and is not so fast that it is difficult to record the peak stress. Stress–strain plots from the test may be of interest but are seldom formally required as, under unconfined conditions, the strain at failure will generally be rather small. For example, the strain at failure is typically in the range 0.2–2%. It is only under confined drained conditions that cut-off materials show failure strains of the order of 5% or greater. The unconfined compressive strength test should be regarded as a quick and cheap test of the repeatability of the material, rather like the cube strength test for concrete. It does not show the stress–strain behaviour that the slurry will exhibit under in situ confined conditions.

6.13.2 *Confined drained triaxial testing*

To investigate the behaviour of the material under the confined conditions of a trench etc., it is necessary to carry out tests under confined drained triaxial conditions. The sample is set up with top and bottom drainage and allowed to equilibrate under the applied cell pressure (which should be matched to the in situ confining stress) for an appropriate time, which will be at least 12 h. It is then subjected to a steadily increasing strain as in a standard drained triaxial test. To ensure satisfactory drainage, a slow strain rate must be used; 1% strain per hour is often satisfactory for bentonite–cement slurries, although this is quite fast compared with the rates used for clays. The stress–strain behaviour will be sensitive to the effective confining stress. The following behaviour is typical:

(a) Effective confining stress < 0.5 × unconfined compressive strength. Brittle type failure at strain of 0.2–2% with a stress–strain curve very little different from that for unconfined compression save that the post-peak behaviour may be less brittle.

(b) Effective confining stress ≈ unconfined compressive strength. The initial part of the loading curve is comparable to (a). This is followed by plastic type behaviour, with the strain at failure increased significantly, typically to > 5%.

(c) Effective confining stress > unconfined compressive strength. Strain hardening observed in post-peak behaviour, and possibly some increase in peak strength. Initial stages of loading curve similar to (a).

Thus, when designing cut-off materials, it is important that the in situ confining stresses are considered and the mix proportions selected accordingly.

6.13.3 *Permeability tests*

For cut-off wall materials, permeability tests must be carried out in a triaxial cell. Samples for permeability testing should not be sealed into test cells with wax etc. This invariably leads to false, high results.

The procedure is therefore to use a standard triaxial cell fitted with top and bottom drainage. An effective confining pressure of at least 100 kPa should be used to ensure satisfactory sealing of the membrane to the sample. The sample may be back-pressured to promote saturation if there has been any drying. Generally a test gradient of 10–20 is satisfactory. The sample should be allowed to equilibrate for at least 12 h (or longer if possible) before flow measurements are made (there may be very little actual consolidation). Measurements should then be made over a period of at least 24 h and preferably 2 days or more. It should be noted that there is usually a significant reduction in permeability with time (see Figure 6.3). This appears to occur irrespective of the age of the sample at the start of the test and appears to be a function of the volume of water permeated through the sample. For critical tests it may be appropriate to maintain permeation for some weeks although this does make for an expensive test.

The permeant from many slurries is markedly alkaline and so aluminium or other alkali sensitive materials should not be used in the test equipment exposed to the slurry or permeant (see section 6.11.1).

6.14 Specifications for slurry trench cut-offs

Typically, a specification for a cut-off wall will include the following components:

(a) Specifications for materials as supplied
(b) Specification of fluid properties
(c) Specifications for mechanical performance of the wall
(d) Durability requirements
(e) Quality control on site during construction

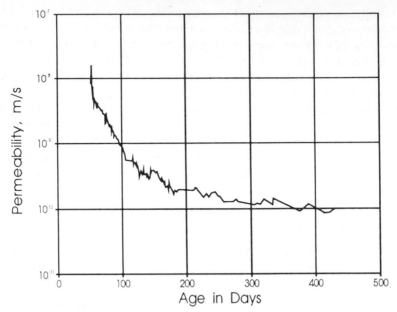

Figure 6.3 Loss of permeability of cut-off wall materials with time.

As the properties of most cut-off slurries develop rather slowly, it is important that the specification separates the on-site quality control testing from the laboratory tests required to confirm the hardened properties. For example, a 90 day strength or permeability is useless as a quality control measure for a job that may be finished before the first of the 90 day results is available. Very often, site control specifications will be concerned with fluid properties whereas hardened properties will be of concern for the mechanical performance and durability aspects of the specification.

6.14.1 *Mechanical performance: permeability*

In geotechnical engineering, for example for earth dams, a cut-off is often required to have a permeability of less than 10^{-8} m/s. As early cut-off walls were required for hydraulic control, most cut-off wall specifications required this level of permeability.

More recently the specifications for clay liners have come to be adopted for cut-off walls and a permeability of 10^{-9} m/s is usually specified; lower values are to be expected if the requirements of the EC Directive are to be applied (see section 6.2).

For a cut-off wall, a permeability of 10^{-8} m/s should give sufficient hydraulic control (i.e. allow minimal water flow) unless there is an unusually high gradient across it. Thus, a permeability $\leqslant 10^{-9}$ m/s is unnecessary for control, although it may give better durability for the wall material. It is significantly more difficult to form a cut-off of permeability 10^{-9} than 10^{-8} m/s, particularly if this permeability is to be achieved at 28 days (the permeability of bentonite–cements drops substantially with time under permeation). Indeed, with current mix designs, a 28 day permeability of 10^{-9} m/s cannot be reliably achieved without sacrificing strain at failure (see sections 6.13.2, 6.14.2).

Structural concrete, despite years of experience, is generally specified only by strength with perhaps minor constraints on cement content and water/cement ratio. Clay–cement materials are much less well understood and yet it is regarded as entirely reasonable to specify strength, permeability and strain at failure. Furthermore, the specified strength may vary between jobs by a factor of more than 30.

6.14.2 *Stress–strain behaviour*

A typical feature of many specifications is that the cut-off wall material should show a strain at failure of not less than 5%. This seems to be an arbitrary figure and it is rare that any justification is offered.

When re-moulded between the fingers, bentonite–cement materials behave as soft clays. However, the materials are cement-based and thus some brittle behaviour must be expected. Investigation has shown that large failure strains (plastic behaviour) cannot be obtained unless the material is subject to a confining stress of at least 50% of the unconfined compressive strength (see section 6.13.2). In situ, the cut-off material will be subjected to imposed stresses from the surrounding ground. Clearly, the large strain at failure is specified to ensure that if there are ground movements, the wall will deform plastically and will not fail as a cut-off. The stresses imposed on the wall are not easily identified and thus for many years it was the practice to specify a maximum strength for the material to ensure plastic behaviour at as low a confining stress as possible. Recently, engineers have refused to accept the concept of a maximum strength and it has become normal practice to specify a minimum strength. It is a popular misconception that strength is always an indicator of quality.

Thus, the effective confining stress is very important. If plastic behaviour is required at low confining stresses, for example near the surface of a cut-off wall, then a low strength material is required unless it can be shown that any ground movements will develop stresses in the wall sufficient to ensure plastic behaviour. The lack of plasticity at low confining stresses is an important reason for the use of a capping layer of clay on top of a cut-off wall (see section 6.15.1)

6.14.3 *Durability*

Bentonite–cement cut-off walls have now been used to contain pollutants for many years and it seems that there have been no reported failures. Laboratory tests show that the material is resistant to many chemicals provided that it is subjected to a confining stress (as will occur in situ due to the lateral pressure of the adjacent ground except in the near surface region). This provides a further reason for the use of a clay capping layer (see section 6.13.2).

A basic discussion of the effects of contained chemicals on cut-off walls materials is given by Jefferis [15]. However, it must be allowed that there is a lack of detailed information on the long-term performance of cut-off materials. Clearly, some interaction between contained chemicals and a cut-off material must be expected although it may not always be damaging and many reactions may lead to a reduction in wall permeability. When considering chemical attack, it must be expected that the general rules for durability of concrete will apply; i.e. the lower the permeability of the wall and the higher the reactive solids content, the better will be the durability. The addition of aggregate (particularly non-reactive aggregate) to form plastic concretes may reduce the durability of the wall under chemical attack. The reduced quantity of the bentonite–cement phase will be more rapidly removed to leave an open aggregate matrix. Without aggregate, the bentonite–cement material may compress continuously under the in situ stresses as attack occurs in order to maintain a homogeneous and relatively low permeability material. Such behaviour has been observed in the laboratory even when chemical attack has led to very large strains.

In addition to external chemical attack, the long-term internal stability of the material must be considered. Clearly, data on this are limited as the history of cut-off walls is still rather short. However, it can be said that there are many surviving examples of ancient construction materials that were mixes of lime, clay and pozzolanic materials. Thus, such mixes appear to have the potential to remain mutually compatible for generations without the formation of new and unstable mineral forms (although it must be allowed that there may have been many unstable mixes and that only the stable have survived).

An important feature of slurry trench cut-off walls is that they are easily renewable. Thus, the wall does not need to be entirely immune to pollutant attack. If the wall is designed to be a sacrificial element that degrades over many decades and is then replaced, the life cycle cost may prove to be lower than if a more complex barrier system is employed.

6.14.4 *The design compromise*

If a high strain at failure is required then a low strength material is necessary. However, a low permeability requires high clay and cement contents. Clay content is limited by the rheology of the slurry. High contents will lead to an unmanageably thick slurry (in contrast cement content has a rather more limited effect on the rheology unless very high concentrations are used). Durability also requires a high solids content and, perhaps particularly, a high clay content. Thus the design constraints are:

(a) Low clay content to give a fluid slurry
(b) Low cement content for high strain at failure
(c) High clay and cement content for low permeability
(d) High clay (and cement) content for durability

Thus, the design of cut-off materials must be a compromise. When a permeability of only 10^{-8} m/s was required it was relatively easy to achieve a 5% strain at failure at low confining stresses. However, the requirement for a permeability of 10^{-9} m/s makes this much more difficult (unless an age at test of the order of 1 year rather than the usual 28 days is specified). If both the early age permeability and strain at failure are to be fundamental then new materials must be developed or membrane liners always used. Thus, the client must be prepared to pay substantially more. Whilst this may be appropriate, it should be remembered that many existing walls designed to 10^{-8} m/s behave satisfactorily. In situ examination of these walls will provide much more useful design data than desk-based refinement of specifications. In particular, it would be most useful to know the actual permeability of these walls as it may be substantially lower than the design value.

6.15 Overtopping and capping

It is important to remember that a cut-off wall can be overtopped (there may be infiltration of rain, etc. even if the site is capped). It is therefore important that drainage is set to remove water from the contained region, which otherwise is effectively a pond. In principle the life of a cut-off wall may be substantially extended if there is always a tendency for inward flow of clean groundwater through the wall rather than a tendency for outward flow of contaminated water.

It seems that in future, the UK regulatory authorities will permit only a very limited build up of water within a landfill and thus in general there will be the tendency for inward flow. However, it should be remembered that a low water level will substantially reduce the microbiological activity within the landfill and hence the rate of breakdown of organic matter.

The water removed by the drains may have to be treated. This will increase the running cost of the site but, if the cut-off and cover is effective, the quantities will be small. Leachate removed by drainage will contribute to a slow clean up of the mobile and thus most hazardous components of the waste, which otherwise may remain as a permanent and unchanging risk.

6.15.1 Capping

A cap or some other form of cover is essential to prevent drying of most mineral cut-off materials (for a landfill this may be a part of the general cover layer or it may be a separate element). Bentonite–cements materials are particularly sensitive to drying and will crack severely if not covered. If the slurry is not protected, cracking may start almost immediately the slurry begins to stiffen. Cracks up to 1 m deep have been found in a wall left uncovered for a few weeks after construction. Figure 6.4 shows the detail of a sound (although seldom used) capping procedure for a bentonite–cement barrier:

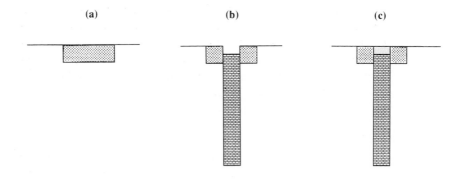

Figure 6.4   Capping procedure for a bentonite–cement barrier.
(▨) clay; (▨) cut-off material.

(a) Excavate a trench along the line of the wall to a width three times that of the cut-off wall. Backfill the trench in layers with compacted clay. The depth of the trench must be selected from consideration of the depth to the groundwater and the potential for drying during extreme weather conditions.

(b) After compaction of the clay install the cut-off in the normal manner.

(c) As soon as any section of the cut-off has set scrape off the surface layer of the cut-off material to a depth of 0.5 m and replace with compacted clay.

If a membrane is included in the barrier, a special capping procedure may need to be developed, in particular to link the membrane to any surface works.

6.16 Future developments

It is clear that the current level of activity with barrier systems will ensure that there is a continuing flow of new developments. Many of these may be fundamental improvements to existing systems. However, it is to be hoped that some will represent lateral approaches to the problem. Two examples of more novel approaches are outlined in the following sections.

6.16.1 *The bio-barrier*

In oil wells, bio-barriers have been developed to clog undesirable permeable regions. An outline of the procedure is as follows. A sample of the natural bacterial population in the formation is obtained, cultured and then starved. On starving, certain bacterial species may remain viable but reduce very substantially in size and develop an electrically neutral cell wall (which reduces attachability). These bacteria are termed ultra-micro bacteria. The formation to be blocked is permeated with a culture of the ultra-micro bacteria and a slow acting feed. The bacteria then slowly develop and expand, and so block the formation. At present the technique is limited in application to materials within a narrow band of permeabilities. It would be most elegant if a system could be developed that selectively clogged a leachate or gas migration path. More detailed discussions of microbiological effects in civil engineering are given in reference [16].

6.16.2 *Active barriers*

The fundamental aim of all barriers systems is to prevent all flow from the contaminated area, repository, etc. This is an ideal requirement and strictly unachievable in a passive system. However, it is possible that active systems may be developed where the flow is dynamically controlled. In many areas of science, if a no-flow condition is required then a guard ring procedure is used. In electrical measuring circuits guarding is well established. For heat experiments, adiabatic conditions may be achieved using a guard dynamically driven to match the sample temperature. Jefferis [17] has developed the guard ring concept for the elimination of edge leakage errors when measuring the permeability of concrete. Thus, the prevention of flow by use of an active guard driven to the sample potential (electrical, thermal, hydrostatic, etc.) is well established. Unfortunately, it is more difficult to develop a guard ring for a contaminated area site, for, as noted in section 6.4, many different potentials may influence flow through a barrier, and many different chemical species may be involved. Mitchell [5] has shown that the migration of cations may be retarded by applying a reverse potential for short periods during a permeation test. However, the procedure must of course accelerate the migration of the anions and it may have little benefit for organics. There may be some scope for the development of electrical sandwich layers designed to retard (or recover) anions and cations separately.

6.17 Conclusions

Vertical barriers are now widely accepted for the containment of gas leachate from contaminated sites, landfills, etc., although the actual number of sites that have been contained is still quite modest. However, the number that pose a threat to groundwater may be of the order of thousands. Thus the potential applications for cut-offs are enormous even if only a fraction of the problem sites are appropriate for such barriers.

To date there have been very few problems reported in the United Kingdom (all relating to installation rather than long-term performance). However, it is important to realize that applications are continually being extended, on occasion beyond the validity of the existing research base — an unsatisfactory situation, which has led to failures in many branches of civil engineering. A systematic investigation of the in situ behaviour of barriers is now urgently required.

There are now slurry trench waste containment walls that have been in the ground for 20 years (or perhaps more). It is important that some sections of these walls are exhumed to: (i) examine the mechanics of waste–wall interaction; (ii) establish whether reaction zones move uniformly through the wall or whether they localize or form fingers, etc.; (iii) investigate any reactions that may have occurred in the cut-off material; and (iv) quantify any change in permeability and pollutant migration mechanisms. Until a number of operational barriers have been exhumed, no guarantee of durability can be credible.

As a result of the wider knowledge and concern about toxic leachates, performance specifications for cut-off walls are becoming more rigorous and it would seem that new cut-off systems will have to be developed. However, it is illogical and potentially very wasteful to develop new systems before the in situ behaviour of the present systems has been properly investigated, particularly as no containment failures have been reported. No amount of laboratory testing or theoretical analysis can reproduce field conditions.

Specifications for barriers are a regular source of problems. Realistic common standards need to be established, addressing the compromises inherent in barrier design. Design procedures for gas barriers also need to be established. Most specifications for these barriers specify only a water permeability (generally 10^{-9} m/s). This is nonsense, as gas and water permeabilities (extrinsic or intrinsic) may be orders of magnitude different. Indeed the most elementary consideration of the microstructure of mineral sealing systems shows that there can be no unique relation between gas and water permeabilities.

7 Reclaiming potentially combustible sites

R.H. CLUCAS and T. CAIRNEY

7.1 Introduction

The reclamation and reuse of derelict and contaminated land for housing and industrial use is increasing, and amongst the sites being reclaimed are many that contain large volumes of potentially combustible materials.

The high incidence of subterranean fires in the United Kingdom in recent years [1] is such that the hazards posed by potentially combustible materials on sites have to be treated as serious. Whilst many of the reported sub-surface fires have been on undeveloped sites, some have occurred on redeveloped land [2, 3] or on waste tips [4]. Typically these sites where fires have occurred have near surface layers containing materials of high calorific value. These are often colliery spoils, tars, wood and paper wastes, and domestic refuse layers, and can occur in such large volumes that their excavation and off-site removal is not practicable. The concern that such material could ignite is compounded by the known costs and complexities of dealing with such combustion incidents. Obviously, preventing combustion is a more attractive solution in such cases and this calls for a clear understanding of the combustion processes and the factors that influence it.

7.2 Combustion processes

7.2.1 *General combustion*

The general processes of combustion are well established. No substance can burn until its temperature is high enough to permit a sufficiently rapid reaction with oxygen. Visible combustion is normally preceded by a gradual rise in temperature, which initiates a slow ignition process. This is a complex process, affected by a variety of factors. Combustion, of course, is the consequence of the oxidation of materials, with the resultant evolution of heat and light. Flame, which features in some definitions, need not be present and is indeed usually absent in subterranean cases. Reactants inevitably are consumed in the combustion process, and thus steady state conditions are not possible. Instead, the temperature rises as the oxidation reaction continues, then peaks and finally decays, as the reaction fades out, and cooling and heat dissipation take over (Figure 7.1). This pattern suggests that the two important stages are:

● The ignition temperature (that at which the temperature build-up process just commences)
● The run-away combustion temperature (that at which the reaction with oxygen becomes very rapid and high temperature rises occur)

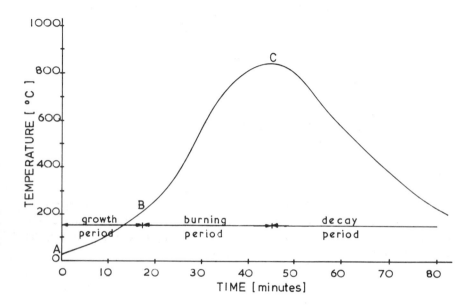

Figure 7.1 The temperature/time graph for a combustible material, e.g. coal. (A–B) growth period, temperature relatively low, heat transfer processes, production of gases and ignition; (B–C) burning period, flashover and fire develops, temperature rises sharply because heat is developing at a higher rate than can be dissipated; (C onwards) decay period as combustible material is burnt out and dissipation rate exceeds the rate of heat production through combustion.

Essentially three conditions have to coexist before any material will burn:

(a) Presence of a fuel source
(b) Availability of enough air or oxygen to initiate and support oxidation
(c) Presence of an ignition source, giving rise to whatever critical minimum temperature is appropriate for the particular fuel/air/inert fill concentrations

This can be summarized graphically in the triangle of fire (Figure 7.2), which makes the point that if any one of these three factors is absent, then runaway combustion cannot occur.

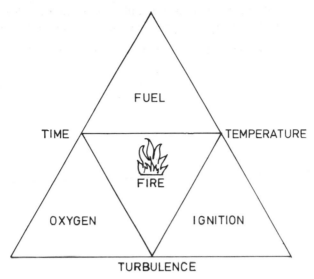

Figure 7.2 Triangle of fire.

This representation is of course simplistic, since other factors can be of importance. For example, the ignition temperature is to some extent dependent on the presence of catalysts or inhibitors, the dimensions of the sample, the pressure on the fuel source, particle size and friability of the fuel source, the surface area open to reaction and the moisture content.

To ensure that combustion occurs, three conditions described as the 3 Ts, have to be met:

● Time
● Temperature
● Turbulence

Time is important simply because a long slow process of gradual temperature increase generally has to occur to permit runaway combustion. Very rapid temperature increases, such as those occurring in laboratory furnaces, need not mirror the actual reaction processes that occur on site and may give misleading indications of what will occur.

Temperature is critical in that it increases the oxidation reaction rate. A 10°C (18°F) temperature increase, for example, will double the rate of a chemical reaction. For any particular mixture of a burnable material and air with inert soil and fill, there will be a well defined temperature at which ignition will occur and reaction rates increase exponentially with temperature rise [5], whilst heat losses only increase linearly as a material temperature rises.

Turbulence is important as the control both on air and oxygen availability to the reaction, and as the mechanism that can cool and stop a reaction by removing heat.

Thus, the generality of combustion processes is well understood, and the normally observed course of events can be predicted (Table 7.1). This shows that a rise in temperature is the necessary first stage. This increases the rate of oxidation and so the temperature rises, which in turn tends to make the process self-accelerating. Finally, the process dies out as the oxidation rate declines when the reactivity of the fuel material declines. If the heating and heat dissipation rates are equivalent, no temperature increase will occur. If the dissipation rate is low, the fuel material will eventually become hot enough to ignite and, in most cases, combustion will ultimately take place.

Table 7.1 General events leading to runway combustion for most organic materials

Temperature (°C)	Result
Above atmospheric	Water vapour is evolved and may be visible as a mist in cold humid weather
< 50	Slow absorption of oxygen
50–80	Oxidation accelerating
50–100[a]	Ignition and spontaneous combustion commences
About 80	A period of steady temperature probably associated with drying out of the material
100–120	Evolution of hydrogen and oxides of carbon begins
100–180	Interaction of oxygen and material accelerates rapidly
180–250	Thermal decomposition begins, evolution of tarry vapours, detectable by their odour and as bluish smoke; thermal runaway likely
250–350	The hottest zones within the material move against air flow, appearing as fires at the surface of the material

[a] At any temperature above 80°C, direct gaseous oxidation generally proceeds at a faster rate in wet materials than in dry materials.

An obvious practical consequence is that a material's liability to oxidize is less if it has been exposed on a site's surface for a long period and its more sensitive constituents have already suffered oxidation. Freshly excavated material will, however, pose a greater risk, as the more unstable constituents will still be present and largely unoxidized.

7.2.2 *Spontaneous combustion*

Whilst the generalities of combustion are well known, the interaction of the various factors that control the detailed progress of the combustion process are much less well understood.

There have been detailed studies of the combustion of coals (although not on coals admixed with inert soils and fills), and data are available on the effects of coal and char particles sizes, coal particle porosities and coal composition. Other workers [6] have studied the effects of self-heating on non-coaly materials, but this research did not extend to the types of combustible fills that occur on derelict and contaminated sites. Thus, only indirect guidance is available for the prevention of subterranean combustion on reclamation sites, although some basic truths seem well established. For example:

- Particle size is important, since this controls the type of reaction that can occur. Smaller particles allow the rapid diffusion of the evolved gases away from the heated fuel particles. Large particles, however, can permit a more complex diffusion process.
- Reaction rates are initially governed by the amount of heat transferred by radiation, convection and conduction from the fuel particle surface.
- Air availability and the temperature of the air as it comes into contact with the fuel particles are known to be the controlling factors.

More practically biased work on coal storage facilities has shown that the carbon monoxide concentration within a coal heap is a good indicator of the onset of ignition, and a level of 25 mg/l has become accepted as a meaningful indicator that ignition has commenced [3]. More refined work on the gaseous precursors to actual ignition (e.g. the presence of hydrogen) has, however, met with little success [7], and so no prior prediction of likely hazardous conditions is yet possible.

Thus, the situation is that no substantial theoretical basis exists that can be used to predict whether combustible materials intermixed with soils, demolition rubble and other inert matter will pose a risk to a reclaimed site.

7.3 Combustion tests

7.3.1 Introduction

Given that no adequate theoretical foundation exists to allow subterranean fire risk prediction, it is necessary to determine if empirical tests are adequate for this purpose. A variety of possible tests exists and each is considered in the following sections.

7.3.2 British Standard tests

There is a British Standard (BS476) [8] on the fire tests appropriate for buildings and structures. This, however, considers the properties only of elements of building and structures, and is not relevant for any assessment of the underlying site materials.

7.3.3 Direct combustion testing

The simplest of the possible tests consists of no more or less than abstracting particles from a site sample, and heating these with a Bunsen burner to some

650°C, for a time of 10–20 min. This gives very obvious data on a material's combustibility and is a useful initial screening test to reduce the number of samples that have to be tested more fully. The data obtained are largely subjective, e.g. 'smokes but does not ignite', 'high emission of volatiles and tars before any ignition occurred', 'material ceased to burn when Bunsen flame withdrawn', etc. The test is, however, not an adequate model of the actual conditions that will occur below a site's surface, since an abundant supply of free air is available and the ambient conditions permit the easy transportation of combustion gases and heat from the tested particles. Equally, the test cannot provide precise numerical data for analysis.

7.3.4 *Loss-on-ignition tests*

This approach has been seen as a useful indication of combustion potential and is commonly carried out on samples taken from contaminated sites. The test certainly provides data that can be used to estimate a material's calorific value [9]. Some authors have gone beyond this correct usage and have suggested that loss-on-ignition data give a cheap and simple indication of the organic content of a sample. This, however, is a false belief, since the measured weight loss includes the loss of free moisture, the loss of water of crystallization, and the weight losses caused by the breakdown of some inorganic materials (i.e. decarboxylation) [10]. Thus, even as a crude indicator of the amount of potentially combustible material present, the loss-on-ignition test cannot be recommended, and its regular use in the analysis of contaminated land samples is unjustified.

7.3.5 *Calorific value testing*

The calorific value (CV) is the heat given off when unit weight of a substance is completely burned. The general consensus is that the most accurate method of measuring calorific value is to burn a 1 g sample in a bomb calorimeter under an oxygen pressure of 3 MPa. With the bomb immersed in a water bath, the total heat evolved during the sample's burning can be determined precisely from the temperature rise in the surrounding water [11]. A measurement accuracy for fuels low in hydrogen (i.e. most British coals) of less than $\pm 0.1\%$ is claimed [12]. The test procedure is not difficult, testing facilities are widely available, and it has become accepted that calorific value data are useful and accurate.

In fact, this is not as certain as is assumed. The measured energy output will include a proportion due to the latent heat of moisture held in the sample, and an error of about 1.2 MJ/kg [13] is likely to arise from this. It also has to be noted that the sample size (1 g) is minute compared to the volumes of potentially combustible materials that occur on most derelict and contaminated sites, and this suggests that a very large number of CV determinations should be carried out to take due account of sample variability. In fact this is seldom done and a spurious accuracy is ascribed to perhaps one CV determination for a site area of 400–600 m^2. Apart from any doubts on the accuracy and repeatability of CV test results, it also has to be noted that the method was devised to measure the energy content of pure coals and not the combustion susceptibility of combustible materials intermixed with inert fills.

There is good evidence that CVs relate well to the rank of a particular coal where rank reflects the carbon, volatile and hydrogen content and the caking properties of different coals [14]. As carbon content increases, so CVs rise and, with carbon contents of 92% or less, calorific values and carbon contents relate very well. A typical British coal can be expected to have a CV of some 32 MJ/kg. When correlation with volatile matter is considered, it becomes obvious that an inverse relationship exists. The higher the hydrogen and volatile content, the lower is a coal's rank and the lower is its CV. This single point emphasizes the most obvious problem in using CV data to predict combustion risks. Low ranking coals are much more volatile rich, generally more reactive, more susceptible to self-heating, and so pose a much greater combustion risk. Yet these are the materials that have the lowest CVs!

Thus, the use of a particular calorific value (e.g. the level of 7000 kJ/kg advocated by the former Greater London Council) as an indicator of combustion risk is intrinsically flawed. Despite this, continued references to the risks posed by particular calorific value levels are made and Smith [15] has recently suggested that any values above 2 MJ/kg could pose a hazard. This would, of course, make most fertile garden soils potentially dangerous, and is, at best, misleading advice.

Apart from the bomb calorimeter tests, use has been made of simpler procedures to estimate CVs. The proximate analysis [9] and Ball's [10] loss-on-ignition tests for estimating organic matter, intermixed with non-calcareous soils, are examples of these approaches. Ball's method has the advantages of being designed for a mixture of materials and of not requiring hand picking of very small sub-samples from the site's samples. Additionally, both methods determine the volatile contents specifically and so have a better foundation for predicting the combustion risks.

A comparison of the CVs given by the bomb calorimeter, proximate analysis and Ball's method is given in Table 7.2. This makes obvious the discrepancies between the bomb calorimeter results and those from the other two methods. The table also indicates, on the accepted wisdom, that samples 3 and 4 could be expected to pose combustion risks. In fact, the use of the direct combustion test (section 7.3.3) proved that neither of these samples could be persuaded to burn, even in the ideal conditions of freely available air and imposed temperatures in excess of 600°C.

Calorific value testing was designed to give the energy output for clean coals. Its use in this context is not in doubt, although the measurement accuracy and repeatability may be less than is often assumed. When used for subterranean fire risk evaluation, particularly where potentially combustible material is intermixed with inert materials, misleading results are inevitable, since the method is being misapplied.

Table 7.2 Calorific values determined by various methods (colliery spoil, former coal stock yard, N.E. England)

Method	Calorific value (kJ/kg)			
	Sample 1	Sample 2	Sample 3	Sample 4
Bomb calorimeter	1000	1000	22 000	18 000
Proximate analysis	4500	3000	7000	3750
Ball's method	3500	2750	6000	3750

7.3.6 *Fire Research Station test method*

The spontaneous ignition test developed by the Fire Research Station (FRS) [16–18] suffers from similar limitations to those of the calorific value test. Grain size and surface area are important factors in combustion, and therefore both methods are affected by the fact that the sample for testing needs to be ground to a fine particle size. The information that both tests provide is an indication of the total available energy rather than of susceptibility to combustion. The sample selected for grinding may also be non-representative of the original material, because larger fragments may have been removed.

In the FRS test, the sample material is ground to its most reactive particle size (less than 2 mm), placed in a wire mesh basket with sides ranging from 20 mm to 200 mm and suspended in a standard convected laboratory oven operating up to 250°C. The furnace is then set at a known temperature and a thermocouple placed in the centre of the sample to monitor the temperature.

The test is continued until the sample temperature has reached its peak level and then fallen back to the set point temperature of the oven (Figure 7.3). Ignition is assessed from a series of tests performed at different oven temperatures and, from this, the material's behaviour under normal conditions is predicted. The FRS method fails to mirror site reality in that:

- The material is ground to a fine particle size to maximize the reactive surface area available
- The test material is loosely packed in the basket and no assessment is attempted of the benefits that compaction could give
- Air is constantly circulating around the sample, in a way that does not mirror actual site conditions.

Thus the FRS method gives a useful indication of the ultimate combustibility in terms of the available energy of the material, but fails to provide any

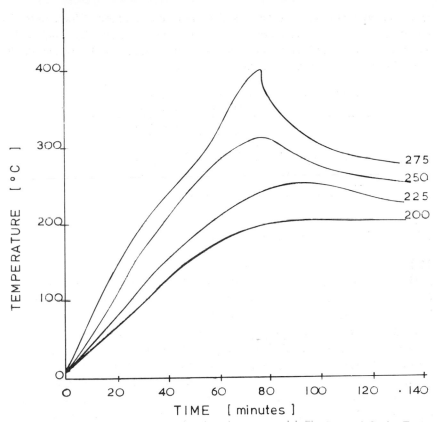

Figure 7.3 Typical result: soil admixed with coal waste material (Fire Research Station Test).

measure of the material's combustion susceptibility in the conditions that will actually occur on a reclaimed site.

7.3.7 *Combustion potential test*

If a test is to be effective then it should mirror as closely as possible the conditions under which it will occur on site. The advantage of the combustion potential test described here is that little or no sample preparation is necessary. Additionally, the method of testing attempts to assess the two critical factors that affect the potential for spontaneous combustion:

(a) The gradual build up of heat, which is retained and not dissipated
(b) The minimum supply of oxygen needed to allow combustion

The test procedure adopted identifies these two properties by:

● Having the test sample in a central tube with a controllable air flow being passed through it (Figure 7.4)

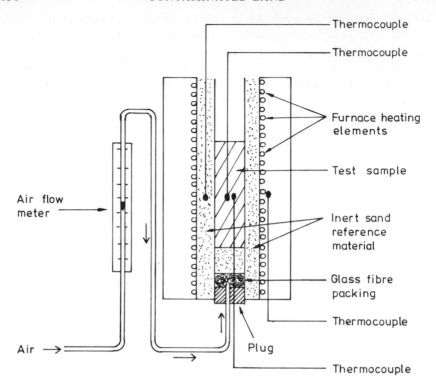

Figure 7.4   A schematic representation of the apparatus for the combustion potential test.

- Surrounding the sample with a cylinder of an inert reference material
- Gradually heating the two concentric cylinders with a wire wound electric furnace up to a temperature of 600°C.

The essential parameters for combustion to occur have been established and are as follows:

(a) There must be sufficient air or oxygen
(b) The air or oxygen stream should be at the same temperature as the material before entering the bed
(c) The heat evolved must be detected immediately by a rise in temperature and not dissipated
(d) The temperature of the material should be uniform when the ignition point is reached
(e) The two curves for the temperature–time of the material bed and the reference material should cross sharply to permit a precise determination of the ignition temperature

From these and previous studies of coals and cokes it has been found that the sharpness of the cross-over point, and hence the precision of the ignition point, could be controlled by regulation of the heating rate. The same result can be obtained with samples composed of soil containing waste and contaminated materials.

In essence, the test employs the time/temperature relationship that is a feature of BS476 [8]. The apparatus and the procedure were adapted from earlier work by Sebastian and Mayers [19] on coke reactivity. In the arrangement used, the turbulence is kept constant and the temperature and time can be varied, although in each determination this is fixed by controlling the rate of heating of the furnace. The point of runaway combustion is identified when the temperature rise of the material is compared with that of the inert reference material (Figure 7.4). In the test, the temperature when the reaction proceeds at a higher rate than the heating rate of the arrangement is taken as the ignition point. This is the point at which the self-heating effect overtakes the external heating rate being supplied, and does not necessarily correspond with the sample starting to glow or with the first appearance of smoke or flame. The resultant temperature–time graph can be examined to establish if ignition and runaway combustion have taken place.

If the test sample is inert, the resultant graphs of the thermocouple outputs from the sample and the reference material will remain parallel (Figure 7.5). This separation of the two curves represents the thermal lag between the two materials and corresponds to the time taken for the heat to pass through the reference material to the sample.

Should the test sample be of a combustible material, the two curves will remain parallel initially but deviate when ignition commences, when the test sample

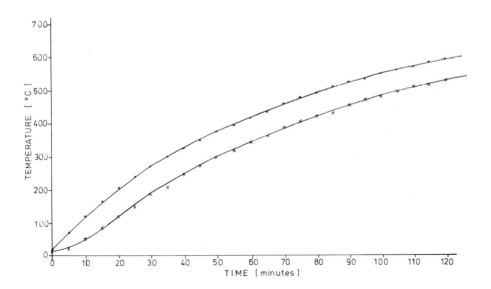

Figure 7.5 Test result for an inert sample. Air flow 2.5 l/min. (●) reference material; (×) shale plus coaling matter.

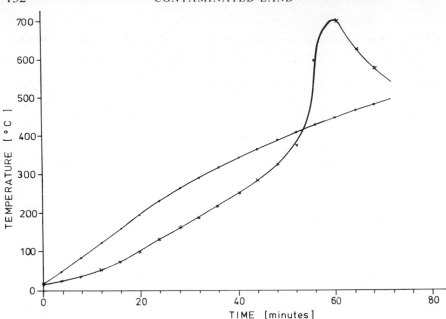

Figure 7.6 Test result for a reactive sample. Air flow 2.5 l/min. (●) reference material; (×) soil with coal admix.

curve's gradient will steepen. The curve will continue to rise and may eventually crossover the reference sample curve if combustion occurs (Figure 7.6).

By using a modified arrangement of the thermocouples it is possible to ascertain the point of ignition and, by making allowance for the thermal lag, it is also possible to estimate the cross-over point, which corresponds to the point of thermal runaway (critical temperature for combustion). Thus, the method affords a means of identifying the point of onset of oxidation (ignition) and the point at which thermal runaway (combustion) occurs (Tables 7.3, 7.4, 7.5 and 7.6). The test procedure generally takes less than 2 h to complete and the analysis is simple to carry out. The value of the test is that combustion susceptibility is directly measured. The lower the temperature at which the test sample's temperature gradient increases, the more susceptible the material.

Table 7.3 Critical air flow to cause runaway combustion (colliery site, N.E. England)

Air flow (l/min)	Ignition (°C) (smouldering)	Combustion (°C) (runaway)
0	165	–
1.5	160	–
2.5	145	–
5.0	120	–
10.0	120	–
15.0	120	200
20.0	120	200

Table 7.4 Ignition and critical air flow to cause runaway combustion for materials with different calorific values (colliery spoil, former coal stock yard, N.E. England)

Air flow (l/min)	Sample 1 Ignt.	Comb.	Sample 2 Ignt.	Comb.	Sample 3 Ignt.	Comb.	Sample 4 Ignt.	Comb.
0	145	–	140	–	160	–	165	–
1.5	115	210	130	215	130	215	160	–
2.5	115	240	125	210	105	230	120	235
5.0	120	250	115	240	125	215	120	220
10.0	115	240	105	225	120	215	135	215
15.0							120	200
20.0							130	200

[a] Ignt., ignition point; Comb., thermal runaway.

Table 7.5 Ignition and combustion test (colliery spoil heap, N.E. England)

Sample type	Calorific value (kJ/kg) average	Air flow 0.5 l/min Ignt.	Comb.	Air flow 2.5 l/min Ignt.	Comb.	Risk factor
BH1 shale, clay, coal	4800 6800 [b]	205	220	115	205	9.7
BH2 shale, sand, coal	4900 8400 [b]	225	–	105	220	6.3
BH3 shale, clay, coal	4100 4700 [b]	220	250	120	235	9.9
BH6 shale, gravel, coal	5700 9500 [b]	190	–	115	180	7.2
BH7 shale, mudstone, coal	4200 6400 [b]	205	245	95	200	7.2
BH8 shale, clay, coal	4400 7500 [b]	215	–	90	185	0.3

[a] Ignt., ignition point; Comb., thermal runaway.
[b] Highest value.

Table 7.6 Ignition and critical air flow (former railway sidings)

Sample	Temperature (°C) [a]					
	Air flow 0 l/min		Air flow 0.5 l/min		Air flow 2.5 l/min	
	Ignt.	Comb.	Ignt.	Comb.	Ignt.	Comb.
TP4 [b]	–	–	135	–	130	230
No. 21	–	–	125	–	110	200
No. 27	130	–	125	290	100	160

[a] Ignt., ignition point; Comb., thermal runaway.
[b] CV = 10 000 kJ/kg.

A second and very practical aspect of the test is that the air flow through the sample can be altered. If the test sample is run at different air flows it is possible to establish the critical air flow at which runaway combustion commences (Table 7.3). This can be compared with the worst possible situation anticipated on site and an assessment of the risk made (see section 7.3.8 and Table 7.5).

7.3.8 *Air permeability test*

Geotechnical research into soil compaction has revealed that at optimum compaction, the air permeability of a soil material is extremely low. The Proctor test [20] can be used to determine a material's dry density value at its optimum compaction level. This density value can then be reproduced in the combustion material and the achievable air flow through the sample at that density can be measured.

This essential supplement to the combustion potential test requires a sample of the material compacted to the optimum density in a cylindrical mould. The bottom is sealed and connected to a variable compressed air supply via a flow meter, with a mercury pressure manometer (or suitable pressure measuring device) in parallel. The arrangement is shown in Figure 7.7 and is similar to that used by Nagata [21] to determine the air permeability of undisturbed soils. The air flow is set to a suitable value and the pressure measured. This procedure is repeated for a range of values. The top of the mould is taken to be at atmospheric pressure and the pressure gradient can therefore be determined. From the measurements, the corrected air flow, and hence the velocity of the air flow through the compacted material, are calculated. A plot of inlet head pressure against velocity of air flow is constructed, from which the air flow velocity at any inlet head pressure can be established (Figure 7.8). The maximum likely velocity of air flow through the material is estimated and compared with the critical air flow velocity for combustion determined from the combustion studies. The results can then be evaluated in two discrete but similar ways. One is based on CP3 (Chapter V, Part 2, 1972, Wind Loads) [22]. This predicts a maximum value for pressure of 52 mmHg on a surface

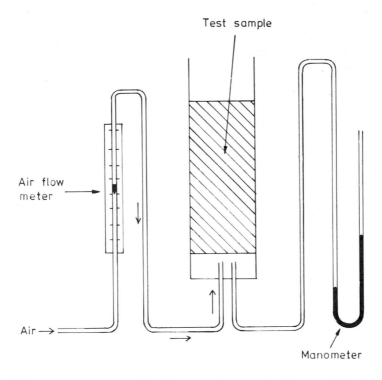

Test sample

Air flow
meter

Air →

Manometer

Figure 7.7 A schematic representation of apparatus for measuring air permeability of samples.

in extreme storms. A similar result is obtained by considering the maximum difference in barometric pressure between high and low pressure atmospheric zones likely to be encountered. Such pressure differences in the atmosphere would be associated with extreme wind conditions and would likely result in the de-stabilization of any embankment itself. From these analyses, a worst possible case can be accepted for the risk assessment analysis of an air pressure of 60 mmHg. The air flow that this extreme wind loading could produce can be measured, and the minimum necessary air flow needed to permit runaway combustion is known from the combustion potential test. By dividing the minimum necessary air flow to permit runaway combustion flow by that which could occur in extreme storms, a risk factor can be determined. The higher the risk factors thus obtained (Table 7.5), the lower is the risk of combustion. Typically, risk factor values in the range of 5–10 are found in stable compacts. In practice, no case has yet been found in which it has been possible to pass the critical air flow to allow combustion to take place through a fully compacted material.

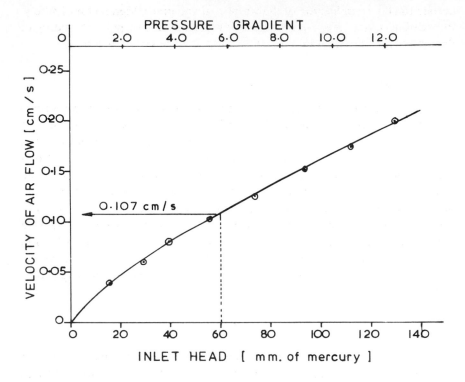

Figure 7.8 Test results for air permeability test (unreactive soil/coal admix). Critical velocity of air flow (from combustion potential test) = 1.04 cm/s. Maximum likely velocity of air flow (from air permeability test) = 0.107 cm/s. Risk factor = 1.04/0.107 = 9.7.

The advantage of the test method is that it tests the material largely as it would exist on site and not after it has been artificially modified by sieving, grinding or other amelioration. The data obtained are thus of practical value in assessing the risk of combustion.

7.4 Use of the combustion potential test

The test has been usefully applied in a number of cases.

7.4.1 *Case study 1*

In the case of a former coal stock yard in the North East of England still underlain by 1.2 m of waste coals, the site was investigated in order to assess its suitability for housing development. The tests showed that the site compaction necessary for the house foundations would prevent the critical air flow needed to allow combustion of the waste coal and give a high factor of safety. Provided adequate steps were taken to avoid accidental or indirect heating of the material, no removal of the waste coal proved to be necessary.

7.4.2 *Case study 2*

In the case of a former coking works stock yard, the site investigation revealed that the area was underlain by 2–4 m of coals, ash and shale, and that the entire site had a calorific value in excess of 10 000 kJ/kg. Combustion tests revealed that even the lowest CV materials would ignite at low air flows and combust at 200–260°C. Thus the site, even if compacted to the optimum possible value, would present a significant hazard, since only very small risk factors were achieved (section 7.3.8). The owners agreed to improve the site material's compaction by a process of double washing to remove larger coal fragments. The washed discard, when returned to the site, still had sizeable calorific values, but could be compacted to higher densities and would not burn even when exposed to the air pressures anticipated in extreme storms.

7.4.3 *Case study 3*

In the case of a tip area around a former colliery at Sunderland, despite the fact that subterranean fires have been found on-site, the initial testing (section 7.3.3) showed that none of the first set of samples taken could be made to undergo runaway combustion, despite being heated to 600°C with air flows up to 2.5 l/min. Further testing, however, showed that some later samples could be ignited at relatively low temperatures, and that, at high air flows, this could result in thermal runaway. This also revealed that the site was much less uniform than had been believed and had on it different coal materials from different mines and seams. A process of coal washing and rigorous site compaction was followed to produce conditions on the site that should prevent any subterranean heating.

7.4.4 *Case study 4*

In the case of a former colliery spoil heap in the North East of England, the spoil heap consisted of approximately $2 \times 10^6 \, m^3$ of colliery spoil, as a conical mound rising some 30 m above the surrounding valley site. The material was variable in density and grading, and consisted of silty sand to gravel sized fragments of shale, sandstone and coal, with soft to firm clayey material and occasional pockets of ash. The site was to be restructured and landscaped for recreational use, and might be developed for holiday flats and leisure activities. Even after restructuring the site would be relatively steep and exposed, and so particularly subject to wind loading conditions, due to the surrounding topography. Testing showed that some of the material could be ignited at low temperature and that at high air flows runaway combustion could occur (Table 7.5). Tests undertaken at site compaction levels, however, revealed that, with the exception of one sample, air flows that would lead to thermal runaway could not be established.

The exceptional sample proved to be an atypically uniform and coarse material that not only was difficult to compact but also was visibly identifiable. Tests were carried out to determine how the material compaction properties could be improved (by the addition of finer inert materials) and how the site owners could ensure that any bands of this atypical material could be identified when the site was excavated. This work showed that even in such a particularly exposed site, subterranean fire risks could be minimized and brought within acceptable limits.

7.4.5 *Case study 5*

The former railway sidings is a level area overlain with 1–4 m of compacted coal ash and boiler waste from the bygone era of steam locomotives. With the exception of one sample, the material showed a reluctance to burn, and thermal runaway could only be achieved weakly at high air flows (Table 7.6). The tests at difference levels of compaction were able to show that site compaction would prevent the critical air flows necessary to allow combustion of the material.

7.4.6 *Case study 6*

The earlier studies refer to colliery discard materials, and enough experience exists to show that such sites can be reclaimed safely. However, when materials are contaminated by fuel oils or reactive chemical wastes, more dangerous conditions can occur. In one case recently investigated, chemically reactive wastes proved to be present. In this case, the risk factors determined that the site material presented a serious hazard and it was agreed that the material would need to be removed. Thus, it is apparent that combustion risks can exist on a wider range of derelict sites than is sometimes believed, and that the decision to develop a site needs to be carefully considered.

7.5 Conclusions

The risk of subterranean heating is a real problem, which can be particularly serious to the infrastructure and fabric of a development. Calorific value determinations of the materials from such a development site have not proved adequate in assessing the potential of materials to undergo combustion, nor proved sensitive enough to identify the conditions under which spontaneous combustion might occur. The use of loss-on-ignition data can be misleading, as can tests that depend on modifying the form of the material under examination.

The combustion potential test described in section 7.3.7 does allow identification of more sensitive materials and assists in establishing which, if any, material needs to be treated or removed. A particularly useful aspect of the test is that it enables estimation of the critical air flow to allow combustion to proceed. This, in turn, provides the necessary information to ensure that the site material can be compacted to prevent this critical value being achievable, and minimizes the risk of spontaneous combustion on the site. Subterranean heating is one of the risks that can occur in the redevelopment of a disused site.

Developers and their advisors need to be able to understand the subterranean heating problems and the solutions offered so that they can judge the necessary level of work, costs and risks involved in order to give a guarantee of security for the site. What the combustion potential test cannot do is give safeguards either against poor workmanship and practice during site compaction, or against later site deterioration that permits large scale admission of air to occur. Only good quality assurance and management can rectify such potential deficiencies (chapter 10).

8 Landfill Gases

M.V. SMITH

8.1 Introduction

Heightened awareness of the hazards from landfill gas production and its potential for migration can be related to development pressures on or near former urban landfill sites. In addition, the pressures to recycle urban and inner city land rather than to utilize 'green-field' sites have led to an awareness of a wider range of situations where methane and other gases may occur. In the light of appropriate desk study and site investigation data, this in turn often leads to remedial measures being required for naturally emitting gases (such as mine gas, marsh gas or radon) and for migrating gases associated with waste disposal sites. In this chapter, emphasis is placed on methane and carbon dioxide gas measurement and assessment, since these are the gases most commonly encountered in site redevelopment. Reference to volatile organic compounds and vapours is made in chapter 3.

8.2 Principal gases and their properties

The principal types and components of gaseous contaminants are given in Table 8.1. Under the heading of chemically contaminated land, features that could give rise to different types of gas include mixed sulphate and organic waste giving off hydrogen sulphide, or hydrocarbon spillages giving off, amongst others, propane, butane or petrol vapour. In addition, where colliery spoil has been used for land-raising, under the right conditions methane, carbon dioxide or carbon monoxide could be given off.

The following brief review looks at various gases in isolation. In practice, however, gases will be present as mixtures, which, in conjunction with temperature, pressure and ignition source effects, will affect flammability limits. Once mixed, gases are unlikely to separate due to density differences.

Methane is a colourless, odourless gas, which can be found where organic decay occurs in anaerobic conditions, such as landfills, peat deposits or associated with coal measures and other carbonaceous material. It is soluble in water and lighter than air (specific gravity 0.55). When mixed with air, methane forms a mixture that can explode, given a source of ignition and suitable conditions, at concentrations from 5% to about 15% methane by volume of air (the lower and upper explosive limits, LEL and UEL, respectively). Methane is not toxic to plants but can cause depletion of oxygen in the root zone, with less than approximately 12% oxygen typically leading to tree death. Methane may be oxidized to carbon dioxide by soil bacteria.

Table 8.1 Principal types and components of gaseous contaminants

Feature	Possible type/source of gas	Principal components
Refuse disposal sites (domestic and non-domestic)	Landfill gas	Methane, carbon dioxide, hydrogen
Peat and organic clay deposits	Marsh gas	Methane, carbon dioxide
Sewer chambers	Sewer gas	Methane, carbon dioxide, hydrogen sulphide
Coal mines and seams	Mine gas	Methane, carbon dioxide, ethane
Gas distribution pipes	Mains gas	Methane, ethane
Coking works	Coal gas	Methane, hydrogen, carbon monoxide
Geological deposits containing radioactivity	Radon gas	Radon
Chemically contaminated land	Various	Hydrogen sulphide, hydrogen cyanide, volatile organic compounds

Carbon dioxide is a non-combustible, colourless, odourless gas, which is very soluble in water and approximately 1.5 times heavier than air (specific gravity 1.53). It has a background level of 0.03% in the atmosphere and there is the risk of asphyxiation wherever the gas is allowed to accumulate in confined spaces. Concentrations of carbon dioxide higher than that of normal air can occur as a result of aerobic and anaerobic decomposition of organic materials, and from respiration. Carbonate-based geological strata may also give off carbon dioxide. Whilst values of more than 3% carbon dioxide have been reported from natural microbiological action in the soil, with higher values in summer than in winter, carbon dioxide can be toxic to plants at levels of more than 5–10% by vol. in the root zone, although different species will vary in their tolerances to the gas. Concentrations of 0.5–1.5% carbon dioxide by vol. in air have been adopted by the Health and Safety Executive (EH40/91) as the maximum acceptable for occupational long-term (8 h) and short-term (10 min) exposure, respectively [1].

The presence of methane, carbon dioxide and other gases can result in oxygen deficiency, which can be a critical condition in its own right. Oxygen is a colourless, odourless gas, which is slightly soluble in water and heavier than air (specific gravity 1.11). Apart from dilution with other gases, oxygen deficient atmospheres can be created by oxidation of wood and other carbonaceous materials and by respiration. Values of less than 18% by vol. are considered significant by the Health and Safety Executive.

Carbon monoxide is a colourless, odourless gas, which is slightly soluble in water with a specific gravity of 0.97. It is produced by incomplete combustion of organic materials. Underground combustion is the most likely source in the context of contaminated land, although it is also associated with carburetted water gas from coking works or gas works sites. Although explosive in the range 12–75% by vol., carbon monoxide is highly toxic by inhalation, with occupational exposure limits of 50 ppm and 300 ppm for 8 h and 10 min, respectively.

Hydrogen is a non-toxic colourless, odourless gas, which has a specific gravity of 0.07 and is explosive with air in the range 4–74% by vol. It can be found during the anaerobic non-methanogenic stage of landfill gas generation, as a constituent of carburetted water gas or as a result of combustion if the temperature is high enough.

Hydrogen sulphide is a colourless, highly toxic gas with a distinctive odour of 'rotten eggs' at low concentrations. It is soluble in water, has a specific gravity of 1.19 and is highly flammable (explosive in the range 4.4–45% by vol. in air). The occupational exposure limit is 10 ppm and 15 ppm for 8 h and 10 min, respectively.

Hydrogen cyanide is a colourless, highly toxic gas with a faint odour of bitter almonds. It is soluble in water, highly flammable (lower explosive limit 6% by vol. in air) and occurs as a white liquid at temperatures below 26.5°C. The maximum exposure limit is 10 ppm (10 min).

8.3 Source identification

In order to develop appropriate remedial measures, the source of the gaseous contaminant needs to be identified. By reference to desk study information (e.g. British Gas pipe distribution records, British Coal mining records and geological/hydrogeological information), in many cases the source of the gas will be known with certainty but, particularly in the case of methane and carbon dioxide, there may be several potential sources as indicated by Table 8.1. The most common sources of methane encountered are derived either from geological sources (mine gas or natural mains gas) or from recent biogenic sources (landfill gas and marsh gas). Carbon dioxide concentrations are generally higher from biogenic sources.

The problem is that portable gas monitoring equipment for methane and carbon dioxide cannot distinguish between different sources of gas and, apart from the need to verify readings from portable instruments, samples for laboratory gas chromatography analysis should also be taken during the monitoring period.

The presence of ethane could indicate mine gas, mains gas or a hydrocarbon leak from a pipeline or tank. The presence of helium, possibly in the range 20–200 volumes per million depending on the particular coalfield, would indicate mine gas, the helium being trapped in gas-bearing deposits after migrating from the earth's crust.

If more than one source of gas is present, the picture is further complicated. Take, for example, the situation of colliery spoil placed over disused mine shafts in an area adjacent to a domestic refuse landfill. In such circumstances, selected ratios of gas concentrations can be useful for comparison. These could include the methane : carbon dioxide, methane : ethane or other ratios.

During the anaerobic methanogenic steady state of gas generation, landfill gas typically has a methane : carbon dioxide ratio of 65:35, although mixing with gases from other sources can modify this. From data contained within *Waste Management Paper No. 26* it can be seen that mains gas typically has a methane : ethane ratio of between 24 and 30, whilst the methane : ethane ratio of mine gas could be between 10 and 12 [2].

However, it should be noted that variations in mains gas and, in particular, variations between different coalfields or even different seams within the same coalfield, could significantly alter these ratios. Consequently, where possible, samples should be taken directly from suspected sources, or the appropriate organizations approached to determine if they hold the relevant data.

An additional problem can be that the further some gases migrate from their source, the more likely they will be subject to chemical and microbiological reactions that alter the soil gas composition. Oxidation of methane to carbon dioxide, for example, can occur in soils giving a greater proportion of carbon dioxide in the sample compared with the source. Conversely, however, carbon dioxide may be removed by dissolution in water, or by alkaline strata.

Radiocarbon dating can establish the age of materials that contain carbon, with mains gas and mine gas being of 'old' radiocarbon age and landfill gas having a 'young' radiocarbon age. However, mixing of gases of different ages from different sources can give an equivocal age.

From the foregoing, it is clear that where there could be several potential sources of gas, it is not always possible to establish with certainty whether one or a combination of sources are being reflected in gas monitoring results.

8.4 Redevelopment guidelines

Advice from central government on the use of gas-contaminated sites, in particular the reuse of landfill sites, is covered by six guidance documents, with revised versions of two of them due to be published in the near future.

(1) Approved Document C of the 1985 Building Regulations [3] states that a 1% concentration of methane by vol. suggests that remedial measures are needed, whilst a 5% concentration by vol. of carbon dioxide suggests remedial measures are needed where people will normally be

present. The Regulations only apply to the area beneath a building and do not consider migrating gases.

(2) Advice in *Waste Management Paper No. 26* [2] suggests that building should only be considered on shallow landfill sites, where tipping ended about 20 years in the past. It is further accepted that, with proper precautions in building design, more recently completed sites may also be acceptable for hard forms of development.

(3) Department of the Environment (DoE) Circular 21/87 [4] provides advice and guidance to local authorities and developers on the identification, assessment and development of contaminated land (defined to include domestic and industrial waste landfills) but does not mention landfill gas specifically.

(4) *ICRCL Guidance Note 17/78* [5] is referred to by Circular 21/87. This document recognizes that flammable gases are produced from biodegradable waste and present a problem for construction, but does not preclude any specific form of development. The seventh edition (May 1988) provided very specific advice with reference to housing. "Housing developments require particularly stringent precautions and should not be located on or near landfill sites which emit gases". The latest edition (December 1990) provides three categories of sites in terms of their gas emission characteristics and these are given in Table 8.2. This edition states "provided that traditional housing development (i.e. houses with gardens) is not envisaged it may be possible to protect the proposed development but it could be unduly expensive to do so". This statement apparently relates to Category 3 sites and not Category 1 sites (e.g. older landfills containing minor amounts of biodegradable material). The problem here is that whilst sites are categorized in terms of emission rates rather than gas concentrations alone (which is an improvement on previous editions), ICRCL 17/78 does not define 'low emission rate' within the text. Significantly, however, in Annex 11 of 17/78 it does refer to Carpenter [6], where the criteria of the former Greater London Council of an emission rate in excess of 0.05 m/s from a 50 mm diameter pipe is taken as significant.

(5) *Waste Management Paper No. 27* [7] states that "Domestic housing with gardens should not be built on gassing landfills. Great care must be taken if development is to take place near such sites. This form of development should not take place near to such sites". This does accept that other forms of development such as commercial and industrial premises and blocks of flats can be built on gassing landfills provided there is sufficient investigation and appropriate precautionary measures.

(6) Circular 17/89 [8] gives advice on the consideration of planning applications for new landfills for developments close to or on existing landfills, and also on the implications for planning of sites generating gas close to existing developments. Planning authorities are warned to

exercise 'due caution' in granting permission for developments or redevelopment on or near landfill, and that "permission should not be
granted unless reliable arrangements can be made to manage migrating
gas and to minimize risks". It also suggests that any land that is included
in the 5 year supply of land for housing which is subject to gas contamination may have to be removed from that classification.

Table 8.2 Characteristics of gas emissions and sites (adapted from *ICRCL Guidance Note 17/78*,
8th edition, December 1990)

Category	Emission characteristics	Examples of sites	Consequences
1	Gas produced continuously, usually released at low rates; concentrations of flammable constituents may exceed lower explosive limit (LEL) values	Natural materials, e.g. silts, peat, coal seams, marshland; older landfills containing minor amounts of biodegradable matter	Building development may be possible with passive protection systems and suitable design allowances (e.g. ventilation of structure and elimination of voids where possible)
2	Production and release may be intermittent but emission rates and concentrations can be high	Some abandoned mine workings	Possibility of short-term hazard: development for non-sensitive uses such as public open space may be possible
3	Gas emitted rapidly in large volumes under pressure, causing vertical and lateral movement; methane often present at high concentrations together with other gases that may modify its behaviour, e.g. CO_2	Landfill sites containing a high proportion of biodegradable material, e.g. household waste, infilled docks or watercourses	Traditional housing development is not suitable; other forms of hard development are best avoided until emission ceases; if such development has to proceed, carry out thorough investigation and then design active protection systems, gas collection/extraction wells, gas detectors and alarm systems; monitor effectiveness of precautions before, during and after development

A revised version of *Waste Management Paper No. 27* (August 1991) is
currently at the consultation stage. Whilst retaining the 'trigger value' for
flammable gas of 1% by vol. (20% LEL) from biodegradation within the
landfill, it is proposed that the trigger value for carbon dioxide is increased to
1.5% by vol. However, the draft goes on to comment that, owing to residual
gas often being trapped within wastes in old landfills, the detection of methane
and carbon dioxide may not indicate continuing gas generation, and that gas
pressure or emission rates should be monitored and pumping carried out to
determine whether the gas source becomes depleted. In terms of development
on landfills, the advice remains that domestic housing should not be built on
landfills that are gassing or have the potential to produce significant quantities
of gas. In terms of proposed housing development adjacent to landfill sites,

the consultation document states that "where landfills are producing large volumes of gas, or have the potential to produce large quantities of gas, no house, garden shed, greenhouse or any domestic extension should be constructed within 50 m of the boundary of the infilled wastes, and no garden should extend to within 10 m of the waste".

Part C of the new Building Regulations are due to come into force within the near future. Throughout the draft stages of the revised document, specific advice was given to deal with different concentrations of methane. However, in line with current central government thinking, Part C of the Building Regulations will now refer to ICRCL 17/78, 8th edition and the revised *Waste Management Paper No. 27*, thus reflecting the importance now placed on emission rates as well as gas concentrations. The revised document indicates that, where methane is unlikely to exceed 1% volume, a suspended floor slab and passively ventilated undercroft should be adopted for housing as outlined in a Building Research Document (yet to be published) on building design and construction techniques where gases are present. For carbon dioxide, 1.5% by vol. is seen as a threshold value and 5% by vol. as an action value requiring specific design measures.

Whether development proceeds on or near to gassing sites very often depends on the planning authority's attitude, the importance of freehold or leasehold developments and whether the 1% methane by vol. is seen as an action level or the level at which no development should take place irrespective of remedial measures or end use. Consequently, planning authorities often have one of three reactions to redevelopment of gassing sites.

(1) Allow the redevelopment but control it by conditions on planning permission.
(2) Allow the redevelopment, but under a Section 106 agreement of the Town and County Planning Act 1990.
(3) Oppose application for the redevelopment as a result of interpretation of advice in *ICRCL Guidance Note 17/78*.

Apart from the *Waste Management Paper No. 27* guidance trigger level of 1% methane by vol. in soil as the maximum acceptable concentration in monitoring points and more recent reference to 'low' emission rates in ICRCL 17/78, there is at present no further guidance in the United Kingdom on appropriate remedial measures for different methane concentrations, flows and/or pressures. In the United States the US Environmental Protection Agency [9] suggested a 25% LEL (i.e. 1.25% by vol. for methane) as an action level for gas venting and other remedial measures. These conservative guidelines are due to the fact that there is a real potential for relatively rapid changes in gas concentrations, possibly on a daily or even hourly basis, and that the presence of gas can be especially hazardous for the proposed use of the land. Consequently, an assessment of the hazards from gassing sites depends on the selection of appropriate investigation techniques, the accurate measurement of gas parameters, and on effective interpretation of site data.

8.5 Gas monitoring

Guidance from central government on gas monitoring methods and techniques is provided by *Waste Management Paper No. 26* (1986), *Waste Management Paper No. 27* (1989) and *ICRCL Guidance Note 17/78* (1990). Other organizations have also published guidance documents. These include the former Greater London Council [10, 11], the Building Research Establishment [12] and the Institute of Waste Management [13]. In spite of the detailed information contained in these references there are still, as yet, no standard methods of gas monitoring.

The purpose of gas monitoring is to ensure that gas emitted from natural or man-made sources does not cause a hazard to the environment on or around a particular site. Before such an assessment can be made five principal areas of gas monitoring need to be considered. These are: the parameters to be measured, investigation techniques, monitoring locations, monitoring equipment and frequency of monitoring.

8.5.1 *Parameters to be measured*

In order to make an assessment of the gas hazard on or around a particular site, a variety of parameters need to be measured. However, all too often gas monitoring from boreholes is seen to be synonymous with methane, carbon dioxide and oxygen concentrations, with no other parameters being considered. This is due in part to *Waste Management Paper No. 27*, which provided a base line for hazard assessment and control by indicating 1% methane by vol. and, in the revised edition, 1.5% carbon dioxide by vol. as trigger values. These are the values on which planning and controlling authorities often base their decisions for site redevelopment. However, with the publication of the latest edition of *ICRCL Guidance Note 17/78* (1990) and the current draft revisions to *Waste Management Paper No. 27* and Part C of the 1985 Building Regulations, there is some indication that central government is recognizing that emission rates are of more importance than concentrations alone.

The parameters to be measured are outlined below.

(a) *Gas concentrations.* The gas concentrations most commonly measured where geologic or biogenic gases are suspected are methane, carbon dioxide and oxygen, although, in the light of desk study information, monitoring of other gases may be appropriate. It is important to measure oxygen concentrations not only because of the asphyxiation hazard in confined spaces but also because portable instruments using catalytic flammable gas detectors require a minimum level of oxygen (typically 15%) to give accurate flammable gas readings. Details of the monitoring instruments, duration of measurement, and peak and steady rate values should also be recorded.

(b) *Gas pressures*. Gas flows from one point to another either because of a differential in gas pressure or by diffusion, which is the result of differences in gas concentration. Gas can therefore move from high pressure to low pressure areas or diffuse from areas of high concentration to low concentration. Measuring gas pressures can therefore be important in assessing gas migration potential or the effects of a sudden drop in atmospheric pressure. Where sensitive differential pressure monitoring apparatus is used, minimum, maximum and average pressures over the duration of the monitoring period should be recorded.

(c) *Gas flow rates*. Data on gas volumes rather than concentrations alone are a crucial factor in determining the risk posed by gas and whether it is safe to build on land affected by landfill and other gases. Techniques to measure the volume of gas include flux boxes, hot-wire anemometers and purging boreholes with nitrogen, although assumptions have to be made regarding the area of influence of the monitoring point.

(d) *Temperature*. Gas temperatures from deeper sites (more than 10 m deep) elevated above the mean annual air temperature can be indicative of gas generation. In shallow sites, gas temperatures can be significant, due to seasonal effects affecting microbial activity and therefore gas production. Elevated ground temperatures across and around a site may indicate landfill gas passing through the ground. The use of false colour aerial photography and aerial thermography using an infrared camera may be of use here.

(e) *Water levels*. The effect of a rise in water level can be to saturate additional degradable material and cause a rise in gas pressure within the degradable material. This, in turn, can lead to increased gas emissions. Difference in water level between the site and surrounding area can also determine if gas will migrate laterally.

(f) *Water quality*. The quality of leachate from a landfill can be indicative of the progress of waste degradation and gas generation potential. For example, a biochemical oxygen demand (BOD)/chemical oxygen demand (COD) ratio of the order of 0.5 suggests that the leachate can be considered to be relatively biodegradable. According to *Waste Management Paper No. 27* a BOD/COD ratio of 0.7 would be anticipated during the initial phases of anaerobic degradation. The proposed EC Landfill Directive [14] suggests that stable methanogenic conditions are established and a sufficiently high level of activity will continue when, taken with minimum pH, temperature and gas production rates, the average BOD/COD ratio is equal to or less than 0.3. As anaerobic degradation proceeds and the landfill stabilizes, this ratio can fall to 0.1 or less, indicating the presence of organics that are hard to degrade (such as humic and fulvic acids typically associated with stabilized tips). However, a BOD/COD ratio of 0.1 or less could also indicate the presence of toxic material inhibiting the BOD results. These and other laboratory methods that can be used to evaluate methane production are reviewed by Harries [15].

(g) *Meteorological data.* Atmospheric pressure, ambient air temperature, rainfall, wind speed and direction, and ground conditions at the time of monitoring should be noted together with records of atmospheric pressure, temperature and rainfall during the previous 24 h. For example, during periods of frozen ground, or after heavy rainfall, gas emission rates could increase as the gas finds alternative ways out of the ground through particular fissures or discontinuities. Similarly, if the extreme atmospheric pressure drop of 40 mbar in 24 h were to occur, gas emission rates could increase, the magnitude of the increase depending on the difference between the gas pressure and the atmospheric pressure, and the efficiency of near surface materials in providing a seal.

8.5.2 *Investigative techniques*

The types of investigative techniques may be related to the staged approach of site investigation (chapter 3). This could comprise sampling of confined spaces and drains during an initial walk-over survey, followed by shallow probing and/or trial pitting, and finally the construction of gas sampling boreholes. Gas monitoring standpipes can be installed in shallow probe holes, trial pits and boreholes.

Drive-in probes and trial pits enable monitoring to shallow depths but cannot incorporate effective seals at ground level. In addition, whilst trial pits enable a good visual assessment of large volumes of material, they also allow large volumes of air into the ground. If low gas concentrations exist it can take a long time for gas atmospheres to re-equilibrate with surrounding materials. Care is therefore needed in the interpretation of gas monitoring data from trial pits.

The use of a dynamic probing rig enables the installation of drive-in standpipes (typically 19–25 mm diameter) to moderate depths and enables a good seal to be installed at ground level. However, problems can be encountered in installation if hard strata are encountered, particularly in made ground.

Cable percussion or rotary rigs are generally used to construct boreholes for monitoring pipe installation, although with the use of an air flush in rotary rigs it can take longer for representative results to be obtained from borehole monitoring pipes installed by this method.

Conventional boreholes with slotted pipe from 1–2 m below ground level can provide mean gas concentration, pressure and flow data, although boreholes can be monitored for gas concentrations every metre as the holes are advanced, to obtain an indication of the depth of gas ingress. In natural ground, it is possible to obtain data on gas concentrations, flows, etc., in specific zones, by installing multiple pipes within a single borehole. In situations of deep fill or landfill sites it is preferable to install pipes to different depths in different boreholes if specific zones of gas are to be identified.

The methane probe construction detail often followed is that originally outlined by Pecksen [11]. This type of construction does not allow gas pressures to be measured and allows air dilution during sampling. In addition, whilst perforated 50 mm diameter UPVC pipe is cheaper than slotted, flush fitting, screw threaded types, it generally requires the use of solvent cements to join the expanded ends or collars. This could lead to traces of flammable gas being recorded in the borehole from that source. The preferred type of borehole standpipe is shown in Figure 8.1. This has a larger percentage of open free area of pipe (typically 6–12% on a 50 mm diameter pipe with slot sizes in the range 0.5–2 mm) and does not require the use of solvent cement. In chemically aggressive ground conditions, polypropylene or HDPE pipe could be considered for long-term gas monitoring. The removable screw threaded gas tap assembly enables flows, temperature, water level and water quality to be monitored after initial monitoring of gas concentrations and pressures. An alternative

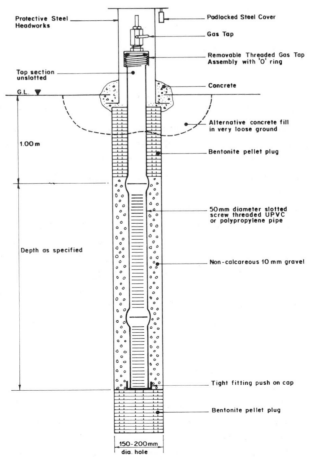

Figure 8.1   Typical borehole gas monitoring standpipe.

arrangement is to have two gas taps so that instead of venting the analysed sample into the atmosphere it is recirculated back into the borehole.

The disadvantage of sealed systems is that they can be relatively unresponsive to transient changes and may be recording pockets of trapped residual gas. By removing the gas tap, emission rates can be measured with a hot-wire anemometer. Alternatively, where flow rates are immeasurably low, the borehole can be purged with nitrogen, and subsequent monitoring in an open borehole can establish if gas recovery occurs.

8.5.3 *Monitoring locations*

The location of monitoring points depends on site specific information obtained from desk study information and will be determined by factors such as the geological and hydrogeological environments and the areas of potential risk.

8.5.4 *Monitoring equipment*

(a) *Gas concentrations.* The main types of portable instrument available for measuring gas concentrations are catalytic flammable gas, thermal conductivity, infrared and flame ionization detectors. A review of the features and limitations of these instruments is presented by Crowhurst [12], *Waste Management Paper No. 27* and *ICRCL Guidance Note 17/78*. Probably the most commonly used instruments are the combined catalytic/thermal conductivity methane instruments (which also measure oxygen concentrations) and carbon dioxide instruments using infrared detectors. Most portable instruments require a minimum volume of 100–200 ml to give an initial reading, with pump operated instruments abstracting gas at between 0.75 and 1 l/min. In situations where there are only small volumes of methane and carbon dioxide in a borehole, this can lead to discrepancies between portable instrument readings and samples subsequently analysed by laboratory gas chromatography. One method of sampling gas concentrations where there are potentially small volumes of gas is to have two sampling points on the gas tap assembly, with small diameter tubing placed inside a borehole at two depths. Gas can be withdrawn from one sampling point by a mechanically pumped infrared carbon dioxide portable instrument, and can then be drawn through a mechanically pumped methane and oxygen portable instrument placed in series, before the sample is returned to the borehole through the second sampling point.

New instrument developments in this area include combined methane, oxygen and carbon dioxide instruments measuring methane and carbon dioxide by infrared gas detectors. Comparisons between these and combined catalytic/thermal conductivity instruments have shown the infrared devices to be extremely accurate and not influenced by the presence of other hydrocarbons, such as diesel oil spillages. Another new instrument works on the catalytic bead principle and is sensitive to gas concentrations of approximately 5 ppm. Its advantage over flame ionization instruments is that it is intrinsically safe, smaller and cheaper. Where a large number of boreholes require frequent monitoring over a number of years, the use of automatic borehole monitoring systems for methane and carbon dioxide concentrations may be appropriate. These use the infrared absorption principle.

(b) *Gas pressures.* In the past, gas pressures, if monitored at all, have typically been measured by micro manometers with results in millimetres of water. There are now instruments available that can measure differential pressures, positive and negative, down to 1 Pa and which can provide minimum, maximum and average pressure readings over the measurement period. One such device contains an internal pressure sensor with two external flexible tubes, the instrument giving the differential gas pressure between the atmospheric pressure and the second tube attached to the gas tap assembly.

(c) *Gas emission rates.* These are typically measured by one of three methods: (i) the hot wire anemometer; (ii) the flux box; and (iii) by purging boreholes with nitrogen and measuring gas recovery.

The hot-wire anemometer operates on the principle of thermal conductivity, and the calibration and safety of the instrument in flammable gas mixtures should be checked. The hot wire is typically sensitive to flows of 0.01 m/s (i.e. 1178 cm^3/min at the point of emission from a 50 mm diameter borehole) and requires an extension piece of the same internal diameter as the borehole to be fitted to provide wind shielding.

At its simplest, a flux box could comprise a 25 l container placed over the top of a borehole or probe, with the concentration of methane, oxygen and carbon dioxide monitored over time to calculate a volume emission rate. Alternatively 200 l drums either cut longitudinally or with the base removed, could be placed on the ground or, in the case of shallow footings, at the proposed foundation level; the concentrations of methane, oxygen and carbon dioxide could then be recorded over time to calculate a volume flow rate in cm^3/min per m^2.

Purging a borehole with nitrogen requires cylinders of nitrogen (typically size B) with regulator and flexible tubing attachments, together with portable methane and carbon dioxide meters. Both the hot-wire anemometer and nitrogen purging require assumptions to be made regarding the radius of influence of the borehole to establish a flow rate in cm^3/min per m^2 (see [11, 16] for discussions on this aspect).

Where emission rates are high, it may not be possible to purge a borehole with nitrogen, in which case the information obtained from a hot-wire anemometer would be sufficient. Where flow rates are immeasurably low using a hot-wire anemometer (i.e. < 0.01 m/s), or where isolated pockets of methane are suspected, purging with nitrogen would be appropriate. Where very low emission rates are anticipated at ground level or shallow depth, a 200 l flux box could be utilized subject to its susceptibility to vandalism and the need to obtain readings over a long time period.

(d) *Gas sampling for laboratory analysis.* In order to confirm the accuracy of portable instruments and to test for a wider range of gases, samples of gas are often taken using a hand held pump with double headed stainless steel sampling tubes. The volume of the borehole needs to be considered when taking gas samples. The sample tube is purged prior to taking the actual sample, which is subsequently analysed by gas chromatography.

(e) *Sampling sequence.* If all the parameters previously outlined were to be measured, the following sampling sequence is suggested from boreholes with a gas tap assembly. Pressure is measured first, followed by gas concentrations and sampling for gas chromatography analysis. A second gas concentration sample is then taken. The gas tap is then removed and flow rates measured with a hot-wire anemometer; this is followed by temperature and water level readings. Finally, the borehole is purged with nitrogen and gas recovery monitored.

8.5.5 *Monitoring frequency*

The more often sampling points are monitored, the greater is the degree of confidence when making an assessment. Initial monitoring of boreholes or probes could follow the sequence of 1, 2, 4, 8 and 12 weeks after installation, although development pressures often curtail the length of this initial monitoring period. Due account should be taken of factors such as variations in atmospheric pressure, groundwater levels and other seasonal changes, which may warrant an increase in the frequency of monitoring. Generally, however, on the basis of the initial monitoring results, recommendations may be made for more detailed monitoring or other immediate actions. *Waste Management Paper No. 27* recommends that monitoring should continue until it can be shown that landfill gas concentrations have fallen below the specific trigger values for methane and carbon dioxide.

8.6 Data assessment and remedial measures

Whilst there is a wide range of gas control measures that could be adopted, the problem arises in assessing the results of gas monitoring against possible options for remedial measures or, indeed, in determining if any remedial measures are required at all. The intention here is not to look at particular gas control measures in detail, but rather to very briefly review the options available for remediation, then to look at the various approaches of assessing gas monitoring information that have been made in relation to the most sensitive use, that of building on or near gassing sites. Finally, various aspects of dealing with landfill and other gases will be developed by reference to case study information.

8.6.1 *Options available for remediation*

The range of remedial measures for dealing with landfill and other gases essentially falls under the headings of excavation, venting, barriers and in-building protection. Gas control measures in general and some of their longer term performance problems are reviewed by Barry [17].

Unless the gassing material is on-site and at very shallow depths, excavation is unlikely to be a viable option but it should nonetheless be considered. Venting could include gravel filled trenches, currently used at 53% of the 192 landfill sites reviewed by Gregory *et al.* [18] as a means of gas control, followed by 45% using pumped systems and 2% using cut-off barriers.

Barriers range from horizontal types such as membranes or clay layers to vertical ones such as grout curtains, diaphragm walls, slurry trenches or a series of pumped wells. Needham [19] provides a review of cut-off barriers, whilst Raybould and Anderson [20] and Ingle and Kavanagh [21] provide case studies of grout curtain and slurry trench barrier installations, respectively.

Depending on site circumstances, three levels of building protection could be considered [22]. These are: (i) removal of the hazard from the vicinity of the buildings (e.g. cut-off barriers or dilution/dispersion of gas); (ii) prevention of gas entry into the building (e.g. membranes and sealed services); (iii) safety procedures and monitoring (e.g. permanent fixed detection and alarm systems).

8.6.2 *Data assessment*

In order to make an assessment of whether gas will be a hazard to the development of a site, the gas monitoring programme should have addressed the following areas:

(1) The existing composition of the gas (e.g. concentrations of methane, carbon dioxide and oxygen) within the site boundary
(2) The existing pressure of the gas within the site boundary
(3) The volume of gas being produced
(4) The potential of organic materials for future degradation and subsequent gas production
(5) The existing water levels within the site
(6) The extent of gas migration from the site (although access problems can restrict this aspect)

The exact scope and scale of the gas monitoring programme will have been determined by, amongst other factors, the end use of any redevelopment proposals, the anticipated nature of the ground (e.g. landfill, peat deposits, etc.), the development surrounding the site, and local geological and hydrogeological conditions. For example, provided gas migration is not taking place off-site to more sensitive areas, the need for, and design of, remedial measures for a playing field or park often presents little problem other than the preservation of trees and plants. However, building on or near old landfill or other sites where hazardous gases may be present requires considerably more caution.

At its most basic, the composition and volume of gas being produced would be required in order to make an assessment. All too often, however, gas monitoring is seen to involve the installation of boreholes with nothing but methane, oxygen and carbon dioxide concentrations subsequently recorded. This situation may in part be due to the perception that gas concentration readings are easy to take and require little experience. Whilst this is certainly not the case (see for example chapter 10, section 10.2), the line is often drawn there owing to the additional equipment and experience required in taking flow or pressure measurements.

Establishing gas concentrations may be sufficient to determine whether a potential hazard exists but it does not necessarily provide information to enable buildings to be suitably designed and ventilated, or measures to be selected to control the movement of gas. Basic flow rate data can, for example, at least establish whether the soil gas regime is dynamic or static.

Although the Building Research Establishment and CIRIA are currently producing guidelines on methane and associated hazards to construction, at the present time no guidelines are laid down by central government or the building industry in linking gas monitoring data with gas control measures and the type of protective measures required in different structures and different site circumstances. This lack of guidance stems from the difficulty of relating the results of gas monitoring to a site specific assessment of risk. Consequently, differing views and assessments have been made in the last 6 or 7 years in resolving the data obtained from gas monitoring. Many of these involve qualitative risk assessments with some data, such as gas concentrations, to back them up. Indeed, as long ago as 1985, Pecksen [11] noted that "the lack of a universally accepted procedure for dealing with the generation of methane gas on sites which are to be developed for housing and other buildings has led to the adoption of fail safe solutions which are expensive and in some cases totally unwarranted". The requirements for a thorough and appropriate site investigation and gas monitoring programme cannot be overstated.

Emberton and Parker [23] chose five criteria, which they felt should normally be met before even considering a landfill site for any form of redevelopment. These are:

(a) The site should have been completed at least 10 years prior to redevelopment
(b) The site should be shallow, with a depth of less than 10 m
(c) The site should have a stable, low water table
(d) The site should not have accepted toxic or hazardous materials, particularly liquid wastes
(e) The development should be appropriate for the site conditions.

In the light of examples of successful developments, Emberton and Parker went on to divide sites, where precautions against gas ingress are necessary, into three main classes:

(1) *Class A*. Deep sites with recently deposited waste producing large quantities of gas; not recommended for redevelopment
(2) *Class B*. Some gas being evolved, containing waste 5–10 m in depth and deposited during the last 10–15 years; buildings on slabs with ventilated undercrofts using extraction fans giving a minimum of 1–2 air changes per hour; plastic membranes incorporated into floor slabs, services should not penetrate slabs, methane detection sensors beneath and above slabs
(3) *Class C*. Small quantities of methane being liberated and lacking the potential to produce large quantities of gas in future years; waste typically at least 15 years old, or sites contain only small proportions of recently deposited domestic waste mixed with 'inert' material; natural ventilation beneath floor slabs of structures, and special care to ensure integrity of the floor slabs

Clark and Warby [24] considered three principal types of protection measures for buildings: a gas proof membrane placed either directly beneath the floor slab or in the top surface of the floor slab; a passively or actively vented undercroft or void beneath the floor slab with a gas proof membrane; a passively or actively vented granular blanket below the floor slab together with a gas proof membrane. They state that if methane concentrations exceed 1% by vol. then they would normally recommend mechanical (active) ventilation.

Parker [25] noted that it could be argued that a passive system (natural ventilation) beneath a floor slab should be installed for methane concentrations up to 2% by vol., a semi-passive system (natural ventilation beneath a floor slab with the addition of vertical stack pipes with rotating or fixed cowls above eaves level) for the range 2–10% methane by vol. and an active system (the use of mechanical pumps to maintain ventilation beneath the floor slab) for higher methane concentrations. However, he recognized that spatial and temporal variations in gas concentrations make it difficult to produce satisfactory guidelines, and that while gas emission rates from the ground surface or boreholes can be measured, there are practical problems in the interpretation of the data.

The assessment of results of gas monitoring has been considered by Pecksen [11] and Carpenter [6, 26] following the work undertaken by the former Greater London Council. An estimate of the potential surface gas emission from the measurement of borehole gas emission rates was made by Pecksen assuming a small radius of influence of $10 \, m^2$ (equivalent to a radius of 1.78 m) in order to incorporate a safety margin for void design. An argument against this, and a far more conservative approach, is that the borehole could have intercepted a fissure or other discrete migration path, and actually already represents the worst case condition.

Gas emission rates were measured by hot-wire anemometer for flows of more than 0.01 m/s in a 50 mm diameter borehole and where flow rates were less than this, by flushing the borehole with nitrogen and then monitoring the build up of methane concentration with time once the flushing had ceased. Carpenter [6] noted that in the redevelopment of London Docklands, although methane concentrations were often in the range 20–30% by vol., the majority of gas emission rates were below 0.01 m/s in a 50 mm diameter borehole. The use of a passively ventilated full size test rig of the housing floor design (blinding at ground level, 225 mm void, membrane above floor slabs) and the use of flow rates in excess of 100 times greater than those measured on site was shown to be satisfactory. Experimental work using a granular blanket beneath a suspended floor slab indicated that methane dispersion was much slower than from an open void, and that there is the problem of gas becoming entrapped between the ground beams on top of piles. Granular blankets were only recommended where gas emission rates were less than 0.01 m/s from a 50 mm diameter borehole. Furthermore, building redevelopment on sites where methane concentrations of more than 1% vol., in association with gas emission rates in excess of 0.05 m/s, had been recorded from a 50 mm diameter borehole, was not advised.

More recently in the case of landfill sites, Sheriff et al. [27] and O'Riordan and Warren [28] have taken the approach of initally calculating gas generation rates and potential surface release from first principles. This essentially involves an estimation of the total carbon and an assumption of a realistic decay rate, although in older landfills this can be particularly difficult. The design philosophy in this approach is outlined by Lord [29] who considers that the difficulty of measuring low gas flow rates cannot at present be remedied, and that a simple mathematical model could be used for the basis of a risk analysis involving the following assumptions:

● The quantity of gas producing material
● Total volume of gas likely to be produced
● Half-life of gas production, to yield a generation rate
● Migration routes taken by the gas
● Permeability of the ground and structure
● Gas solubility

This leads to a prediction of ventilation rates and, from there, to a gas control system to ensure that methane concentrations are maintained below an agreed safe level.

This approach, however, includes assumptions about the total volume of gas available from biodegradation (theoretically taken as $460\,\mathrm{m}^3$ of methane per tonne of dry refuse if the original total organic carbon content of the refuse is assumed to be 50%). Given the heterogeneous nature both of landfill material as a whole, and of the variation of types of biodegradable material within it, it is doubtful if generation rates can be accurately estimated. Practical yields, for example, are often in the range 30–$180\,\mathrm{m}^3$ of methane per tonne of dry refuse owing to much of the organic carbon being washed out in the leachate [30].

The change of gas production with time can be represented mathematically as

$$G = L_0\,(1 - \mathrm{e}^{-kt})$$

with the rate of gas production as

$$\frac{\mathrm{d}G}{\mathrm{d}t} = kL_0\,\mathrm{e}^{-kt}$$

where G is the volume of methane produced up to time t, t is the time in years since the tip has become anaerobic, L_0 is the ultimate volume of methane to be produced and k is a gas production rate constant, which, from published sources, could vary from 0.02 to 0.2 depending on site conditions. If it is assumed that the half-life of landfill gas production is around 10 years, then k is equal to 0.07. In other words

$$G = L_0(1 - 10^{-0.03t})$$

and the rate of gas generation would be

$$\frac{\mathrm{d}G}{\mathrm{d}t} = 0.07\,L_0\,10^{-0.03t}$$

Thus, for example, if the worst case situation of L_0 is taken as $460\,\mathrm{m}^3$ of methane per tonne of dry refuse, some $16\,\mathrm{m}^3$ of methane per tonne of dry refuse would be anticipated after 10 years, compared with some $6\,\mathrm{m}^3$ of methane per tonne of dry refuse after 10 years if L_0 were taken as $180\,\mathrm{m}^3$.

Whilst a theoretical gas generation rate can be estimated, the evolution rate from an actual site is more complex to calculate, since it depends on factors such as: (i) the difference between the landfill gas pressure and the atmospheric pressure; (ii) the temperature; (iii) the permeability of the fill; (iv) the water level and (v) pathways for the gas to escape. With respect to gas escape, a borehole provides a vent for gas release and could thus be considered as the worst case situation. In practice, therefore, given the variable nature of bio-

degradable material encountered and variations within and between different
sites, there is no substitute for appropriate field measurements. Indeed, Pecksen [11] originally considered a mathematical approach based on total organic
carbon contents and a realistic decay rate to establish gas generation rates but,
particularly in the case of old landfill or infilled docks, concluded that efforts
were better channelled through field measurements of gas concentrations and
emission rates.

At this point, it is worth considering fairly typical cases where shallow filled
sites containing either clay fill with occasional brick and timber fragments and
small pockets of organic matter, or landfill sites of some age with clay, ash,
glass, metal and timber fragments. In these situations methane and carbon
dioxide concentrations of some 1–5% by vol. are often recorded in some of
the monitoring boreholes, in conjunction with negligible gas pressures
(< 10 Pa), immeasurably low flow rates when using a hot-wire anemometer,
and limited or variable recovery after purging boreholes with nitrogen. Gas
associated with these types of site, together with cases of methane associated
with peat or organic alluvial clay, whilst being extremely difficult to model
mathematically, are often susceptible to having expensive fail safe gas control
solutions adopted in the building design. The situation of applying mechanical
ventilation for any methane concentrations recorded above 1% volume during
gas monitoring is one such example. This point also relates back to what extent
the 1% methane volume value is seen as an action level, or indeed whether
development should proceed at all if methane concentrations are above 1%
volume.

Very often any value over 1% methane by vol., whether 2% or 20% and
irrespective of source, is automatically seen as requiring a whole host of
building gas protection measures so that, in the absence of any gas flow and
gas pressure information, the assessor becomes the prisoner of the gas concentration monitoring results. If gas concentration readings were taken in
conjunction with gas pressures and gas flows as a matter of course, together
with a consideration of the source of the gas, this situation would not always
be the case.

Whilst the presence of landfill and other gases does not necessarily imply
a static soil-gas regime, and methods of assessing potential long-term post-
development effects need to be considered, it is the initial site assessment
(including desk study, site investigation, gas monitoring programme and
laboratory testing) which forms the basis of an engineering judgement and
determines to what extent gas protection measures need to be incorporated
into the design. The assessment therefore needs to consider if the results of
the initial site investigation represent typical conditions and, in the light of
uncertainty regarding post-development conditions, if the worst case conditions have been considered. Finally, the practicality or 'buildability' of the
proposed remedial measures needs to be considered. For example, the use of
a 450 mm void beneath a cast suspended floor slab means that man entry after

casting, to remove the formwork and support system, is not possible. Alternatives could include a larger void or the use of hollow block 'sleeper walls', as described by Rys and Johns [31] in an industrial development on a landfill site in West London. In addition, any proposed remedial measures should be robust enough to withstand the installation process on site.

8.6.3 *Case studies*

Industrial development, Southern England. The site is located in an area of former sand and gravel extraction. Desk study information indicated that much of the gravel workings had been backfilled for more than 30 years. However, relatively recent fly tipping meant that site levels were now some 4–5 m AOD (above Ordinate Datum) with surrounding ground levels lower at 1.5 m AOD. A preliminary borehole investigation indicated a sandy clay/clayey sand fill with a large proportion of demolition rubble and ash extending to 8 m below ground level. Domestic refuse containing timber, cloth, wire and glass was locally present. The development proposal, for a single large industrial unit which, owing to the nature of its operation, would in part be continuously internally ventilated, was to excavate to +0.25 m AOD, and then raise the ground level to +2.0 m AOD with controlled filling of granular material.

Gas concentrations were monitored every metre as the 200 mm diameter cable percussion holes were advanced with the results shown in Figure 8.2. Two 50 mm diameter monitoring pipes were then installed in each hole, one with a response zone from −4.0 m OD to 0.5 m OD (i.e. the material to be retained) and the other with a response zone from 1 m below ground to +1.15 m OD (i.e. the material to be excavated). Where methane and carbon dioxide concentrations were recorded they were generally higher from the lower response zone and typically in the range 0–5% methane volume and 0–10% carbon dioxide volume, the carbon dioxide readings often being higher than the methane readings by a factor of 2 or 3. Gas emission rates were immeasurably low (< 0.01 m/s) when using a hot-wire anemometer, whilst groundwater levels were depressed (range −1.95 to −1.5 m OD) across the whole site and gas temperatures were at, or a few degrees centigrade below ambient temperatures depending on depth. During initial visits, caps were left off boreholes for some 2.5 h after monitoring and gas concentrations were measured again. Methane and carbon dioxide concentrations fell during this period and oxygen values increased, suggesting that only small volumes of methane and carbon dioxide had collected in the standpipes, and that these were being diluted with oxygen rather than being replenished.

Flushing some of the boreholes with nitrogen during subsequent visits confirmed the very low emission rates. However, other factors also needed to be considered in the risk assessment. These were:

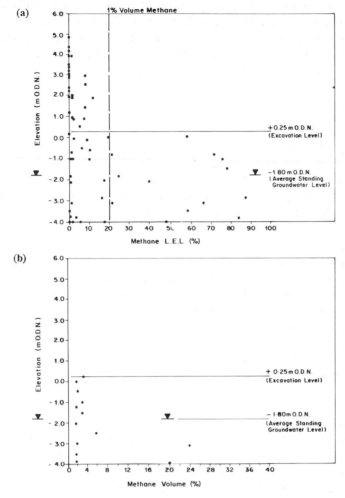

Figure 8.2 Methane concentrations versus elevation taken as the borehole was advanced. (a) LEL versus elevation during drilling; (b) gas (CH_4) volume versus elevation during drilling.

(a) Virtually the whole site would be 'capped' by a piled structure, with the piles themselves possibly acting as a migration path for landfill gas.

(b) The average standing water level of $-1.80\,$m OD could not be assumed to represent the natural long-term water level in the area. Whether pumping of water in the gravel or chalk was taking place was not clear, but from other sources of information a long-term water level of $+0.5\,$m OD was appropriate. A rapid rise in water level could, of course, significantly change the rate of gas emissions.

(c) A number of boreholes encountered domestic refuse below $+0.25\,$m OD, which could decompose further over the design life of the structure so that, whilst some areas of the site may have entrapped pockets of gas, other parts could have more serious areas of active decomposing material.

The option to excavate all the material off site was not economic, particularly with fill below the standing water level. On the basis of worst case conditions a gas impermeable membrane was built in below the ground bearing floor slab together with a passively vented granular blanket to provide a high permeability dispersion layer (granular fill was in any case extending to + 0.25 m OD). The membrane was designed to be placed above pile caps thereby avoiding having to seal around individual piles and, in the light of detailed geotechnical assessment, settlements of the ground bearing floor slab would be minimal so that membrane rupture due to ground movement was not seen as a problem.

Proposed commercial development, London Docklands. Previous geotechnical investigations typically proved 3 m of made ground overlying 3 m of alluvial clay, 1.5 m of peat, 1 m of sandy clay and, at 8.50 m below ground level, flood plain gravel across the northern and eastern parts of this site. Made ground extended to 8.50 m below ground level in areas where former finger wharves had been infilled. Due to the presence of infilled docks and alluvial clays, borehole gas monitoring comprised 100 mm diameter standpipes taken into the flood plain gravel and multipoint gas monitoring pipes with specific response zones. Long-term monitoring of these installations has taken place since 1989 and methane concentrations in the range 20–50% by vol. were found with carbon dioxide concentrations typically up to 5% by vol. in a well defined linear strip of land at the eastern end of the site on, and just beyond, the eastern end of the proposed building line.

The multipoint installations comprised 19 mm diameter pipes with 1 m response zones within the peat, organic clay and made ground, typically with a 2 m bentonite seal between each gravel filled response zone. Monitoring of these boreholes at the eastern end of the site, as they were advanced, indicated 2–3% methane by vol. and 0–1.5% carbon dioxide at 5–6 m below ground level (i.e. towards the base of the organic clay). Subsequent monitoring of the multipoint wells at the eastern end of the site indicated peak values up to 50% methane by vol. in the organic clay response zone, with up to 1% carbon dioxide and oxygen concentrations generally above 17%. The water level within these small diameter pipes was above the response zone and monitoring small volumes of methane quickly depleted the methane concentrations from a peak value. Gas chromatography analysis confirmed the portable instruments and showed no ethane, hydrogen sulphide or hydrogen to be present.

Flushing of the boreholes with nitrogen indicated a variable but low emission rate from the standpipes, no recovery whatsoever of methane and carbon dioxide, and normal oxygen atmospheres in the multipoint wells when monitored for the rest of the day. Subsequent visits (initially weekly but then monthly) indicated methane concentrations returned to 40–50% vol.

The linear distribution of low flow rate methane concentrations, apparently from an organic clay layer, led to a more detailed historical search. Local information indicated that sewage from surrounding areas used to be tipped directly into a tidal ditch that flowed into the River Thames at the eastern end of the site. Historical records also showed a 'sea wall' running on a line immediately west of the gassing area. Trial pits of 7–8 m depth proved the existence of this wall and that it had been constructed of a mixture of clays, silts and peats, probably excavated from the tidal ditch, into which the sewage had been tipped. Local data indicated that methane normally occurs on the east of the wall but not on the west side, for a mile or more inland.

A deep trial pit excavated in the centre of the gassing area was backfilled with brick rubble. Subsequent monitoring of two of the 'worst' boreholes closest to the pit over a 6 month period, showed a significant reduction in gas concentrations, with methane concentrations now in the range 0–3% by vol., carbon dioxide less than 1% and oxygen concentrations increased. A vent trench solution on this site currently looks a realistic remedial option.

One further point of interest here concerns the use of nitrogen to purge boreholes and estimate a free flow emission rate. Figure 8.3 shows two plots of methane recovery against time. Figure 8.3(a) shows a rapid initial recovery with the rate of change reducing with time. The water level in this 100 mm diameter borehole was 5.35 m below ground level giving a borehole volume of some 42 000 cm^3. The initial methane concentration in the borehole prior to purging was 42% by vol. Taking the steepest straight line portion of the recovery curve suggests that a maximum free flow emission rate of 3.6% methane by vol. per minute may be extrapolated, which would equate to some 1500 cm^3 of methane. More typically, however, it is not always possible to completely purge the borehole of methane and, after an initial rise, methane concentrations often rapidly reduce with time, as indicated by Figure 8.3(b). This situation may arise as a result of the nitrogen flowing out of the top of the borehole creating a differential pressure gradient or suction effect and drawing methane immediately around the borehole into the pipe. Ten minutes or so after flushing, this effect is not felt. The initial methane concentration in this borehole prior to purging was 27% by vol. A third situation can arise where methane, carbon dioxide and oxygen concentration recovery into a borehole is very variable, suggesting discrete pockets of gas entering the borehole rather than a continuous flow. For example, from a situation of no methane or carbon dioxide and 21% oxygen at the measuring point 33 min after flushing, 2% methane and 0.5% carbon dioxide was recorded 25 min later. Twenty minutes after this, the situation was normal again.

It can be seen that calculation of volume flow rates from nitrogen purging as outlined above is somewhat different to the hot-wire anemometer. A flow of 0.01 m/s from a 50 mm diameter borehole equates to a volume flow of some 1200 cm^3/min. The worst case situation often adopted here is that this volume flow rate is 100% methane instead of, say, 10% of this value if 10% methane volume was initially recorded in the borehole.

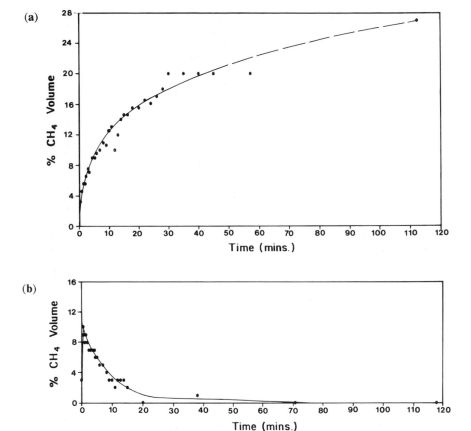

Figure 8.3 Plots of post nitrogen flush methane concentrations against time. For explanation of parts (a) and (b) see text.

Business park and public open space development, Southern England. The site comprises a number of worked out gravel pits backfilled with refuse between the early 1950s and mid-1980s to a depth of 6–10 m. The site has been surrounded by a bentonite slurry cut-off wall excavated through terrace gravels and keyed into the underlying tertiary clay in order to control landfill gas and leachate migration. Gas monitoring boreholes had been installed around the perimeter of the site in natural ground two years prior to the installation of the bentonite trench. Figure 8.4 indicates the gas concentrations taken on a weekly basis before and after trench completion at one of the

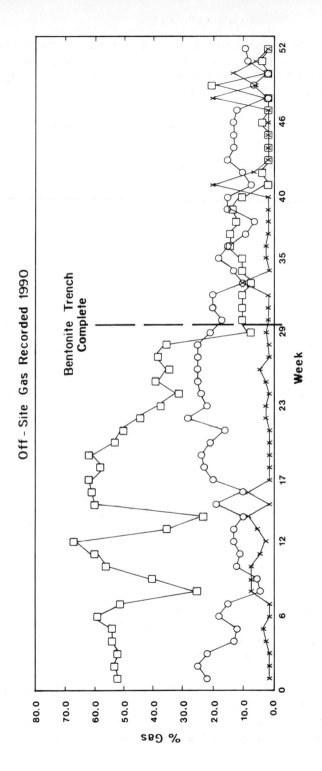

Figure 8.4 Gas concentrations recorded in an off-site borehole before and after completion of the bentonite cut-off wall. (□) Methane; (×) oxygen; (○) carbon dioxide.

monitoring boreholes. Whilst it shows a considerable time lag for the methane and carbon dioxide concentrations to reduce after trench installation (probably due to these gases in the soil pore space between the trench and the borehole slowly dissipating), the oxygen concentrations have generally remained depressed in the 20 or so weeks following trench installation.

Proposed redevelopment of deep landfill site, West Midlands. The site is a former brick pit excavated into sloping ground, with the deepest part of the workings extending below the lowest part of the original ground level by some 15 m. Cable percussion drilling proved 25 m of waste overlying Mercia Mudstone, with much of the tip effectively above ground. The brick pit was run as a refuse tip and civic amenity tip in the period 1960–1980, and redevelopment options were being considered. Much of the fill comprised building waste, with ash and clinker and some domestic refuse. Some of the results of the initial gas monitoring programme are shown in Table 8.3. In the light of these data, together with chemical contamination and geotechnical considerations, a 'soft' form of after-use was proposed comprising playing fields. The neglible gas pressures, profile of the tip, loose density and permeable nature of much of the waste, and general absence of perched water (leachate) during drilling suggested that vertical gas migration rates should not be inhibited and that the movement of gas laterally to the boundaries of the site would be limited. This was confirmed by monitoring boreholes placed immediately off site adjacent to sensitive areas of dwellings where a normal soil gas regime was recorded.

Table 8.3 Results of initial gas monitoring programme on a deep landfall site, West Midlands

BH. no.	Date	Ground water level [a]	Methane		Oxygen vol. (%)	Carbon dioxide vol. (%)	Remarks
			Vol (%)	LEL (%)			
2	05/11/90	18.68	3	–	2.4	5	
	19/11/90	17.80	6	–	0.5	6	Water sample taken; gas
	26/11/90	17.83	3	–	0.7	4	sample taken (GE 9)
	03/12/90	17.78	5	–	0.8	7	
	17/12/90	17.91	5	–	0.7	5	
	14/01/91	16.38	2	–	6.5	3.5	Gas sample taken (GE 15)
	18/02/91	–	2	–	0	4	Unable to remove gas tap
3	05/1190	20.73	3	–	4.3	7	
	19/11/90	20.87	19	–	0.5	10	Gas sample taken (GE 19)
	26/11/90	20.60	1	–	2.2	10	
	03/12/90	20.59	14	–	0.7	6	
	17/12/90	20.70	18	–	0.4	11	
	14/01/91	19.56	14	–	1.0	8	Gas sample taken (GE 6)
	18/02/91	19.45	13	–	0	10	

[a] Metres below ground level.

BH. no.	Date	Gas pressure (Pa)			Gas velocity (m/s)			Gas temperature (°C)	Ambient air temperature (°C)	Atmospheric pressure (mbar)
		Min	Max.	Average	Min.	Max.	Average			
2	05/11/90				0	0.64	0.09	15.7	8.6	1026
	19/11/90				0	0.60	0.37	18.1	9.4	1007
	26/11/90				0	0.96	0.38	17.5	8.9	1007
	03/12/90	2	5	3	0	1.93	0.61	18.0	8.5	1029
	17/12/90				0.36	0.44	0.40	18.2	1.2	1031
	14/01/91	−1	0	0	0	0.19	0.33	18.3	1.5	1036
	18/02/91	−3	7	2	−	−	−	17.7	5.2	1017
3	05/11/90				0	0.28	0.02	16.5	8.6	1026
	19/11/90				0	1.80	1.13	17.3	9.4	1007
	26/11/90				0	1.69	0.87	17.7	8.9	1007
	03/12/90	5	9	7	0.53	1.59	1.26	17.5	8.5	1029
	17/12/90				0	1.37	1.33	17.9	1.2	1031
	14/01/91	7	9	8	0	1.79	1.41	19.0	1.5	1036
	18/02/91	7	10	8	0	1.72	1.55	18.9	5.2	1017

An interesting feature of the gas monitoring is the negligible gas pressures (less than 10 Pa) compared with the emission rates subsequently measured by hot-wire anemometer. Whilst these were variable, with zero (i.e. < 0.01 m/s) often recorded, the temperature difference between the gas (measured at 10 m below ground level and clearly elevated above the mean annual air temperature expected at this depth) and ambient air temperature was significant, with convection currents probably being induced and an enhanced flow rate resulting. Indeed on 14 January 1991, a plume of 'steam' was clearly in evidence leaving the extension pipe on borehole 3 as the gas condensed under clear, calm atmospheric conditions and ambient air temperatures approaching freezing.

Industrial development, Southern Coast of England. A previous geotechnical investigation on this low lying reclaimed coastal site proved up to 5 m of sea dredged sand overlying alluvial clays, with a peat layer at some 6–7 m below ground level. The presence of loose sand deposits enabled the use of 19 mm diameter gas piezometers, which were installed using a dynamic probing rig at a fraction of the expense of large diameter pipe installed by cable percussion methods.

A number of probes were driven to 3 m below ground level with response zones from 1 to 3 m, whilst a number of others had response zones from 4 to 6 m or 6 to 8 m below ground level. No methane or carbon dioxide was recorded in the shallow probes together with normal oxygen concentrations. Two of the deeper probes recorded peak values of between 2% and 4% methane by vol., although 0% methane was recorded on some visits during the inital 3 month monitoring period. Water levels on the site were very high, typically between 0.3 and 0.7 m below ground level, and were tidally controlled.

Gas pressures were very variable, with both large positive and negative pressures recorded. However, after an inital peak reading (in the range + 300 to

+4000 Pa or −3500 to −100 Pa), pressures consistently fell towards 0 Pa if positive, or rose towards 0 Pa if negative. The largest positive and negative pressures were recorded nearest to the coast, in probes where methane and carbon dioxide were absent. The large pressure variation was considered to be caused by tidal variations either pressuring soil gas to cause positive pressures or creating a suction effect (negative pressure). Not surprisingly, due to the small volumes of gas, once the gas tap assembly was removed, immeasurably low gas flows were recorded (< 0.01 m/s).

Prior to the site investigation, Building Control was insisting on a passively vented granular blanket with perforated pipes below the floor slab and the incorporation of a gas proof membrane into the slab for any development on peaty ground. Whilst recommendations were made for a well constructed slab with services entering above slab level together with ventilation of all spaces above the slab, Building Control only relented on the type of membrane. This was reduced from a 'gas proof' membrane (with an indicative price of £4000–8500 per 100 m² for membranes tested as having permeability values of less than 0.2 ml/m² per day per atm) to a double sheet of 1000 gauge polythene, so considerable savings were nevertheless made.

Residential development, South East England.   The site is located adjacent to a 30 m deep former chalk quarry, which has been landfilled and incorporates a pumped gas extraction system around its perimeter. In 1987 a prepurchase investigation of the site adjacent to the landfill found gas to be present, with methane concentrations of up to 30% by vol. at low emission rates in boreholes and trial pits. In spite of these findings, the land was purchased and the leasehold development went ahead.

The site has a cover of sand and chalk was typically encountered at 1–1.5 m below ground level. In the vicinity of proposed housing units, plain 100 mm diameter pipes were placed on the top of the chalk via small trial pits. During the course of 1989 and early 1990 these showed maximum isolated methane concentrations up to 18% by vol., with carbon dioxide concentrations generally less than 5% by vol. Purging the pipes with nitrogen typically showed a 2% methane volume recovery after 90 min, although on occasions there was no recovery after 120 min.

Two trial floor slabs were constructed and continuously monitored; one had a 450 mm high passively vented void beneath a screeded precast suspended floor slab. UPVC vent pipes (100 mm diameter) led from the underside of the slab to the edge of the slab. In the final design, these led up outside of the building to a fixed type venting cowl above eaves level (fixed rather than rotating to avoid long-term maintenance) to augment the efficiency of the system. One gas proof membrane was laid on a 75 mm lean mix concrete base at ground level and thereby formed the underside of the void, with a second membrane placed above the floor slab. The second slab had a similar pipe arrangement but a 150 mm void, a membrane above the floor slab and no membrane at ground level.

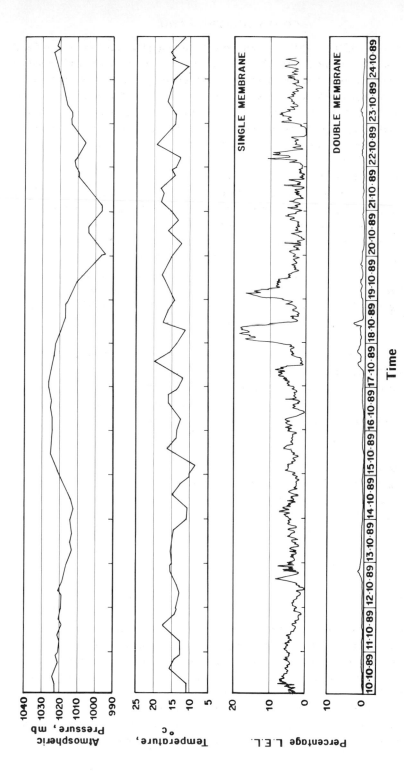

Figure 8.5 Continuous LEL readings from fixed point data loggers beneath trial floor slabs.

Fixed continuous read out gas monitors were placed in each void space and the data stored by the instruments until periodically downloaded. Manual readings were also made on a weekly basis using portable instruments and these corresponded very well with the data logger. The slabs were monitored from 1989 to the first part of 1991 and indicated the efficiency of the double membrane system. The higher methane concentrations in Figure 8.5 relate to the single membrane system, the lower values to the double membrane system. Careful installation of the double membrane system was subsequently adopted across the site and no gas alarms were installed.

The double membrane system is a safeguard against methane migration from the adjacent landfill. Close liaison with the landfill operator (who originally owned the building land) ensured that he will continue to operate the pumped extraction system. Indeed, during the latter part of 1990 an improvement in the efficiency of the pumped system was seen, with methane concentrations in the monitoring points on the residential site falling and staying at zero from that time on.

The development, which has completed the first phases and is occupied, is a leasehold one and no ancillary buildings, e.g. greenhouses or extensions, are allowed on site.

9 Establishing new landscapes

G.S. BEAUCHAMP

9.1 Introduction

Derelict land until recently formed a greater and greater part of our environment, as old industries ceased operation and new businesses decided that establishing their operations on a greenfield site was preferable to the possible inconvenience and additional expenditure involved in the reclamation of derelict, contaminated or despoiled sites.

Nevertheless, as environmental consciousness and legislation increased, land prices rose and available space became more and more at a premium, the onus has shifted back onto landowners to restore sites to an acceptable condition. It is no longer acceptable merely to provide a tall screen, hiding the site from the outside world; now the site must reflect the surrounding area, with planting appropriate to its location and setting, linking with existing, or providing additional, wildlife corridors, habitats and ecosystems to the satisfaction of the local authority.

Previous chapters have described the complexities of land contamination and the variety of techniques available for the reclamation of such areas. Removal of, or treatment to negate the effects of the contaminants is, of course, only the first stage of the reclamation process. It then becomes necessary to establish a new and viable use of the land. Sites may become available for housing, commercial or industrial redevelopment, they may be required to support agriculture, forestry or recreational uses. Whatever end use is proposed, however, it is likely that some, if not all, of the site will require to be landscaped.

In order to ensure that the scheme is successful, the nature of the restoration, end use and landscaping have to be determined at an early stage of the reclamation process, and should involve the landscape architect in the reclamation team from the outset. This will enable the resources available within the site to be recognized and safeguarded for use at a later stage. Topsoil, for instance, is a valuable commodity, which can easily be damaged or destroyed by careless handling or contamination; its presence on site therefore needs to be recognized at an early stage and the resource safeguarded for use at the appropriate time. The lack of topsoil on a site need not restrict planting options, however, as will be discussed later.

No two sites are ever identical; each site will require to be appraised on its own merits and a specific methodology determined, which in every case will take the following into account:

- Site location: latitude, altitude, aspect
- Surroundings: urban/rural, visual quality
- Soil quality, content, alternative materials
- Vegetation type and cover, ecology, amenity
- Physical constraints: slope, drainage
- End use
- Financial constraints
- Timescales available

Sites can be considered as falling into two broad categories: (i) those where the contaminants have been eliminated; and (ii) others where the contaminants have either been treated in order to render them inert or encapsulated by means of a covering, or containing layer. In this latter instance, it is important to ensure that planting does not penetrate through the containment layer to the contaminated soil below; equally adverse would be the ensuing uptake of contaminants by the plants themselves, particularly if there was any possibility of these plants forming a part of the food chain.

In each case, it is necessary to consider the requirements of the types of vegetation (trees, shrubs, grasses or other herb layer plants) proposed on the site, to ensure that these needs can be fully satisfied in all respects, whilst still maintaining the integrity of the reclamation scheme.

9.2 Plant requirements

The basic requirements of all plant forms are well known:

- Sunlight, for photosynthesis
- The means of achieving some sort of anchorage in the ground
- Water and oxygen
- Nutrients

9.2.1 Sunlight

Needless to say, it is unreasonable to expect plants to grow well in areas where sunlight is excluded by extensive overshadowing, from walls, under bridges or canopies. These sorts of locations are better served by treatments such as hard paving, or a shade-tolerant grass mix, although species of ivy (*Hedera*), elder (*Sambucus nigra*) and laurel (*Prunus laurocerasus*) will tolerate heavy shade.

9.2.2 *Anchorage*

Plants obtain support by anchoring their roots in the growing medium. The greater the ultimate size of the plant, the more anchorage it will require. The plant's root system is a continually growing and changing mechanism, and varies with plant species, age, season, soil structure and type. However, all root systems expand in order to maintain the supplies of essential nutrients, water and oxygen to the plant cells. To achieve this, the soil in which the plant is growing needs to have a good structure, with particle spacings of sufficient size to hold both oxygen and water, yet not so large that the soil is entirely free-draining. A very free-draining soil will fail to retain sufficient water for the plant to use, and will also give poor anchorage. It is therefore important to ensure that any soil, or soil substitute, which is loose in structure and which is proposed for use on site, has its structure and density improved by means of some moderate compaction or rolling; water retention can also be improved by the addition of substantial quantities of composts, mulches or similar products in advance of any planting taking place.

Soils are traditionally classified as either clay, sand or silt, or a combination of two or more of these categories. The texture of a soil is determined by the relative proportions of the mineral particles present (Figure 9.1); the soil structure relates to the accumulation of soil particles into larger compound units. Either subjective assessment or physical testing can determine these aspects and thereby the need for improvement to the soil prior to its use on

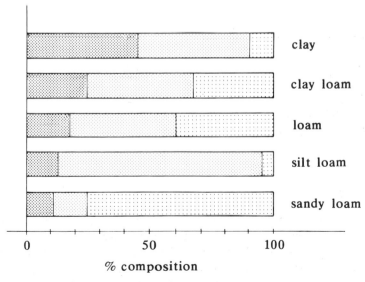

Figure 9.1 Soil texture and composition. (▨) clay; (☐) silt; (☐) sand. Redrawn and adapted from Strahler [1].

site. Figure 9.2 indicates the range of soil types that are generally found, and their relative suitability for different end uses. From this it can be seen that the most suitable soils are those that combine two or more of the three different categories, and which are not exclusively one or other of the principal soil classifications.

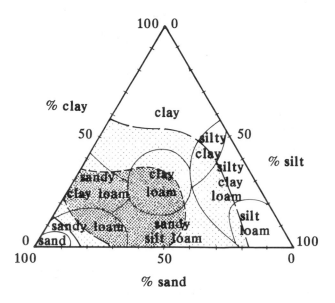

Figure 9.2 Soil texture and land use. Soil types suitable for: agriculture (▨); amenity planting (▢). Redrawn and adapted from Roberts and Roberts [2].

Compaction of soils, through trafficking by machinery or by the placement of an overburden, even temporarily, will damage the soil structure and restrict the diffusion of oxygen to the roots. There may also be implications for drainage, depending on the type of soil in question: sandy soils may continue to drain, as these soils have a higher percentage of macropores, whereas in clay soils the particles will coalesce on compaction and the number of larger pore spaces will be reduced. The opportunity for root growth will be restricted, therefore. Methods to reduce or remove compaction are discussed in more detail later. It is clearly most advantageous, however, if the compaction of soils that are to be used in landscaping can be avoided at the outset, by judicious site planning and careful routing of machinery.

The depth to which tree roots penetrate, and the depth therefore that is required for anchorage, is, for a wide variety of species, only the top 1 m or so of soil; very rarely will tree roots penetrate to below 2 m (Table 9.1). It is this sort of maximum depth of soil, or soil-forming material, that is required as a rule, within areas to be planted.

Table 9.1 The rooting depths of different tree species [a]

< 1.5 m	1.5–2.0 m	> 2.0 m
Ash	Cedar	Lime[b]
Beech	Fir	
Birch	Oak[b]	
Cherry	Poplar	
Crab apple		
Hawthorn		
Hazel		
Holly		
Horse chestnut		
Larch		
Maple[b]		
Pine		
Robinia		
Rowan		
Spruce		
Whitebeam		
Willow		

[a] Information taken from Gasson and Cutler, Tree Root
Plate Morphology, Arboricultural Journal 1990 [3].
[b] Species with tap root.

9.2.3 *Water availability*

All plants require water, which is absorbed by the root hairs, to enable photo-synthesis to occur. The root system is constantly expanding through the soil in search of water; the water storage capacity of a soil is therefore important in ensuring the satisfactory development of the plant. This capacity is largely determined by the soil structure, and the presence of pores of a suitable size, namely macropores (> 50 μm diameter), which allow unimpeded penetration of roots, water and air; and mesopores (5–50 μm diameter), which hold a reservoir of moisture available to the roots [4]. Sandy soils are, by their nature, free-draining and will store relatively little water (Figure 9.3), whilst clay soils are capable of holding considerable quantities and can be prone to waterlogging. Too great a quantity of water in a soil can be as great a problem as too little, as it will result in the deprivation of oxygen to the root system, which will ultimately cause the plant to die. Too little water is, of course, equally detrimental to plant growth, and very free-draining soils, as previously stated, should be avoided unless large quantities of water-retaining mulches can be incorporated.

Soil water availability is related to the quantity of water held in the soil and the texture of the soil. The soil suction, or the pressure required to remove water from the soil, can be expressed in bars, and ranges from just above 0 bar, when the soil is fully saturated, to 15 bars, at which point plants are unable to remove any more water from the soil and wilting occurs. The field capacity of a soil occurs after a saturated soil has been allowed to drain (under gravity) for some 48 h and only the water held in the pore spaces remains. This will be the

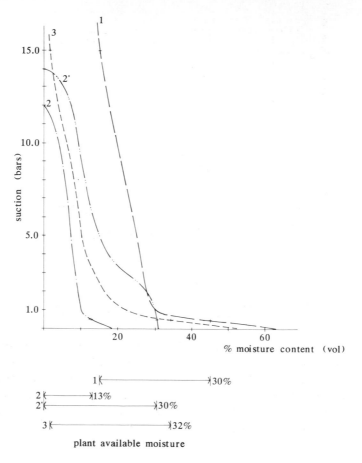

Figure 9.3　The moisture characteristics of certain soil types. Material 1: heavy London clay, low organic content, low permeability; material 2: crushed brick/rubble forming a sandy gravel, no organic content, very high permeability; material 2′: material 2 with 20% organic content added, lower permeability; material 3: clean river dredging, moderate organic content, low permeability.

maximum amount of water that the soil can hold for any duration and, at this point, the soil suction is 0.5 bars. Figures 9.3 and 9.4 indicate the various characteristics of different soil types in relation to soil moisture levels, and demonstrate that whilst clay soils can retain more water than sandy soils, wilting can occur sooner in clay soils because some of the moisture contained in the soil's smallest pore spaces is under such high suction that it is unavailable to the plants.

It is therefore important to ensure that any soil-forming material that is proposed for use as a planting medium is capable of both storing and releasing soil water to the plants themselves. Soils with a greater range of available water, i.e. which are not largely sand or largely clay, are therefore to be preferred.

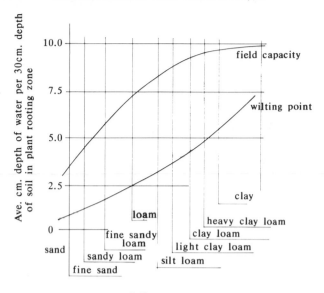

Soil textures

Figure 9.4 The relationship between soil texture and its moisture levels of field capacity and the wilting point. Redrawn and adapted from Strahler [1].

The texture and structure of the soil, the potential for drainage provision, and the gradients of finished levels must therefore all be considered to ensure that adequate water supplies will be available. Irrigation systems and even manual watering of planting are likely to be uneconomical, therefore full consideration must be given to the movement of water across and through the site, at an early stage, to enable the scheme to be successful.

9.2.4 *Plant nutrients*

All plants need a regular and continuous supply of nutrients. Nitrogen, phosphorus, potassium and magnesium are the principal elements, together with certain positively charged ions such as calcium and sulphur and other trace elements including zinc, copper, nickel and water-soluble boron, which are only required in minute quantities.

Topsoil generally contains the majority of these nutrients by the very nature of its composition, which will include naturally degenerating plant and animal tissue. Subsoils and other generally inorganic materials, however, are much less likely to contain all of the elements necessary for plant growth. It is therefore essential that soils are chemically analysed at an early stage of the project, to determine their suitability to support plant growth and to provide an indication of the treatment necessary to rectify any deficiencies that may occur.

9.3 Soil cover

As the medium in which trees, shrubs and grasses are expected to grow, the importance of having a thorough understanding of the soil proposed for use on site, cannot be overemphasized. Evaluation at an early stage of the project will enable a variety of different options to be examined and the correct decisions to be made, taking into account factors such as end use, maintenance, financial circumstances and so on.

9.3.1 *Evaluation*

A thorough site survey, including excavations by borehole or auger, is essential at the outset of any project, to determine the nature of the substrate, which may well vary across the site. This should enable a picture to be built up of the extent and depth of the various materials, suitable and unsuitable, present on the site. Any materials that appear to have potential for reuse can then be analysed more fully, in terms of their texture, structure and chemical analysis (pH, nutrient content, salinity and so on). This should complete the picture and enable firm decisions to be taken regarding the ability of on-site materials to meet the requirements of the reclamation team.

Any material that appears to have some potential for use as a growing medium must always be considered in the light of the ameliorative work necessary to bring that material up to the standard suitable for the type of planting or other end use, for which it is required. Substances such as crushed concrete or brick rubble will be of little or no use unless they are improved by the addition of a water-retaining, nutrient-rich product. This aspect of soil amelioratives is discussed in more detail in section 9.3.3.

9.3.2 *Depth of cover*

There are three main considerations to be taken into account when determining the depth of soil cover required on a site.

Firstly, the presence (or otherwise) of contaminated materials which have been retained on site, for whatever reasons. Chapter 5 has described techniques for ensuring that there is no upward movement of these contaminants, by means of capillary break layers placed over the toxic substances. These layers are then themselves covered with a layer of subsoil, and occasionally topsoil, into which planting can be expected to take place. It will be important to ensure that there is no possibility of the plant roots penetrating through the break layer and absorbing the contaminants through their root systems. Not only would this be harmful to the plant but there could also be implications for the food chain, in addition to the loss of integrity of the break layer itself.

Secondly, the type and nature of the planting proposed, which will itself relate to the proposed end use of the site, and the landscape into which the development has to fit. Areas that are merely to be restored as grassland will require a shallower depth of soil cover than will areas of tree planting, although agricultural grassland will require a better quality of soil than, for instance, rough grass for informal open space. Soil depths can also be reduced if only shallow-rooting species of trees and shrubs are proposed for the site, although it is then necessary to ensure that future management of the planting does not allow deeper-rooted species to encroach or invade the planted areas. This is assuming, of course, that the contaminants in question will actually be harmful to the planting. It is worth pointing out that the break layer itself may well be inhospitable to the plant roots, in which case it can be assumed that the root zone will be confined to the layer of soil above. This could then result in the plant's ultimate growth being restricted; there may also be implications for the wind-firmness of the planting.

Nevertheless, an optimum overall depth of approximately 1 m of soil cover is generally accepted as being adequate for most planting; this can be reduced to 200–300 mm where grassed areas only are proposed. Of these overall depths of soil, at least 30% should comprise a proper topsoil that has the necessary humus, or moisture-retentive matter, providing a supply of nutrients to the plant roots. Where good grass cover is needed, on playing fields, agricultural land, golf courses and so on, a minimum topsoil depth of 150 mm is essential, and more will be required if arable crops are expected to be produced.

If the site cannot supply a full 1 m depth of soil cover (topsoil and subsoil, or subsoil only) then the species and size of planting proposed requires consideration. If the ground conditions below the soil layer are not expected to be harmful to plant growth, small whips and transplants can be planted (in pits if necessary). If the reverse is the case, then shallow-rooting species are to be preferred. Table 9.1 indicates the different rooting depths of some of the principal species of trees likely to be used in any planting scheme.

The third consideration is financial; where budgets are limited, then there is likely to be a need for minimal depths of soil cover. This should not, however, be at the expense of the overall integrity of the scheme. It would be preferable to revise the end requirements, if finances are constrained, rather than to produce a second-rate environment that is likely to require further restoration at some later stage.

9.3.3 *Soil ameliorants*

Because of the expense and general scarcity of good quality topsoil, all landscape schemes are having to examine ways in which a suitable growing medium can be obtained for use on site. It is clearly most cost-effective to make use of materials that are available, either on site or close at hand, or which can be imported at little or no cost. As previously stated, the ability of a substitute material to reflect good soil characteristics is all important: soil texture, structure, drainage capacity, pH and nutrient content all need to be considered.

Most soils found on reclamation sites will contain many of the general requirements stated above. The areas in which deficiencies generally occur are in the soil pH, its nutrient status, and occasionally in its water-retentive capacity.

The acidity or alkalinity of a soil can drastically affect plant growth as well as the microbial activity within the soil. Soils with a low pH are acid soils, which will support gorse and heathland vegetation; extremities of acidity can often be rectified by the addition of lime, although it is essential to ensure (through the process of soil evaluation) that there are no ongoing chemical processes (such as the oxidization of iron pyrites in a colliery spoil material) that would give rise to continuing acidity.

More frequently, however, soils found on contaminated sites are alkaline in nature. There is no easy solution to excessive alkalinity. Where the pH is around 8, alkaline- or lime-tolerant plants can be grown, and can often form rich communities containing otherwise rare species such as orchids. Where soils exhibit such extremes of pH that they cannot be readily treated, the simplest solution is often to bury them.

The improvement of a soil's nutrient content is generally the most critical aspect in ensuring good plant growth. It is most important that the soil contains materials that will retain water, and which contain the necessary nutrients for good plant growth. Several alternative options are available, which may well be worth considering. The quantities of ameliorants to be incorporated will vary with the precise requirements of the particular soil: on average, a quantity of 25–30% of the total soil mass should be incorporated (equivalent to a 50 mm thick layer on top of a 200 mm thickness of soil); where soils are particularly poor then this figure might be increased to 40–50% of the total mass.

Sewage sludge. The addition of composted sewage sludge to generally infertile soil can be of benefit, as it is a good source of organic matter, nitrogen and phosphorus. The nutrients are generally released slowly and will provide an improved soil structure, with good water retention. The use of sewage sludge has been developed over recent years, with much improved techniques now available, making the handling of the material more acceptable.

To be of benefit as a mulch or water-retentive additive, the sludge needs to be incorporated in a dry form, to prevent the development of anaerobic conditions, and needs to be well mixed with either the infertile soil, or a binder such as straw, which will assist in improving the handling qualities of the material. The mixed sludge is stored under cover to dry, for a period of 6 months, which improves its spreading ability.

Composted sewage sludge is now available as a commercial product [5] and is a good means of recycling an otherwise unacceptable waste product. Where a site has an infertile soil medium, the use of such a product should always be considered after discussion with the relevant control bodies. However, in urban areas the sewage may well be contaminated with heavy metals and will therefore not be suitable for use. It is essential that a full chemical analysis is carried out to confirm the acceptability of the sewage product and its compatibility with the soil with which it is to be incorporated. There is strict legislation in place nowadays that controls the use of sewage sludge.

Sludges are more usually available as a wet product than a dry one, simply because they are costly to dry out. However, when dried, they are significantly reduced in volume, and are accordingly cheaper to transport. Consequently, the two costs may balance each other out. Wet sludges may be difficult to incorporate into soils and will not add the water-retentive bulk of dried or composted sewage. This must be borne in mind when specifying this type of material.

Spent mushroom compost. Another waste product that may be available locally is spent mushroom compost. This material is also a good source of nutrients, particularly nitrogen, and can either be used as a surface mulch or can be mixed into the soil to improve the nutrient content. It is a very lightweight material, however, and is not recommended for spreading as a surface mulch in windy conditions!

Shredded bark. This is another recycled product that is widely available. As a soil ameliorant, it is of limited nutrient value, but will improve the structure of clay soils, is a useful mulch for suppressing weed growth and gives valuable protection against moisture losses, if laid as a 50 mm thick surface dressing.

Farmyard manure. This is a nitrogen-rich material which, like sewage sludge, is a good source of nutrients for plant growth. However, its use may be limited in that it is unlikely to be available in sufficiently large quantities to be a viable option.

River dredgings. These can be a useful soil substitute if they are available locally at a minimal cost, and if they are free from toxic metals. It will, however, be necessary to analyse the dredgings at the outset as they could have a high saline and/or metals content, if the river is tidal or has been contaminated from another source; in this case they may be unsuitable or uneconomical (or both). As with sewage sludges, the material will be highly saturated and will need to be allowed to drain for a few months before it is able to be readily handled.

Whichever type of soil ameliorant is used, it is essential to ensure that its qualities are suited to the type of soil that is to be improved. Sandy soils will require a different formulation to clay soils, for example, and this may require careful consideration to ensure that the correct balance of soil properties is achieved.

9.4 Soil fertility

It is important to consider soil fertility in both the long-term and the short-term. Soil ameliorants will provide the necessary nutrients over a specific period of time (18 months or so) but if conditions are largely hostile to the encouragement of microbial activity, then the nutrient supply could eventually be seriously depleted. Depending on the proposed end use of the site, it will be necessary to assess future plant needs in order to ensure that these can be adequately met.

The use of land for agriculture or forestry will be the most demanding in terms of plant nutrient requirement. Grassland established for pasture or silage will require regular and frequent applications of fertilizer, particularly nitrogen. The number and size of applications will depend to a large extent on the quality of the land in question, and will be determined by analysis. Forestry plantations will also benefit from annual applications of nitrogen, unless nitrogen-fixing plants such as alder, gorse, broom, or lupins are planted or sown to assist in raising the nutrient levels of the soil. It is important to note that the presence of specific organisms is essential for lupins to carry out nitrogen-fixation; these organisms may not always be present in poorer soils.

Conversely, the presence of fertile soils is disadvantageous when establishing 'wild flower' meadows or grasslands. These species-rich herb layers are best established on poorer soils, where the more competitive weeds and grasses will be less inclined to overpower and smother out the slower-growing varieties. Applications of fertilizer will not, therefore, be necessary, unless these are applied solely to any trees or shrubs planted within the wild flower areas. With this in mind it is important to remember that grass cuttings from areas sown with a wild flower mix should be raked off and removed from site, to maintain low fertility levels within the soil.

Soil fertility is not only related to its nutrient content, however. Equally important is the establishment of good microbial activity within the soil, to restore a more natural balance to the lifecycles therein. The earthworm is particularly important in this respect, as it will mix the soil, aerate it and provide root and drainage channels to the benefit of any plant life, as well as being of importance in the wildlife chain generally.

British Coal's Opencast Executive have carried out research which has demonstrated that soils with a good earthworm population are vastly superior to soils where there are few, or no earthworms present. Typically, such soils are found on newly reclaimed land [6]. This is often because the earthworms normally present in the topsoil are largely killed off when this soil is stockpiled in large mounds; the only worms to survive will be those present in the topmost layers of these mounds. When this soil is eventually spread back over the area, the remaining earthworms are likely to all be concentrated in one place, leaving the rest of the site barren. It can take years for the worms to migrate across a large area. In the absence of a good earthworm population such areas will probably drain badly, and may be anaerobic.

British Coal's solution to this problem is to take thin slices of the top layers of stockpiled topsoil, and lay them in narrow strips, up to 50 m apart, across the restored land. This allows the earthworm population to multiply naturally, to the benefit of a wider area; numbers reach a near normal level more rapidly than would otherwise be the case.

The research described has also demonstrated that land management practices have an influence on the development of earthworm populations. Drainage by subsoiling is important for deep-burrowing species; food supply is also a controlling factor and, where the grass sward is grazed rather than cut (with organic manure applied rather than mineral fertilizers), and cultivation is avoided, then conditions are generally more favourable to the establishment of an earthworm population.

9.5 Site preparation

The design of any reclamation proposal may well include the creation of new levels and contours across the site area. This is often the case where contaminated materials have to be buried within the site. It is important to create natural-looking mounds wherever possible, and the gradients and heights of such features should always reflect the topography of the wider area.

The reclamation of contaminated sites will inevitably involve the movement of machinery across much, if not all of the area concerned. This can be particularly damaging to the soil structure as the weight of the machinery will cause severe compaction of the surface layers of the soil, reducing the size of pore spaces and leaving the land more susceptible to waterlogging. The use of machinery across wet soils is likely to worsen the problem and this should therefore be avoided as far as possible.

In order to relieve compaction, and also to improve the soil's drainage capacity, the technique of subsoiling, or ripping, is used. This involves the use of single, double or winged tines, pulled through the subsoil layer (in advance of any topsoil being spread) in a down-slope direction at a depth of between 450 and 750 mm, and at centres of between 1.5 and 3 m, depending on the type of soil in question. Surface water drainage is also improved when the gradient of the surface is at about 6° or 1 in 10. The length of slope that can be adequately drained in this way varies with the soil type but 50 m has been found to be acceptable, with a ripped channel, a gravel or rubble toe drain, or a ditch installed to carry the water off the site.

Erosion and slope stability are problems that can often arise on slopes where the soils have recently been replaced. The establishment of a vegetation cover is useful in preventing surface erosion; ensuring that slopes do not exceed gradients of 20° or 1 in 3 is also helpful in this respect. Problems associated with slope stability are generally water-related, so it is advisable to ensure that water drains away at the surface rather than below ground. It is particularly important to maintain the soil cover at the desired gradient where this is overlaying and protecting contaminated materials beneath.

A number of commercially available mattings and geotextiles can be useful in assisting vegetation establishment whilst ensuring slope stability. These can consist of natural materials such as coconut fibre, coir or straw combined with a polymer mesh, which will degrade slowly over a period of years, or of more permanent materials such as polypropylene, which will add durability to surfaces whilst still allowing vegetation to grow through. The cost implications of the use of these types of materials need to be set against the potential cost of any reconstruction of slopes and vegetation, should erosion prove to be a problem. This is particularly important where the incorporation of steep slopes is unavoidable.

9.6 Establishing grass cover

Having decided on the ultimate end use of the site, it is likely that some, if not most of the site will require to be grassed. The type of grassland to be established will vary, from agricultural pasture to golf course fairway or playing field, to rough grassland and open space or wild flower meadow. Each type of sward will require different establishment and management techniques, if it is to be successful.

Cultivation of the surface layer is essential in all instances to create a tilth in which the grass seed can germinate. Stone picking may also be necessary, to a greater or lesser degree, depending again on the end use. Areas where the grass will be closely mown should have all stones larger than 30 mm removed, whereas with areas of rough grass this figure can be increased to 50 mm. The quality of the cultivation can also be varied according to the requisite end-product. A rough tilth is adequate for wild flower grasses whilst a fine tilth is necessary for playing fields and other quality grassed surfaces.

A pre-seeding fertilizer should be incorporated where good and rapid establishment and growth is required. The principal exception to this use of fertilizers, both pre- and post-seeding, is with wild flower mixtures and their associated grasses, where any sort of fertilizer will encourage the incursion of the coarser, more competitive grasses, which will ultimately eradicate the original species and reduce the diversity of the sward.

The selection of an appropriate grass seed mix will depend on the desired end use, the site location and the soil condition. Agricultural swards must contain productive and nutritious cultivars, whilst playing fields or fairways must be durable and have a good colour. Mixes are available for the entire spectrum of soil and site conditions, from acid to alkaline and from damp to shady; low-growing, low-maintenance cultivars are increasingly popular where maintenance costs need to be kept to a minimum. Where soil fertility levels are low, or where the soil pH is low, then species such as flattened meadowgrass, browntop bent, sheeps fescue and hard fescue can be grown.

There are also cultivars of the various grass species commercially available that are tolerant of a variety of contaminated soils; the use of these may be appropriate in some instances. Cultivars of perennial rye grass, such as Wendy, are tolerant of the leachates produced by controlled waste; cultivars of slender creeping red fescue, such as Merlin, will grow on soils with a wide range of pH values and particularly on alkaline soils contaminated with heavy metals such as lead, zinc or copper. Browntop bent cultivars such as Parys Mountain, on the other hand, grow well on contaminated neutral to acidic soils. As research and development continues, these cultivars will be continually refined; it is important therefore to seek qualified advice to obtain the most suitable grass seed mix for any one site, particularly where the soils are atypical.

Sowing rates will also vary according to the location and the requisite end-product. Agricultural land should be sown at around 50 kg/ha; playing fields and closely mown swards at around 300 kg/ha; and wild flower mixes at around 50 kg/ha. Rates may need to be increased by up to 50% where ground conditions are poorer than normal, and increased levels of clover can also be included to improve the soil's nitrogen content.

The timing of operations can be critical to the good establishment of grassed areas. In the majority of instances, sowing should take place in spring (April–June) or autumn (September). These periods are preferred because soil conditions will be warm, without being too wet or too dry. Sowing in late September can delay the first growth until the following spring, by which time excessive damp, snow and so on may well have reduced the chances for good establishment. The exception to this is again for the wild flower seed mixes, which are likely to contain species that require a cold period for vernalization prior to germination the following spring.

The correct management of grasslands is essential to the success of any scheme. It will inevitably take some years for substitute soils to reach normal levels of fertility and organic content, and returns from agricultural land will be correspondingly low for the first few years. The continuing application of organic manures and fertilizers is therefore essential, as is the ongoing analysis of the soil to monitor its improvement and enable the management techniques to be adjusted accordingly.

Where a close sward is required, for playing fields and so on, regular cutting is essential to encourage tillering. Grass cuttings can be left on the surface for the benefit of the soil where the cuttings are not so long that they smother the grass underneath, and where improving the soil fertility is important. As grass cutting can be an expensive maintenance item, the use of low-growing and low-maintenance species should always be considered. Areas of wild flower grasses, after the initial establishment period when up to ten cuts may be required in the first year, are generally only cut twice annually, in mid-July and late September, and the cuttings removed to maintain low fertility levels.

9.7 Establishing trees and shrubs

Areas of tree and shrub planting within a reclamation scheme will add height, colour and pattern to the environment, with seasonal changes contributing to this variety and interest. The choice of species will vary according to the location, soil condition, and the function of the planting — for screening, protection, amenity and nature conservation value. Planting will ideally reflect any existing vegetation so that the reclaimed area will ultimately blend in with its surroundings.

The use of native, indigenous species of trees and shrubs is increasingly popular as a result of a greater awareness of the need to protect our environment. Such species are generally of greater value in providing food and shelter for birds and insects, as well as providing autumn colour and fruits. Nevertheless, ornamental and amenity-type planting can be important in certain locations to add more varied colour, flowers and so on, around buildings and in an urban context.

If woodland planting is proposed then it will be necessary to consider the natural succession of the vegetation, and to plant species that reflect all the layers of plants found therein, from shrub understorey up to the forest canopy. Careful planning at the design stage can reduce the amount of management ultimately needed by allowing nature to take her course. It may also be preferable to phase planting over a number of years, to give a mixed age structure to the woodland.

Successful planting begins with the use of good quality plant material, although careful handling and planting is also essential. The size of plants to be used should also be given due consideration, as smaller material will ultimately catch up in growth with larger plants, and can often adapt better to any inhospitable site conditions. Small plants will also be cheaper and can therefore be planted in greater numbers, so that any early losses will be less noticeable. For this reason, whips and transplants are the most commonly planted sizes of trees, although where a more immediate impact is required then light or selected standard trees can be specified.

Whips and transplants are generally planted either into pits, or by 'notching' into the prepared ground. In every instance it is important that the root system is spread out fully and that the plant is set into the ground at the same level that it was growing in the nursery. If a pit is excavated for the planting hole then it is essential that the sides and base of the hole are loosened, to enable the root system to expand outwards. The pit should be backfilled with a mixture of soil, compost and slow-release fertilizer to help establishment; often a water-storing polymer, in granular form, is also incorporated to provide water to the root system during dry periods, avoiding the need for watering during the maintenance period, which can otherwise be costly.

Depending on the size of plant material to be used and the finances available, whips and transplants can be planted at between 1 m and 2 m centres; shrubs are planted at between 1 and 2 per m^2, depending on their ultimate height and spread. Planting at 3 per m^2 will ensure a rapid covering of an area by the shrubs, although thinning may be required ultimately, depending on the nature and form of the material planted.

Stakes are usually only required for plants taller than 1200 mm, and on exposed, windy sites. It has been generally established that the use of stakes for longer than the first year can be detrimental to the planting, as the plant stem fails to thicken at the base. Short stakes are preferred, as they permit the crown of the tree to flex in the wind.

The timing of planting operations is also important. Although the planting season is usually from November to late March, planting in November or early December is often more successful as the soil is still fairly warm and the plant roots can establish somewhat prior to the post-December rain and snow. Soil conditions often deteriorate into January and beyond, and can become totally saturated by the end of the planting season, which again will limit working on the land due to the greater risk of compaction. Planting should not be carried out in frosty conditions, or when ground temperatures are below 3°C.

Any plants grown in containers can be planted at virtually any time of the year, although they will need to be well watered through any dry periods if they are planted in the summer, unless a water-storing polymer is incorporated into the soil around the base of the plant. Evergreen species are usually best planted between April and October, and again will require regular watering to ensure good establishment, if water-storing polymers are not used. It is good practice to water in areas of new planting at the time of planting, particularly as the soil should be dry for the planting to be able to take place. Water-storing polymers also require a thorough soaking at the outset if they are to work efficiently.

In many areas, it will be essential to protect new planting from possible damage caused by rodents or small mammals, or by vandalism if the area is prone to trespass or adjacent to public rights of way. Large areas of planting are best protected by enclosing them with fencing, to which can be attached rabbit-proof mesh if required; deer will need to be controlled by higher fencing. Individual plants can also be protected by the use of guards or tubes, where this is more economical than fencing.

9.8 Maintenance

The correct maintenance, or ongoing management of any planting scheme is essential to the success, or otherwise, of the project. Without this, all monies expended on the planting itself are likely to have been wasted. The first 3–5 years of any new planting are critical to the future development of the landscape; it is during this period that the majority of plants will fail if they are not adequately maintained. This is particularly important when the planting comprises smaller material such as whips and transplants.

The young plants, as stated earlier, require a good supply of water and nutrients to promote growth. Any other vegetation in the vicinity of the plant will therefore be competing for the soil's water and nutrients, and reducing the quantities available to the plant itself. Regular weeding, to remove such competing vegetation from around the base of the plant, is therefore essential; a 300 mm diameter of clear soil is the usual recommended area. The application of a layer of mulch or compost can be effective in keeping weed growth to a minimum. Where plants are spaced at 1 m centres (or less) there could be a conflict between the need to clear the soil around the base of the plant, and the need to maintain surface vegetation cover, particularly on sloping sites, to reduce the risk of erosion. With plants spaced at 1.5 or 2.0 m centres, however, this becomes less of a problem.

If conditions during the summer are dry, then ample quantities of water will need to be applied across the planting area. Twenty-five litres per square metre or per plant is recommended where a water-storing polymer has not been incorporated into the soil during planting. Additionally, plants should be inspected to make certain that they have not lifted out of the ground, and firmed back in if necessary. Where soils are likely to be generally infertile, then the regular application of compost, fertilizer or manure, once or twice yearly, is advisable.

Depending on the density of the planted material and the ultimate form of the planted area, any significant numbers of dead trees and shrubs should be replaced during the subsequent planting seasons. Should widespread losses be encountered in any particular area, then it may be necessary to investigate the reasons for this, as it may indicate problems within the soil related to the original contamination.

As time goes by and the planting becomes better established, provided that the soil fertility levels have improved, then regular maintenance should become less important. Areas may only require the occasional visit, to remove litter or other debris, strim unwanted weed growth and so on. If the scheme has been designed to provide a natural succession of plant growth, using native species, then the planting will develop naturally without a significant need for human input. Areas planted for forestry may require thinning to enable the ultimate tree crop to flourish. More ornamental planting may, on the other hand, require annual pruning to promote stem colour or flowering. Whatever the particular needs of any one area, a long-term management scheme can be drawn up to highlight the necessary work and evaluate the financial implications.

9.9 Species selection

When designing the landscape of a site, the type and form of planting will be determined by such factors as end use, the nature conservation value of the site, its relationship with its surrounding environment, amenity and visual qualities, and the degree of management to which the site will be subjected.

The actual species selected for use on the site will also depend on a variety of different factors: the location of the site (latitude and altitude) and the degree of exposure to wind and sun; the soil type, pH and drainage; visual interest such as autumn colour, flowering and berrying; value for nature conservation as food sources or habitat; low or high maintenance requirements and so on.

Exposed coastal sites in the north of Scotland, for example, will be greatly restricted as to the species of trees and shrubs that may be successfully specified. These could be restricted to the forestry-types of conifers such as pine and spruce, native broadleaves such as ash, whitebeam and poplar, and to gorse, broom and the Scotch roses (spinosissima). A site in the Midlands or South of England, on the other hand, which is sheltered from the prevailing winds, will have a much wider range of species to choose from, including the major native hardwoods such as oak and beech. The planting of elm is still not recommended, as although this tree is starting to regenerate naturally in the southern half of the country, it is still too early to be certain that the trees will be resistant to the dutch elm disease.

9.10 Natural regeneration

The methods for establishing vegetation on contaminated or reclaimed sites, outlined in the previous sections, assume that there is a positive requirement for a particular end result on the site in question, within a short timescale. However, it is relevant to mention that nature will generally take a hand in the revegetating of land, if the site is left undisturbed. Depending on the particular location of the site, the soil type and nutrient content, regrowth can be expected to commence very quickly, with a low, herb layer forming the initial vegetation cover. All sites below the tree line will ultimately revert to woodland, the original vegetation cover across the United Kingdom, provided there is no interference from outside agencies.

The natural succession of vegetation within an area can be of great significance to nature conservation, supporting a wide variety of plant and animal life. Contaminated sites in particular, where the soils can often be very different to the norm, frequently give rise to rare and unusual plant habitats [7].

Alkaline soils, such as result from sodium waste, power station ash or other lime-rich waste products (e.g. those produced by iron smelting and steel-making), give rise to a species-rich flora that can often contain a large number of rare and unusual species, particularly orchids. These plants, in turn, are host to a wide range of insects, particularly butterflies and moths, which feed and breed on the flora. The time span involved in the creation of these sorts of habitats is usually in the order of 10–20 years, however.

Acidic soils are generally the result of colliery works and mining spoil heaps, which may frequently contain high levels of such metals as lead, copper and zinc. The natural regeneration of these types of soil is often more difficult as the wastes are generally toxic to plant growth until the metals have been broken down by the gradual process of soil weathering and leaching. Any plant growth will generally resemble heath or moorland, with plants such as heathers and gorse establishing. Some species of grasses and other plants are tolerant of particular metals and may also establish across the area.

Regeneration is less of a problem on soils with more normal pH levels, and soils with a higher nutrient status will rapidly establish a sward of coarser, more competitive rough grasses, such as are regularly found on roadside verges or beneath hedgerows, and so on.

Where drainage problems create permanently wet or damp areas, a different habitat will establish, with reeds, rushes and other marsh or aquatic vegetation developing; this can also attract a variety of interesting fauna. Wetland areas that lie above the water table and which do not contain a natural lining of impervious clay soil will eventually dry up, however, as the growth of rushes develops across the area and the quantity of dead vegetation builds up within the pond or marsh, unless management techniques are exercised to remove this material at regular intervals.

9.11 Conclusion

The establishment of a successful landscape across a previously contaminated, reclaimed site is one of the principal keys to ensuring a successful future for the site in question, both environmentally and economically.

To ensure that any proposals are successful, site planning must be undertaken from the outset in order to pay full regard to the materials and other qualities available within the site, the surrounding environment, the nature of the proposed end use, and any financial constraints that may exist. This will allow the preparation of solutions that fit both site and budget.

The flexible nature of landscape techniques will allow for a number of alternative solutions to be prepared, to fit the range of budgetary or other constraints, using a variety of materials, plant sizes and species and planting methods as necessary. In all instances, however, the ultimate impact on the local environment must be the most important factor to be considered. If the new environment fails to blend with its surroundings or to suit the type of use to which it is to be put, it will not be successful.

Landscaping is a process that takes all of the environmental features and rearranges them into a shape and form appropriate to the location. To enable this to be achieved, the landscape architect should be involved in the reclamation process from an early stage; otherwise the end treatment is likely to be merely cosmetic: a mask that may, in time, slip and then require additional, and possibly costly, reparative treatment.

10   Quality assurance

T. CAIRNEY

10.1 Introduction

There are obvious advantages in ensuring that contaminated land reclamations are carried out to clearly defined quality levels. In some cases, the planning consent, permitting the reclamation to go ahead, might be subject to conditions that particular problems have to be resolved to the satisfaction of the planning authority, or of such statutory consultees as the National Rivers Authority or the Waste Disposal Authority. Failure to achieve the required end quality can, in extreme cases, lead to the planning consent being withdrawn, although it is more common for financial penalties, arising from having to recheck or even to redo already completed work to more stringent standards, to result.

Even if difficulties with the statutory bodies do not arise, most contaminated sites are reclaimed for resale to other parties. Whether it is the case of a large site being reclaimed for sale to one large developer, or the more common situation of a building company selling off individual house plots, on a reclaimed site, makes little practical difference. In both instances, there is a vendor and at least one buyer, and these parties are bound to have very different interests and priorities. The prime differences invariably are over the financial value of the reclaimed land, whether it is fit for its proposed reuse, and whether any remnant liabilities still persist, despite the reclamation work.

Potential buyers of reclaimed sites often obtain the services of specialist advisers, who expect to be given full details on such issues as:

- The site's pre-reclamation condition
- The site investigations that were carried out
- The contamination and other problems identified
- The reclamation strategy selected to deal with the site's problems
- The detailed reclamation specifications
- The reclamation records and post-reclamation testing that prove the specifications were achieved
- The infrastructure works installed, their design capacities and design methodologies
- The necessary permissions and endorsements that should have been obtained
- Whether the land is now fit for its planned reuse

● Whether any remnant liabilities still persist that could adversely affect the potential purchaser

Anticipating these predictable demands is obviously sensible, and it should be normal practice for an end-of-reclamation summary report (Table 10.1) to be drawn up and related to the relevant supporting documentation (site investigation reports, monitoring results, quality control checks, etc.). In fact, such summary reports are still quite rare, and reclamation data still tend to be produced (often in very time limited circumstances) only to answer particular queries.

Table 10.1 Typical contents of an end-of-reclamation summary report

1.	Instructions from client
2.	Desk study ● site description
	● past history
	● likely difficulties and hazards
3.	Site investigations and contamination testing (supported by site investigation reports)
4.	Land reclamation options and the chosen reclamation method
5.	Specifications established for the site reclamation, including required end quality levels
6.	Schedule of reclamation activities and timings
7.	Summary of records showing the achievement of the selected specifications (supported by monitoring and quality control records)
8.	Validation of the standard of the reclamation works
9.	Topographic surveys, pre-reclamation and post-reclamation
10.	Listing of all planning approvals and endorsements from control and insurance bodies

This approach all too often leads to disputes over the quality of the reclamation work, which can be resolved only by initiating additional site exploration and monitoring. The net result is usually a reduction in the site valuation. This type of conflict situation is difficult to justify, given that land valuation disputes are entirely predictable and invariably expensive both in time and when extra quality validations have to be carried out.

10.2 Quality systems

Avoiding time-consuming and expensive debates, re-investigations, and probable land value reductions indicates that the reclamation works should be carried out to whatever defined and provable quality standards are appropriate for that site. Quality systems are, of course, far from new. The Ministry of Defence introduced quality assurance systems in the 1940s for the war-time munitions industries, and the British Standards Institute produced its standard BS5750: Quality Systems as early as 1978.

The publication, in 1981, of a guide for the use of BS5750, allowed interpretations that are clearly relevant to the construction industry. This guide [1] poses those critical questions that have to be addressed in any meaningful quality standard:

(a) *Quality system*
  - Has a system been specified for the necessary inspection and testing?
  - Does the system include the criteria for acceptance and rejection?

(b) *Inspection representative*
  - Has a named individual(s) been made clearly responsible for quality inspection?
  - Does he/she have the required authority, time, staffing, and facilities to carry out the specified duties?

(c) *Control of inspection and testing equipment*
  - Are the accuracies of all inspection and testing equipment known?
  - Have regular and routine calibration checks taken place?
  - Are these available to the inspection representative?

(d) *Records*
  - Are the collected test records adequate to prove that regular and adequate testing has taken place?
  - Do all records include the agreed acceptance standards?
  - Do the records list any corrective action that should have taken place?

(e) *Sampling procedures*
  - Have recognized sampling methods been used?
  - Are these the same as those defined in the reclamation specifications?
  - Do the chosen sampling methods give enough confidence that the required quality levels have been achieved?

(f) *Control of non-conforming materials*
  - Are non-conforming materials clearly specified?
  - If repair work has had to be carried out, is it in accordance with the original specifications?

(g) *Training*
  - Have all appropriate personnel been trained to the required standards?

Interpreting the British Standards guidance for use in contaminated land reclamations is not difficult.

Quality system.   Whilst the essential information for defining an appropriate quality system has invariably been defined, i.e.

- The chosen reclamation strategy
- The different elements and activities that have to be included
- The programme of reclamation activities
- The standards that have to be achieved in each of the reclamation elements (e.g. compaction levels, allowable remnant chemical concentrations, etc.)

it is surprising how seldom this information is collected into a definitive quality system document, available for use by all the supervisory personnel.

The fact that this often is not done is probably the most significant reason for failure to achieve the required quality product. Site reclamations can last for 2 or more years, and so site supervisors can move on and be replaced. These newer staff can lack a full appreciation of what is important, particularly since they invariably have to react to short-term emergencies (e.g. dealing with the public reaction to closing a right of way rendered unsafe by the reclamation work, disputing a contractor's interpretation of the quantities detailed in the reclamation contract, obtaining a river water abstraction licence to wash gravel on-site, reacting to a Health and Safety Executive complaint, etc.). Thus it is far too commonplace for routine environmental monitoring to be delayed, or completed to only a partial level, simply because this work is routine and so appears to lack any real priority.

Inspection representative.   Appointing an individual to supervise the quality of particular activities is usually not difficult. The normal resident engineer arrangements generally ensure that the civil engineering inspections are carried out to a high standard. It is still quite common for consulting engineers to lack the in-house expertise to supervise the quality of the contamination and environmental aspects of reclamation work, but this can be covered by appointing an outside specialist. There is no special reason why the use of more than one inspection representative should be other than satisfactory, provided that a senior member of the consulting engineers' organization is able to coordinate the activities of these different inspectors, and support them appropriately, when circumstances require this.

An example of a failure in such support came to light when a former metal smelting works was to be reclaimed for a large retail complex, surrounded by public access landscaped areas. Planning approval to encapsulate some copper and zinc slags below the landscaped areas had been allowed. To avoid any contamination risks to the required vegetation cover, the reclamation design included a 1.5 m thick cover of soil, with high organic content and good water retaining properties, to encourage plant roots to remain in the clean materials.

A landscape architect was appointed to the resident engineer's team to supervise the landscaping. Part of this individual's duty was to test all imported soils. These tests soon revealed that no locally available soils matched the design specifications, and that all had very poor water-retaining characteristics. The landscape architect called for a halt in the landscaping work, until necessary redesign and plant species selection could be completed.

This, however, would have caused an overrun of the contract period, and the resident engineer (who saw environmental issues as very secondary to the contract problem) insisted on work continuing as planned.

The reclamation was completed to time and appeared initially satisfactory. However, the developer had had to agree to maintain the landscaped area for 5 years, before the local authority would accept ownership. Two very dry years occurred and abnormally high proportions (> 40%) of the planted species died off.

The local authority became concerned that plant roots might have migrated into the encapsulated slags, and insisted on contaminated studies of grass and leaf samples. These surveys showed no phytotoxic hazards. Further testing then revealed that the granular sandy soils, which had been used, dried out very easily and were inhospitable for the plant species selected. After a long and contentious dispute, the developer had to improve the soil's properties and replace the bulk of the already planted trees and shrubs.

In retrospect, this unnecessary expense resulted from an obvious failure to recognize the valid concerns of a specialist inspection representative.

Control of inspection and test equipment.   Contaminated site reclamations usually require two quite different systems of inspection and testing. One, for the physical properties of soils and construction materials is well defined in various British Standards and in the specifications issued by governmental departments (e.g. [2]), presents few if any problems, given the industry-wide expertise available in this area.

The other, for the chemical analysis of soil materials, liquids and gases, poses greater difficulties, partly because analytical expertise is not so wide-spread, and partly because use often has to be made of off-site analytical laboratories. Lord [3] considers this latter point and makes clear that since no officially approved catalogue of analytical methods for contaminated land reclamations yet exists, different laboratories can use methods of very variable sensitivity and accuracy. He also notes the very significant differences that can arise from the various methods adopted in laboratories to prepare samples for the analytical process. Thus, it is not possible to accept chemical analytical results as invariably and consistently accurate. Smith [4] confirms this view, and advocates that all analytical reports should include:

- A description of all analytical methods used
- A listing of each method's accuracy and precision
- A statement on the laboratory's quality control and quality assurance systems

Since few, if any, analytical reports list of all the information that Smith rightly sees as essential, it is prudent to use analytical firms that can demonstrate a high level of analytical quality assurance, preferably by having been accredited by such organisations as NAMAS. It also is a sensible policy to ensure that analytical laboratories are supplied with the relevant data on a sample's origins, so that the most appropriate analytical methods can be employed. Giving a laboratory a sample of builder's rubble, which has a content of bitumen road surface debris, will lead to analytical results indicating high toluene extract levels, which then can be misinterpreted as due to dangerous tarry contamination. If the laboratory had been advised of the presence of the road surface debris, analytical costs might have been reduced, and concerns over a contamination hazard avoided.

Even if the laboratory service is fully advised and operates to a high analytical quality assurance, difficulties can very easily occur in testing for environmental hazards. A Midlands site had been a set of marl pits, which were later filled largely with inert wastes from the local pottery industries. Landfill gases were not thought to be a particular hazard, since biodegradable wastes were not encountered in the site exploration, but gas monitoring boreholes were installed around the perimeter of the site for routine gas monitoring over the 2 year reclamation period. Monitoring results were taken monthly in the autumn and winter of 1988/1989 and in the winter of 1989/1990, by a junior member of staff using field portable gas monitoring meters. No results were available in the summer of 1989, since site stripping had destroyed the moni-

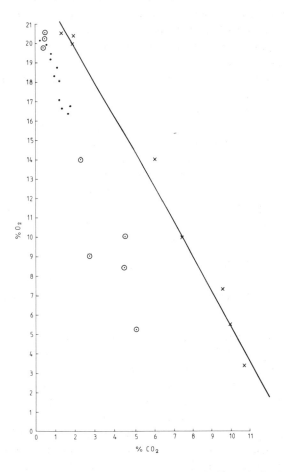

Figure 10.1 Oxygen/carbon dioxide relationships. Borehole 7, Marl pit site. (●) Gas meter reading, winter 1988/1989 and 1989/1990; (○) gas meter reading, summer 1990 (developer's equipment); (×) results from gas chromatographic analysis, summer 1990 (purchaser's advisor's data).

toring boreholes, which were not replaced until that autumn. When the site came to be sold on to a house builder in the summer of 1990, the buyer had his own advisers check the landfill gas conditions (Figure 10.1). The differences between the gas monitoring data collected during the reclamation work, and those obtained by the buyer's advisers, proved to result from a failure to calibrate the field portable gas meters properly and to ensure that gas monitoring took place in the hotter months, when biodegradation of the small proportion of wastes, that could produce landfill gases, was at its highest. The result of this very avoidable situation (which should have been prevented by giving the junior site supervisor a better level of training) was a very large reduction in the price obtained for the site.

Records. Test records can be extremely numerous on any site of a reasonable size, and whilst these records are generally collected to a reasonable standard, dealing with them is often seen as a bureaucratic and boring task. The problem is really to store, check and analyse and present the records in a way that is not too time-consuming, and on sites where computer storage and graphical plotting systems are available, this should be simple. The increasing use of such systems as MOSS, for example, allows regular production of plans showing sampling locations, appropriate contouring and presentation.

Where the greatest failure often occurs is in the completeness of the records. It is still uncommon to find records that list the required quality standard and the actions that have been initiated to correct any unsatisfactory situations. A simple example occurred on a site in the north of England where oily contamination from a coal products works had polluted a near surface aquifer, which, in turn, was in hydraulic continuity with the adjacent river. As part of the planning approval, the requirement for works to prevent any off-site migration of oily contamination was included, and regular groundwater quality monitoring on the adjoining land had to be carried out at monthly intervals. One sample taken early in the reclamation project, on land immediately beside the river, proved to have an extremely high toluene extract value, which was not confirmed by any more specific analysis. Later sampling from the same location, however, indicated that no oily pollution had moved off-site. Late in the reclamation project, the planning authority suggested that the protection works, to prevent off-site migration of the oils, might not be fully satisfactory, and used the single off-site high toluene extract value to justify this concern. Since the monitoring records included no indication of any corrective or check action that had been taken when the high toluene figure was found, it proved extremely difficult to convince the statutory body that the protection works were in fact satisfactory, and that this single high pollution concentration was due to other causes, as in fact proved to be the case.

Sampling procedures. This is perhaps the greatest source of confusion when contaminated land work is undertaken. Smith [4] has shown that minor position changes on a contaminated site can lead to very large variations in contamination levels, and it is also obvious that contamination levels vary with depth. Despite this, it is still all too common to read site investigation reports that list, for example, "ashes and interbanded ashy clays — from 1.5 m to 3.0 m" and to have a single sample of this band analysed to produce cadmium, zinc and copper concentrations far higher than any safe levels. The obvious question that arises is whether the entire ashy band is contaminated to this level, or do the analytical results perhaps relate to a thin and atypical horizon, which attracted the sampler's attention.

Obviously, this sort of doubt is easily avoided (chapter 3, section 3.7.4) by increasing the clarity of the site investigation reports and indicating clearly the level at which a sample was taken, and whether or not is was felt to typify the entire horizon. Internal quality control of this type is spreading in the site investigation industry, although a good deal more improvement has still to be achieved.

The example cited above led to quite serious problems, since the local planning authority had formed the view from the site investigation data provided to them, that the site was extremely contaminated and could only be safely developed if stringent and expensive analytical controls were exercised when the site was opened up. In fact, excavations indicated that the site investigation data had not properly typified the site, which proved to be only marginally contaminated. Convincing the planning authority of this proved difficult.

Control of non-conforming materials. This is usually not a difficult matter, at least in the initial design phases of a reclamation. The civil engineering industry is familiar with the concept of identifying non-conforming materials and requiring contractors to deal with these in particular ways.

When repairs are needed later in the reclamation work after the debates and concerns that dominated the planning approval process have been forgotten, the situation need not be so foolproof. A former tar works in South West England had heavy oil soaking of the demolition rubble and waste products surface layer, and its reclamation called for drainage and other works to reduce the oil content, after which the reduced (mainly metallic) contamination of the site would be counteracted by the installation of a clean cover. The local planning authority required considerable demonstration and testing to prove that the clean cover would be effective for at least 100 years and insisted that the entire site was capped before any redevelopment took place. This was to ensure that no airborne contamination could result. Whilst this initial part of the work proved successful, redevelopment of the site was significantly delayed because of a downturn in the demand for housing, and parts of the site were sold on to different building contractors who had little understanding of the contamination risks that could arise if the clean cover were not maintained at its design quality. One of these contractors found it necessary to alter the previously installed drainage services, which had been included with the clean cover, trenched through the cover, and then backfilled his excavations with granular material obtained from a disused railway embankment.

Unfortunately this backfill material not only failed to meet the design physical properties of the original clean cover, but also proved to be contaminated with high concentrations of heavy metals. Obviously a failure in proper communication led to a very significant loss in quality assurance, which, in turn, had heavy financial penalties.

Training.   Deficiencies in training are least apparent in those activities (e.g. land surveying or soil mechanics testing for civil engineering staff, and trial pitting for environmental scientists) that appeared as practical exercises in the site supervisor's formal education. This indicates that most employers still prefer to limit the professional development training they themselves provide, and expect newer staff to pick up necessary expertise by observation and experience.

In many cases (e.g. the landfill gas monitoring of the marl pit site and the sampling of the metal-contaminated ash site, noted above) this is not a cost-effective system, and greater note has to be taken of the 'abnormal' conditions that regularly occur on contaminated sites and the need to train staff to react to these.

A similar view is expressed (in chapter 13) where the need to ensure safe working practices is discussed.

10.3 Appropriate quality systems

The required quality standards will not always be the same and the extent of quality assurance work will be much less on some reclamations than on others. The obvious point here is that priorities have to be identified — some conditions are potentially more hazardous than others, and some quality monitoring activities can be carried out to less stringent standards. Simple examples of actual site reclamations should make this clear.

10.3.1 *Housing redevelopment, Central Scotland*

This small area of land (about 1 ha) is fairly typical of the marginally contaminated sites that occur within larger towns. Stiff clays originally surfaced the site and still extend to 7 m depths. No groundwater table is present in the upper 5 m of the site. Small scale excavations in the centre of the site had taken place about 1940 and had resulted in a shallow depression up to 2.5 m deep, which was later infilled with post-war demolition rubble. At some later stage, a surfacing (up to 350 mm thick) of domestic ashes was tipped over most of the site.

Site investigation was by means of 24 trial pits, and chemical analyses of the samples revealed that, whilst the ashes contained the expected low to medium contamination of copper, nickel, zinc, lead and cadmium (Table 10.2), the underlying clays were uncontaminated, except where oil spills (from a car breaking operation) had taken place.

Table 10.2 Phytotoxic contaminants in the ashy band of a site in Central Scotland

Contaminant	Concentration found (mg/kg)	
	Range	Mean
Copper	53–310	140
Nickel	20–105	68
Zinc	110–4170	400
Lead	155–1020	490
Cadmium	2.0–5.7	2.4
Chromium	20–45	31
Arsenic	1–8	4
Phenol	<1	
Toluene extract (oils)	1100–5000	1670
pH	6.21–7.81	
Sulphates	300–1500	645

The presence of phytotoxic contaminants (mainly in low concentrations) was seen as unhelpful to the house sales, and the developer chose not to cap the site but to strip the top 400 mm, and redeposit the ashy material as a linear sound reducing mound, between the housing and an adjacent busy road. This sound mound was then capped with a clean cover, to prevent leaching of the metallic contamination (chapter 5, section 5.5).

Quality standards for this reclamation proved to be extremely simple and consisted of no more than the following:

(1) *Site stripping process.* Ensuring that all stripped materials were removed to the sound reduction mound location. Recording the physical types of materials stripped off each area.

(2) *Site inspection post-stripping.* Visual confirmation that all ashy materials had been removed and that each housing plot was underlain by clean materials. Photographic records were taken to prove that ashes had been removed. Visual identification of any oil soaked ground that existed. If encountered, oil soaked materials were excavated, and the arisings removed to the sound reduction mound area. Oil soaked materials were tipped separately, from the ashy strippings, to allow evaporation to occur. Testing of each house plot was limited to:

- one clay sample from each garden area analysed for phytotoxic contaminants; confirmation was noted that every sample was in fact chemically uncontaminated
- spike tests were conducted under each house foot print, to ensure that no landfill gases or oil vapours existed

- occasional plate bearing tests were made on any clay materials that visually seemed less dense than normal
- plate bearing tests were completed on exposed demolition rubble fills to confirm their suitability for foundation loads

(3) *Sound reduction mound.* Oily materials were mixed with the granular ashy fills, and then rolled prior to being capped with a clay surface capping, below which a drainage blanket was installed.

The staffing needs were limited to the normal site foreman, supplemented by periodic inspection visits by the consulting engineer's staff.

The final records consisted only of site plans showing final and initial ground levels, all test locations, photographic records and a summary of the test results attained. These proved acceptable to the buyers' legal representatives and to the National House Builders Council. No dispute over the quality of the finished reclamation arose.

10.3.2 *Housing redevelopment, South West England*

Where a site contains mobile contamination, in addition to relatively immobile metals, reclamation inevitably becomes more complex, and achieving the necessary quality standards is a rather more significant task.

This particular site had been the location of oil and creosote storage tanks for several decades, before which it was a tip for copper smelter slags. Leakages from the storage tanks had resulted in a high degree of oil saturation of the fill deposits (demolition debris, ashes and the slags from a former copper smelter), which averaged some 3 m in thickness. The existing site surface sloped very steeply to a river, used for amenity and water supply purposes, and whose water quality had to be ensured throughout the reclamation.

Site investigation by several sequences of trial pits, and deeper boreholes, revealed the expected high metal levels, a very pervasive oil/tar/phenol contamination, and very high soil sulphate levels. The bearing capacities of the site materials were such that piled foundations for the housing units proved necessary.

The reclamation strategy adopted was:

(a) To break out the concrete slabs, which surfaced almost the entire site.
(b) To install land drainage at 10 m spacings, to draw off the free oily liquids. These drains were designed to act as permanent structures, to hold the site's groundwater level 1.5 m below the average level at which the site materials were oil contaminated. A central sump then collected the oily liquids, which were taken off-site for treatment and final disposal. No construction works took place until the drains had been in operation for an entire winter, to ensure that a maximum possible flushing out of the oil spillages was achieved (Table 10.3).

(c) Foundation piles were then installed, after all exposed tarry contamination was removed for on-site encapsulation.

(d) A designed clean cover, and its overlying filter blanket, were laid to very tight particle size distribution and density specifications.

(e) The final ground levels were achieved by importing up to 4 m thicknesses of inert building rubble. This large-scale raising of the site levels was necessary to achieve acceptable road gradients and the required landscape topography.

Table 10.3 Variation in oily contamination in one of the drainage trenches on a site in South West England

Week	pH	Phenols (mg/l)	Oils (mg/l)	
1	7.7	283	251	site breaking
4	7.2	96	100	out in
8	7.3	114	80	progress
16	7.4	197	135	
20	7.3	170	100	
24	7.5	50	60	
28	7.1	38	43	
32	7.1	15.5	25	
36	7.2	11.4	6.1	
40	7.1	11.0	6.0	
44	7.2	<10	5.5	

The quality system needed for this project included:

(1) *Drainage works*
- Recording the levels, gradients and positions of all drainage works; ensuring that these would not intersect the planned locations of pile foundations.
- Monitoring the qualities and flow rates of liquids entering the central sump at weekly frequencies, and recording when particular zones of the site appeared to have drained.
- Associated with the drainage monitoring, the river water quality at the upstream and downstream boundaries of the site were monitored weekly, to check if any oily pollution was occurring.

(2) *Off-site removal of oily liquids*
- The volumes of liquids removed for off-site treatment and disposal were recorded as a requirement of the waste disposal licence.

(3) *Tarry wastes*
- Locations of all exposed tarry wastes were recorded, as was the depth of excavation needed to remove these. Whenever tarry wastes occurred close to proposed pile locations, particular care was taken to remove the deposits. Where this proved impossible, note was made of the necessity to provide appropriate external protection for the particular concrete piles.

- All tarry wastes were stockpiled, air-dried, and then rolled, to high-ways specification, in a concrete tank base, which ultimately was covered to provide the on-site encapsulation cell.

(4) *Clean cover*

- All materials for use in the clean cover were derived from crushing concrete demolition waste. The particle size distribution curve of the crushed concrete was checked before each load was taken for use in the cover. Specific criteria for acceptability/rejection were devised and enforced.
- The cover was laid in 200 mm thick layers and the compaction of each layer was confirmed, by nuclear densiometer readings at 10 m centres.
- The filter blanket required above the clean cover was checked in an identical manner.

(5) *Imported inert fill*

- Each load of imported fill was visually checked to ensure that no biodegradable materials were included.
- A sample of each load was taken and analysed for its metal, toluene extract, and phenol contents. Criteria were enforced to ensure rejection of any material that failed to meet the necessary chemical quality.

These quality controls were such that a representative of the consulting engineer was permanently located on the site, during these initial reclamation stages, and this individual was supported by on-site and off-site analytical services. A full end-of-reclamation report was produced as detailed in Table 10.1.

10.3.3 *Industrial development, West Midlands*

When a site has landfill gas problems in addition to mobile contamination, the quality system becomes even more demanding. This West Midlands site had been in part an iron foundry, manufacturing cast iron baths, and in part a refuse tipping area, which had been in intermittent operation for some 80 years. Coal mining occurred directly beneath the site, on which several mine shafts had existed. Whether these shafts had been properly filled and capped was not known. An infilled canal arm was also known to have existed.

Because of the coal mining and the possibility of land subsidence, the site exploration was particularly thorough. This exploration revealed that areas of high phenol concentrations occurred, together with oil soaking from demolished oil storage tanks, but the major contamination appeared to be from the lead and arsenic pigments, which had been used to glaze the cast iron baths. Apparently during the site demolition, the stores that had contained these hydroscopic salts had been bulldozed. Colliery wastes containing potentially combustible material were also widespread around the locations of the former mine shafts. Although the tipped wastes (up to 4 m thick) were all of a considerable age, landfill gases with methane concentrations of some 40% by vol. and carbon dioxide levels of up to 13% were frequently encountered in some gas monitoring boreholes.

The physical condition of the site made it necessary to improve its bearing capacity by vibro-compaction and the reclamation strategy was devised around this. The main reclamation phases were as follows:

(a) Zone the site into areas affected by particular potential hazards and deal with each in a specific controlled way.

(b) Install vent trenches at 10 m centres across the zone affected by landfill gases. Several months' monitoring and experimentation had revealed that passive venting and excavation of discrete areas of wastes would resolve the gas hazards.

(c) Collect all exposed patches of the lead/arsenic salts and remove these for off-site treatment. Excavations to about 2 m depths proved essential and this area was reinstated with a clean granular cover.

(d) Locate and remove all oil soaked ground. Stockpiling of this material allowed enough evaporation to give a marked reduction in the oily contamination, and only the more heavily contaminated material had to be removed to a licensed tip.

(e) Compacting the site surface to the levels needed to give the required bearing capacities and prevent enough air entering to allow any subterranean smouldering or burning in the more susceptible colliery spoil materials (chapter 7).

Quality controls had to include:

(a) Weekly landfill gas monitoring (of both gas concentration and flow rates) to prove the effectiveness of the passive venting trenches, and to prove that the 10 m trench spacing was appropriate

(b) Chemical analysis to prove that the lead/arsenic levels had been reduced, to below the ICRCL guideline values, by the removal of the visible pigmentation wastes

(c) Chemical analysis on the oily soaked area to show that the remnant phenol, and oil concentrations were at the allowable final levels, once excavation of patches of oily ground had been completed

(d) Routine monitoring of the chemical qualities of the stockpiled oily arisings to determine if they could safely be introduced back into the site, or whether they would have to be removed to a licensed tip

(e) Regular monitoring of the site compaction, to prove that no subterranean smouldering hazard would be likely

(f) Routine confirmation that the necessary ground bearing capacities had been achieved

(g) Routine updating of site plans, sample location plans and analytical records

This proved to be very successful. The work load, however, called for almost the full time attention of a senior member of the consulting engineer's organization, supported by the scientific personnel from the analytical firm.

The examples noted above all are derived from the engineering-biased reclamations that have so far dominated UK reclamation practices. With the availability of chemical and microbial treatment methods offered by specialist contractors who are now able to enter into commercial contracts that include end-quality achievement levels, it is obvious that a rather different pattern of land reclamation strategies will soon become the norm. These are likely to include appropriate mixes of various on-site treatments, on-site encapsulations, and a reduced emphasis on clean covers.

In quality assurance terms this inevitably will lead to more complex situations. Different organizations are likely to be involved on a single site, at much the same time, and the reclamation techniques each is employing will have very different requirements and achievable qualities.

One such reclamation that is still at the planning stage seems likely to include:

(a) Benzene vapour stripping from alluvial sand deposits that have been contaminated by spillages from a coke works' effluent. Given the known health risks that benzene vapours will pose to site personnel, this work would obviously have to be monitored to a particularly high standard and completed before it would be safe to excavate deeper contamination.

(b) Microbial treatment of free and emulsified oil wastes that have entered the site's near surface groundwater table. As this groundwater could flow into an adjacent river, regular monitoring and confirmation of the efficiency of the microbial treatment will be important, before the cleansed water can be returned to the aquifer.

(c) Excavation of the alluvial sands, after their benzene levels have been reduced, to remove tarry wastes deposited at the groundwater level from the coke works spillages. Separation of the excavation arisings into more and less contaminated soils will probably be necessary to reduce the volume of tarry wastes that will have to be encapsulated on-site or treated. Regular sampling and analysis of stockpile materials to a high quality standard is thus implied.

(d) The inclusion of temporary cut-off works to insulate selected areas of the site to allow the vapour stripping, groundwater treatment and tar excavation activities. Monitoring the effectiveness of these temporary walls, to ensure that no polluted waters migrate out of the working areas, will be a necessity.

(e) The usual need to check that the reclaimed site compaction is to an appropriate high quality and that remnant contamination and vapour levels are safe and acceptable.

(f) Regular off-site monitoring of river water, groundwater and air qualities.

The likely complexity of these different treatment methods, and the very large body of monitoring data that inevitably will arise, suggests that an overall quality audit system will be necessary in such a case. This obviously will impose a higher quality assurance cost than has hitherto been normal, but the benefits (of totally cleaning a site rather than merely reducing its hazards) are likely to outweigh the extra reclamation costs.

10.4 Summary

It sometimes is said that quality assurance is no more than normal best practice; equally common is the view that quality assurance implies bureaucracy, delays and extra costs. There obviously is an element of truth in both viewpoints. On marginally contaminated sites, quality systems need not be particularly onerous and the main difficulty is to convince the parties concerned that it is worth while to spend the initial time identifying the appropriate quality system. On more complex sites, particularly where the contamination is in a mobile form, quality systems will be expensive to implement, but their lack is very likely to mean that the value of the reclaimed site will be reduced.

When the newer reclamation technologies bring different specialized treatment methods on to a single site, it is difficult to avoid the belief that this must call for a high degree of quality audit, probably carried out by an organization independent of those applying the reclamation technologies, and so a sizeable extra cost is likely.

Quality standards will become more commonplace when the clients for contaminated land reclamations insist on them. This process is already becoming apparent as foreign investors come into the UK reclamation market and require the assurances that are more often provided in the United States or elsewhere in Europe.

11 UK legal framework

G.J. LONGBOTTOM

11.1 Introduction

Since the late 19th century when Victorian society began to take an interest in the health of the populace, the body of law relating to public health and in particular to contaminated land has steadily increased.

Concern over the well-being of the environment and a wider appreciation of the world ecosystem in our present century, has given added impetus to legal controls with the result that the contaminated land owner and remediation practitioner are now faced with a complex legal framework. Not only is that framework complex, it is continually changing as the law evolves. As a result it cannot readily be presented in simple check-list form for use as an accessible guide by landowners and their consultants.

Such a basic guide, however, is needed and it is the objective of this chapter to provide a guide to the major aspects of the relevant law relating to England and Wales. Readers will appreciate the need to obtain their own legal advice when dealing with specific circumstances, as this chapter is not intended to be more than a general and necessarily much simplified guide to the subject as a whole. The law is stated as at December 1991.

11.2 Sources of law relating to contaminated land

In the United Kingdom, there are two primary sources of law, Statute and Common Law, and two secondary sources, delegated (or subordinate legislation) and case law. Non-statutory Guidance Notes and Codes of Practice also exist; whilst not law in themselves, these often have a bearing on the outcome of court action when the question arises as to whether or not reasonable steps had been taken by the defendant.

11.2.1 *Legislation: primary and subordinate*

11.2.1.1 *Primary legislation.* The primary source of law is legislation by Parliament in the form of Acts of Parliament. Law arising from this source is also referred to as Statute Law and is said to be 'on the Statute Book'. Acts of Parliament generally arise at the instigation of the Government of the day, which will lay a Bill before Parliament for its consideration. That Bill is essentially a draft of the proposed law and, following debate and perhaps modification during its passage through Parliament, it will eventually emerge as an Act of Parliament. Once 'on the Statute Book' an Act of Parliament may become law immediately or may be implemented in sections over a period of time. Bringing it into effect may occur over several years and parts of the Act may never be brought into force. This process makes the law difficult to discern as it is no simple matter to determine at any one time which parts of an Act of Parliament are in force and which are not.

11.2.1.2 *Subordinate or delegated legislation.* Acts of Parliament commonly include provision for Ministers to provide the detailed requirements of parts of the Act or indeed to bring into effect parts of the Act itself at some time in the future. These powers are exercised through subordinate legislation and are used to provide the many detailed rules and requirements that are necessary to allow the Act to function as intended. Subordinate legislation takes the form of orders or Statutory Instruments.

Statutory Instrument 1980 No. 1709, the Control of Pollution (Special Waste) Regulations 1980, is an example pertinent to contaminated land. Made by the Secretary of State for the Environment, the Secretary of State for Wales and the Secretary of State for Scotland for their respective countries under powers conferred upon them by the Control of Pollution Act 1974, this subordinate legislation sets out the definition of special waste and the detailed rules to be followed with regard to the administrative aspects of its disposal as follows:

Part I: Definitions, particularly of special waste (Section 2(1))
Part II: Prescribes the Consignment Note system to be used when transferring special waste
Part III: Exceptions for regular consignments
Part IV: Requirements for the keeping of regulations and site records
Part V: Provision for the giving of directions as to the disposal of special waste
Part VI: Deals with enforcement specifying penalties and in Section 16(5), a defence to prosecution

11.2.2 *Case law*

Case law is the description of that body of law that arises as a result of the decisions handed down by judges in individual court trials. (These decisions interpret both Statute and the Common Law and 'fill in the gaps'). For example, the rules for negligence were set down by Lord Atkin in the famous negligence case of *Donoghue v Stevenson* and similarly the rules that apply to 'things kept on land' arise from the decision in the nuisance case of *Rylands v Fletcher*.

Decided cases (i.e. case law) are then used by the Courts as precedents for determining the outcome of subsequent cases of a similar factual background. The process provides some certainty in the pretrial prediction of the decision likely to be delivered by the courts in individual cases.

11.2.3 *Common law*

Common law is the name given to that body of law that sets out those rights and remedies that are founded in long-established principles and are not embodied in Acts of Parliament. This area of the law may be considered to be static in that new rights and remedies are not constantly being created or removed. It is not, however, stagnant since, in applying the common law, the Courts have a limited ability to adapt existing rights and remedies to accommodate new circumstances. Actions at common law relating to contaminated land generally arise in nuisance (see section 11.3.3).

11.2.4 *Guidance Notes and Codes of Practice*

Guidance Notes, Practice Notes and Codes of Practice are issued by Central Government in relation to virtually all aspects of contaminated land. Well-known examples are the Waste Management Paper series issued by the Department of the Environment and Her Majesty's Inspectorate of Pollution and the Guidance Notes issued by the Interdepartmental Committee on the Redevelopment of Contaminated Land (ICRCL).

These documents have no legal standing except that they may be adduced as evidence to establish whether or not a party exercised due care. It should be noted, however, that compliance with these documents does not necessarily ensure discharge of a legal obligation. This principle was clearly stated in the introduction for the Draft Code of Practice on Duty of Care issued in September 1991 by the Department of the Environment, The Scottish Office and the Welsh Office:

> The purpose of this code is to set out practical guidance for waste holders subject to the duty of care. It recommends a series of steps which would normally be enough to meet the duty. The code cannot, however, cover every contingency; the legal obligation is to comply with the duty of care itself rather than the code. Breach of the duty of care is an offence.

11.2.5 *European Community regulations, decisions and directives*

Membership of the European Economic Community (EC) by the United Kingdom carries with it the obligation to conform with law made by the Community. The Community utilizes three types of legislation:

(a) *Regulations*. These are directly applicable law in Member States and are therefore enforced in the National Courts of each State. Regulations are principally used to give effect to uncontroversial or pre-cise measures that require no further legislation to give them practical effect.

(b) *Decisions*. These again are directly binding but, unlike regulations, are generally addressed to specific bodies and not the Member States at large. Decisions are often used in environmental policy with regard to international conventions and procedural measures.

(c) *Directives*. These are the main vehicle for the implementation of Com-munity policy and are binding on Member States. They specify the results to be achieved and time-scale for compliance whilst leaving the precise form and method to the Member State. In the United King-dom, most EC Directives are given effect by Ministers using Statutory Instruments under delegated powers derived from Statute. A re-cent example is the EC Industrial Plant Emissions Directive 84/360 brought into the UK legal framework by Statutory Instrument SI 1991 No. 472, the Environmental Protection (Prescribed Processes and Substances) Regulations 1991, made under powers conferred on the Secretary of State by Section 2 of the Environmental Protection Act 1990.

Figure 11.1 gives a simplified illustration of the process by which European Community Policy becomes law in the UK.

It is apparent from the above that whilst an EC Directive is not law at the time of issue, it provides a direct pointer to what the law will be within the ensuing 18 months to 3 years. All those involved with contaminated land thus need to be able to keep abreast of EC policy to enable them to make decisions in the present and thus avoid expense and/or liability in the future.

11.3 The law relating to contaminated land

11.3.1 *General outline*

In sections 11.3.2 and 11.3.3, the principal areas of law that affect contami-nated land are identified and briefly reviewed. Specific aspects are examined in greater detail in section 11.4 placing emphasis on practical guidance rather than legal analysis. It will be appreciated that in a book of this nature, it is not possible to deal with every detailed legal requirement and facet, and when particular proposals or transactions are contemplated it will be necessary to seek specific legal advice. Cognizance of the areas covered, however, should assist in identifying that need.

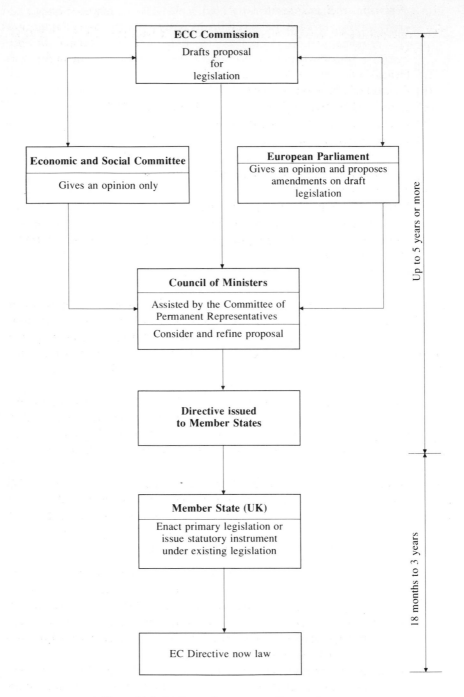

Figure 11.1 Implementing European Community policy.

11.3.2 *Principal legislation*

The principal legislation relevant to contaminated land may be identified as follows:

(a) Health and Safety at Work Act 1974
(b) Control of Pollution Act 1974
(c) Control of Pollution (Special Waste) Regulations 1980
(d) Occupiers Liability Acts 1957 and 1984
(e) Building Regulations 1985
(f) Collection and Disposal of Waste Regulations 1988
(g) Control of Pollution (Amendment) Act 1989
(h) Water Act 1989
(i) Town and Country Planning Act 1990
(j) Environmental Protection Act 1990
(k) Controlled Waste (Registration of Carriers and Seizure of Vehicles) Regulations 1991
(l) Water Resources Act 1991

(a) *Health and Safety at Work Act 1974.* This Act requires employers to ensure that their workplace is as safe as is reasonably practical for both employees and visitors. In the context of contaminated sites this may not prove to be of direct relevance to landowners but it is obviously important for developers who will need to carry out reclamation works on the land.

(b) *Control of Pollution Act 1974 (COPA 1974).* This was the major enabling Act for the control of waste disposal in the United Kingdom and it provided the administrative framework for the collection and disposal of waste materials. It is now largely superseded by the Environmental Protection Act.

(c) *Control of Pollution (Special Waste) Regulations 1980. SI No. 1709.* These regulations, made by the Secretary of State under powers conferred by the Control of Pollution Act 1974, define certain wastes considered difficult to deal with as 'special waste'. Part I of the regulations states that the term 'special waste' shall apply to any controlled waste which:

(1) consists of or contains any of the substances listed in Part I of Schedule 1 and by reason of the presence of such substance,
  (i) is dangerous to life within the meaning of Part II of Schedule 1, or
  (ii) has a flash point of 21°C or less as determined by the methods and with the apparatus laid down by the British Standards Institution in BS3900: Part A, 8: 1976 (EN53), or
(2) is a medicinal product, as defined in section 130 of the Medicines Act 1968(b), which is available only in accordance with a prescription given by an appropriate practitioner as defined in section 58(1) of that Act.

Part II of the regulations requires the use of a consignment note system to provide an audit trail for the disposal of the waste, and Part IV requires the producer, carrier and disposer each to maintain, respectively, a register of special waste produced, carried and disposed of.

Provision is made in Section 62 of the Environmental Protection Act for the making of new Regulations on special waste. The reader is directed to section 11.4.12.2 for additional information regarding special waste.

(d) *Occupiers Liability Acts 1957 and 1984*.   A general duty of care is placed upon occupiers of land by these Acts to ensure that visitors are kept as reasonably safe from injury as practicable. It is to be noted that the duty applies to 'occupiers' and not simply landowners and that the duty, which is one of reasonable care, extends in a qualified form to trespassers.

(e) *The Building Regulations 1985*.   As subordinate or delegated legislation under the Building Act 1984, the Building Regulations describe in detail standards of construction to be adhered to in the construction of buildings. They are principally aimed at securing reasonable standards of health and safety for persons in or about a building and others who may be affected by any failure to comply. It is therefore at the time of application for Building Regulation approval that the limited requirements would affect contaminated land. One would normally anticipate that the amelioration of contamination had been achieved by means of planning conditions, but in the event of permitted development, for example, the Building Regulations provide a further opportunity for control.

(f) *Collection and Disposal of Waste Regulations 1988 SI No. 819*.   These Regulations were made on 3 May 1988 under powers conferred by the Control of Pollution Act 1974, and came into effect fully on 3 October 1988. They provide guidance as to the classification of waste, and the circumstances when a disposal licence is not required but are not applicable in Scotland. Of particular interest to owners and practitioners in land restoration is Schedule 6 to the regulations, which itemizes instances in which a waste disposal licence is not required for the deposit of specified controlled waste. The reader is directed to Section 11.4.11.2 for additional discussion on this matter.

Power to prescribe cases when controlled waste may be deposited, treated, kept or disposed of without a licence is conferred upon the Secretary of State by S33(3) Environmental Protection Act. It can therefore be anticipated that regulations similar to those described above will be made or indeed may be retained in the form of the current regulations.

(g) *Control of Pollution (Amendment) Act 1989*.   This Act provides in Section 1 for it to be an offence "for any person who is not a registered carrier of controlled waste, in the course of any business of his or otherwise with a

view to profit, to transport any controlled waste to or from any place in Great
Britain". Sections 2 to 4 of the Act make provision for the registration of
carriers and Section 6 makes provision for the seizure of vehicles used for the
unlicensed disposal of waste. The Act was given effect on the 14 October 1991
by Statutory Instrument 1991 No. 1624. The Controlled Waste (Regulation of
Carriers and Seizure of Vehicles) Regulations 1991 require all carriers to be
registered on 1 April 1992.

(h) *Water Act 1989 (WA 1989)*.   This Act created for the first time a single
authority (The National Rivers Authority) responsible for water quality
in England and Wales. Section 107 of the Water Act 1989 (soon S85 of Water
Resources Act 1991) makes it an offence to cause or knowingly permit the
pollution of ground or surface water and the National Rivers Authority
is charged with enforcing this and other such provisions. The NRA is given
an important power under S115 of the 1989 Act (S161 of the 1991 Act) to
require the clean-up of the sites that are causing water pollution (see section
11.4.8).

(i) *Town and Country Planning Act 1990*.   With certain exceptions, 'development' (as defined by TCPA) requires the prior approval of the local planning authority. Development is often construed as including remedial work on
contaminated land, whether or not subsequent built development occurs. When
called upon to determine a planning application, the authority is required by
S70(2) TCPA 1990 to have regard to the development plan, so far as material
to the application, and to any other 'material considerations'. The Planning
and Compensation Act 1991 has inserted a new S54A into the TCPA, which
requires the Authority to make the determination, in such cases, in accordance
with the Development Plan unless material considerations indicate otherwise.
It is under the head of 'material considerations' that the presence and effect
of contaminated land must be addressed. Circular 21/87 and Planning Policy
Guidance Note 1, January 1988 refers. Four decisions are possible:

- Refusal
- Unconditional permission
- Permission subject to conditions
- Permission (with or without conditions) subject to the applicant first
  entering into an agreement with the authority under S106 TCPA 1990

The Grant of Permission for development of a contaminated site usually
involves the imposition of a condition that no construction work commences
until the local Planning Authority is satisfied with the remedial treatment.
Circular 21/87 gives guidance to authorities on this point.
    In the event that a refusal is issued on the grounds that the site is contami-
nated the owner may serve a purchase notice on the authority under Sections
137–8 of the TCPA 1990 on the grounds that the site has become incapable

of reasonably beneficial use in its present state. Acceptance of the notice or its confirmation by the Secretary of State would result in the authority being obliged to purchase the land.

(j) *Environmental Protection Act 1990 (EPA 1990).*   Enacted to make provision for the improved control of pollution arising from certain industrial and other processes, the Environmental Protection Act 1990 contains three sets of provisions of relevance to contaminated land. Part II introduces a new regime for waste management to replace Part I of the Control of Pollution Act 1974. The most important new provision imposes a duty of care on all those who handle waste at each stage of the journey from production to disposal (see section 11.4.9.1). Table 11.1 shows when these provisions are likely to come into force. Part III codifies the existing provisions relating to statutory nuisances (see section 11.4.7.3). S143 provides for the introduction of registers of land which may be contaminated.

Table 11.1   The Environmental Protection Act 1990: the implementation of Part II (Waste)

In June 1991 the Government announced the following timetable for implementing the provisions relating to waste regulation and disposal set out in Part II of the 1990 Act (and also in the Control of Pollution (Amendment) Act 1989).

Target date	Task
Summer 1991	Lay regulations and issue circular on the registration of waste carriers and seizure of vehicles; waste regulatory authority powers to seize vehicles used for fly-tipping come into effect Consult on draft regulations to implement a Duty of Care on waste producers and holders
November 1991	Issue circular·and Code of Practice giving guidance on compliance with the Duty of Care; lay regulations implementing Duty of Care
April 1992	Duty of Care regulations and provisions on the registered carrying of waste come into effect
Early summer 1992	Consult on regulations and statutory guidance for the licensing of waste disposal
Summer 1992	Consult on a scheme of charges for licensing
December 1992	Lay regulations, announce charging scheme and statutory guidance, issue circular on licensing
April 1993	Licensing regulations and charging scheme came into effect

Note : the Government's recently announced plans for an integrated Environment Agency casts some doubt on the implementation of the waste licensing regime in April 1993.

(k) *Controlled Waste (Registration of Carriers and Seizure of Vehicles) Regulations 1991 SI 1991 No. 1624.*   Made by the Secretary of State on 17 July 1991 under powers conferred by the Control of Pollution (Amendment) Act 1989, these regulations came into force on 14 October 1991. The Regulations bring into effect the 1989 Act and set out the requirements and mechanics of registration of carriers and seizure of vehicles. All carriers of controlled waste are required to be registered on 1 April 1992. The reader is directed to sections 11.4.12.4 and 11.4.12.5.

(l) *Water Resources Act 1991.*   This consolidation Act, which draws together existing legislation, replaced the Water Act 1989 on 1 December 1991. The corresponding sections of both Acts of particular relevance to contaminated land are set out Figure 11.2.

11.3.3 *Common law*

The most important common law duty arising from the ownership of a contaminated site is based on the tort of nuisance. There is private nuisance (which involves the unlawful interference with another person's use or enjoyment of his or her property) and public nuisance (which means an unlawful act that endangers the lives, safety, health or comfort of the public). There are some fairly subtle distinctions, all of which need not be pursued here. Of more immediate relevance is the species of nuisance represented by the rule in *Rylands v Fletcher*. This involves strict (i.e. no fault) liability and therefore warrants special attention. The rule was stated thus:

> The person who for his own purposes brings on his land and collects and keeps there anything likely to do mischief if it escapes must keep it at his peril and if he does not do so is prima facie answerable for all the damage which is the natural consequence of its escape.

This has obvious applications to land contamination, although it is arguable that conventional industrial use lies outside the ambit of the case. It has been asserted that for *Rylands v Fletcher* to apply: "it must be some special use bringing with it increased danger to others and not merely to be the ordinary use of the land or such use as is proper for the general benefit of the community." However, a relatively recent Canadian case held that the escape of landfill gas (which caused an explosion in an adjoining housing development) arose from the 'non-natural' use of the landfill site.

This suggests that *Rylands v Fletcher* may have wider application to environmental hazards than at first appears. If this is true a plaintiff will only need to establish that loss or damage was caused by the non-natural use. Strict liability means that it will avail the landowner nothing to show that it took all prudent steps to prevent the loss or damage occurring. One could, on the other hand, consider *Cambridge Water Company v Eastern Counties Leather plc and v Hutchings & Hardings Ltd* (Queens Bench Division, 31 July 1991) in which it was held that the storage of solvents was not an unnatural use of the land in question and therefore the rule in *Rylands v Fletcher* had no application. It would appear that this area of law is in a state of flux.

The other important common law duty arises from the tort of negligence. Since the 'neighbours' to whom a duty of care is owed have been widely

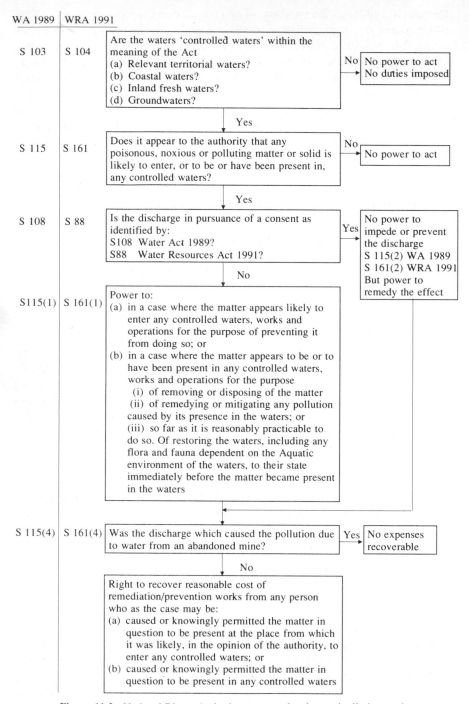

Figure 11.2 National Rivers Authority powers related to antipollution works.

construed, and since the presence of dangerous contaminants demands a high standard of care, negligence might perhaps be thought to offer plaintiffs more scope than nuisance. However, it is essential to establish fault (i.e. a failure to act with reasonable prudence) and although the principle of 'res ipsa loquitur' (the thing speaks for itself) can be invoked in an environmental context (i.e. the escape of toxic material would not have occurred but for a failure to take appropriate precautions), the adherence to technical and professional advice should afford the landowner an effective defence. Even if negligence could be established, there has to be some physical damage for the plaintiff to succeed. Economic loss alone is not sufficient.

11.4 Working with the law: effect on the various parties

In sections 11.1 to 11.3, the source and current state of the law have been outlined. How, in practice, the law affects contaminated land will now be explored in response to a series of questions that might well be asked by any of the parties involved.

11.4.1 *What legal framework is in force?*

Section 1 makes it plain that the law is constantly changing and that for the purpose of assessment it is necessary to fix a point in time. For the purposes of this guide it is assumed that the Environmental Protection Act 1990 (EPA) is fully implemented and has replaced Parts I and III of the Control of Pollution Act 1974 (COPA), and that the Water Resources Act 1991 (WRA) is also in force.

11.4.2 *When will that be the case?*

This condition of the law should be reached by April 1993. In the meantime the steps outlined below represent good practice and should generally ensure compliance with Part I the Control of Pollution Act 1974.

11.4.3 *How does the actual law today fit in?*

At the time of writing (Autumn 1991) the Environmental Protection Act 1990 is not fully in force and the Duty of Care to which constant reference will be made does not yet exist. The majority of the law relating to waste is still contained in the Control of Pollution Act 1974 and associated regulations, and the powers of the National Rivers Authority described are contained in the Water Act 1989. However, if one takes steps to comply with the forthcoming provisions of the EPA and Water Resources Act 1991, obligations under the existing law will generally be fulfilled. With regard to the law at large, in general terms, if one takes steps to avoid prosecution under Statute, these same steps will for the most part prevent one from falling foul of common law actions in nuisance. The Government consults widely on proposed legislation. Table 11.1 shows when the different elements of Part II (Waste) of the Environmental Protection Act 1990 are expected by the Government to come into force. Trade and industry associations will be among the Government's consultees and will keep their members informed. Solicitors and other consultants take care to keep their clients apprised of significant developments.

11.4.4 *How will the various parties interact with the law?*

It may be stated generally that:

(i) *Public authorities* will be acting under Statute:
- To discharge general duties placed upon them to protect the environment, e.g. enforce water quality standards for drinking water
- To carry out specific tasks, e.g. to compile a register of land which may be contaminated
- To prosecute persons who have committed offences.

(ii) *Landowners and developers* will be generally acting to ensure compliance with their legal obligations, in particular the Duty of Care, although, of course, they may be exercising their rights to ensure that Government and Statutory Bodies and other landowners discharge their duties.

(iii) *Consultants and Contractors* may have a number of roles:
- Advising clients how to discharge their legal obligations
- Arranging for the discharge of their clients' legal obligations
- Discharging their own legal obligations

(iv) *Members of the public, or groups representing various interests* must not be forgotten since they may:
- Prompt action by local or central authorities to deal with alleged pollution
- Take action themselves against neighbours (whether individuals or companies) for loss or damage caused to persons or property

11.4.5 *Where will an action originate?*

The owner of contaminated land can expect the law to be brought to bear by either a regulating body, probably the local authority or the National Rivers Authority, using its statutory powers, or a neighbour using the common law. The costs and risks of legal action makes the former more likely; at present the legal system makes it difficult for members of the public to take action to safeguard the environment, as personal loss is generally required. However, prosecutions by national and local pressure groups can be expected to increase. Greenpeace has already had some success.

11.4.6 *A scenario for analysis*

Let us imagine that a parcel of land exists, which is contaminated by substances on and buried within the ground. What is the extent of the landowner's liability? What legal provisions might be brought into play and by whom?

11.4.7 *How might a problem arise?*

Contaminated land often lies undiscovered until some major works or a pollution incident brings it to light. Three particular provisions of the Environmental Protection Act may, however, change this and we will assume it is via those provisions that our parcel of contaminated land is located and remediation required.

11.4.7.1 *The register of contaminated land, S143(2) EPA.* Section 143(2) of the EPA requires local authorities to maintain registers of land subject to contamination within their area. 'Subject to contamination' is defined in extremely wide terms as land that is being or has been put to contaminative use, and 'contaminative use' in turn means any use of land that may cause it to be contaminated with noxious substances. It is not yet clear how these registers will operate in practice but, as they are to be available for public inspection, they obviously make it much more likely that contaminated sites will come to the attention of both local authorities and third parties.

11.4.7.2 *A duty to inspect, S61(1) EPA.* Section 61(1) of the EPA places a duty on every waste regulation authority "to cause its area to be inspected from time to time to detect whether any land is in such a condition, by reason of relevant matters affecting the land, that it may cause pollution of the environment or harm to human health". Relevant matters are in turn defined in subsection 61(2) of the EPA as "the concentration or accumulation in, and emission or discharge from, the land of noxious gases or noxious liquids caused by deposits of controlled waste on the land". The duty to inspect will therefore seek out some parcels of contaminated land and bring them to the notice of the Waste Regulation Authority.

11.4.7.3 *Statutory Nuisance, S79 Environmental Protection Act.* Part III of the Environmental Protection Act codifies and makes more effective a number of existing provisions relating to so-called 'statutory nuisances'. S79 of the Environmental Pollution Act specifies a number of matters which constitute 'statutory nuisances'. These include "any premises in such a state as to be prejudicial to health or a nuisance". Since 'premises' is defined as to land, it is clear that contaminated land can qualify as a statutory nuisance. S79(i) states that "it shall be the duty of every local authority to cause its area to be inspected from time to time to detect any statutory nuisances . . . and, where

a complaint of a statutory nuisance is made to it by a person living within its area, to take such steps as are reasonably practicable to investigate the complaint". Again, the duty to inspect will tend to bring contaminated sites to the notice of the local authority.

11.4.8 *What action may be taken by the authorities?*

Let us assume that the land has been identified as both contaminated and in need of remedial treatment. What action can the regulatory authorities take?

S61(5) EPA imposes a duty on the Waste Regulation Authority (WRA) to consult the National Rivers Authority (NRA) and S61(7) of the EPA imposes the duty ". . . to do such works and take such other steps . . . as appear to the authority to be reasonable to avoid such pollution or harm".

Consultation with the NRA is most important since S161 of the Water Resources Act 1991 gives the NRA significant clean-up powers:

> where it appears to the Authority that any poisonous, noxious or polluting matter or any solid waste matter is likely to enter, or to be or to have been present in, any controlled waters, the Authority shall be entitled to carry out the following works and operations, that is to say:
> (a) in a case where matter appears likely to enter any controlled waters, works and operations for the purpose of preventing it doing so; or
> (b) in a case where the matter appears to be or to have been present in any controlled waters. Works and operations for the purpose:
>   (i) of removing or disposing of the matter;
>   (ii) of remedying or mitigating any pollution caused by its presence in the waters; or
> (iii) so far as it is reasonably practicable to do so, of restoring the waters, including any flora and fauna dependent on the aquatic environment of the waters, to their state immediately before the matter became present in the waters.

When a local authority identifies contaminated land that constitutes a statutory nuisance under S80 of the EPA, the local authority shall serve notice:

> (a) Requiring the abatement of the nuisance or prohibiting or restricting its occurence or recurrence
> (b) Requiring the execution of such works, and the taking of such other steps, as may be necessary for any of those purposes

11.4.8.1 *Power of entry.*

Waste Regulation Authority. Powers of entry for inspection under S61 of the Environmental Protection Act are given to the Waste Regulation Authority by subsection 3. The power is limited to three particular instances. Two specifically concern waste disposal but the third requires simply that the authority ". . . has reason to believe there may be concentrations or accumulations of noxious gases or noxious liquids . . .". Power of entry to carry out remedial works is not given expressly but S61(7) empowers Waste Regulation Authority ". . . to carry out work and take other appropriate steps whether 'or not the land is affected or on adjacent land". A flow chart setting out the powers and duties of Waste Regulation Authorities in respect of S61 is presented in Figure 11.3.

National Rivers Authority. S161 of the Water Resources Act 1991 does not provide an express power of entry for inspection or carrying out abatement or restoration works. It is, however, implied in that the section states that the authority 'shall be entitled' to carry out specified works. The reader is referred to the flow chart presented in Figure 11.2.

Local Authority. In relation to the Statutory Nuisance provisions of S79 EPA, the local authority is given an implied right of entry to carry out abatement measures under S81(3) of the EPA, in the event of any failure to comply with the local authorities abatement notice.

11.4.8.2 *Recovery of costs.*

Waste Regulation Authority. Power for the recovery of costs associated with works undertaken pursuant to S61 EPA is provided in subsection 61(8). Costs are recovered from the then 'owner' of the land. The reader is referred to the flow chart presented in Figure 11.3.

National Rivers Authority. Power for the recovery of costs associated with works undertaken pursuant to S161 of the Water Resources Act 1991 is provided in subsection 161(3). Costs will be recovered from the person who caused or knowingly permitted the polluting matter to be present in the controlled waters or to be likely to be so. The reader is referred to the flow chart presented in Figure 11.2.

Local Authority. In abating or preventing the occurrence of a statutory nuisance under S79 of the EPA, costs are recoverable from the person who caused the nuisance and, if that person is not the owner of the premises, then expenses can be recovered from the current owner. Power for the recovery of any expenses reasonably incurred is provided in subsection 81(4).

11.4.9 *What if immediate remedial action is not necessary?*

Even if the WRA, the NRA or the local authority do not consider action necessary under the provisions analysed in section 11.4.8, the mere existence on the land of 'controlled waste' (definition in S75 EPA) brings into effect the duty of care imposed by S34(1) of the EPA. Since 'controlled waste' means "household, industrial and commercial waste or any such waste" and is subject only to a few limited exceptions, it will be apparent that the duty of care has wide application.

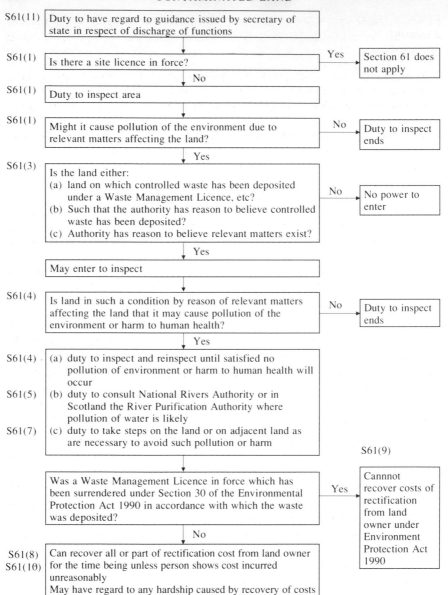

Relevant matters affecting land section 61(2) Environmental Protection Act 1990
The concentration or accumulation in, and emission or discharge from, the land of noxious gases or liquids caused by deposits of controlled wastes in the land

Controlled Waste S74(4) Environmental Protection Act 1990
Household, industrial or commercial waste as defined in Section 75(5), (6) and (7) Environmental Protection Act 1990 (subject, if the regulations so provide, to regulations under Section 83(1) or 75(8) Environmental Act 1990)

Figure 11.3 Waste Regulation Authority duties and powers in respect of closed landfills.

11.4.9.1 *The Duty of Care.*   S34(1) of the Environmental Protection Act 1990 states that save in relation to a householder's disposal of his or her own domestic waste:

> it shall be the duty of any person who imports, produces, carries, keeps, treats or disposes of controlled waste, or, as a broker, has control of such waste, to take all such measures applicable to him in that capacity as are reasonable in the circumstances
> (a) to prevent any contravention by any other person of [the provisions relating to the unlicensed disposal of waste] set out in section 33 above;
> (b) to prevent the escape of the waste from his control or that of any other person; and
> (c) on the transfer of the waste, to secure:
>   (i) that the transfer is only to an authorised person [in particular a licensed waste disposal manager or a registered waste carrier] or to a person for authorised transport purposes [in particular direct imports and exports]; and
>   (ii) that there is transferred such a written description of the waste as will enable other persons to avoid a contravention of that section and to comply with the duty under this subsection as respects the escape of waste.

It will be seen that S34(1) EPA takes effect in three parts. It imposes the duty in the first place to prevent others disposing of the waste illegally; secondly to contain the waste at source; and finally to ensure that in the event that the waste has to be transported to another location, it is properly transferred.

A code of practice to assist in the discharge of the duty is being prepared by the Department of the Environment, the Scottish Office and the Welsh Office as directed by Section 34(7). Section 34(10) states that the code "shall be admissible in evidence and if any provision of such code appears to the court to be relevant to any question arising in the proceedings it shall be taken into account in determining that question".

It is important to note that a person's duty of care under S34 of the EPA cannot be avoided by delegating it to another. The duty may of course be discharged by another, but the liability remains.

If a landowner delegates the task of dealing with a contaminated site to a consultant or contractor, the landowner will still be liable for prosecution in event of a breach of the duty of care. The landowner merely has a possible claim in contract against his consultant or contractor. However, a landowner cannot obtain an indemnity for any criminal sanctions he suffers that are contrary to public policy. There is obviously no substitute for close control and supervision when dealing with the treatment of contaminated land through consultants or contractors.

11.4.9.2 *How does the duty of care affect the landowner?*   If the waste is to be retained on the land, the landowner must ensure that he takes the appropriate measures to contains the waste and prevent:

(a) The escape of the waste itself, e.g. by means of dust blow, landslip or unauthorized removal
(b) The escape of noxious liquids that may cause pollution of the environment or harm to human health
(c) The escape of noxious gases that may cause pollution of the environment or harm to human health
(d) The entry into controlled waters of any liquid or substance likely to cause pollution

A benefit of compliance with the Duty of Care is that it will go a long way to prevent conditions arising that would give rise to an action at common law in nuisance or negligence.

Table 11.2 presents a check-list of the steps to be taken to comply with the duty of care.

11.4.9.3 *Is there anything else?*   In addition to discharging the Duty of Care the landowner also needs to ensure that his land is in a 'safe' condition for the entry of persons (including trespassers) in order to discharge the Duties of Care imposed by the Occupiers Liability Acts (section 11.3.2(d)) and the Health and Safety at Work Act.

11.4.10 *What if remedial action is necessary?*

We will assume that the risk is such that the NRA does not need to take immediate action, but that the WRA considers action necessary. Under S61(5) of the EPA, the WRA must consult with the NRA as to the necessary steps to safeguard controlled waters, thus ensuring the NRA's involvement in the restoration process. The WRA may take action itself, or under S61(7) EPA ". . . take such other steps as appear to the authority to be reasonable . . ."; these other steps could allow the landowner to carry out works himself in agreement with the authority. One would anticipate that this latter course will be preferred by the WRA where co-operation is forthcoming from the landowner, since it places the least strain on the authority in terms of resources and finance. In addition, this course of action will allow the landowner to retain some control over what is done and how much is spent.

11.4.10.1 *What must be done now?*   The basic steps and safeguards required in order to carry out any remediation programme are:

● Steps will need to be taken to ensure proper containment of actual or potential pollutants as described in sections 11.4.9.2 and 11.4.9.3.
● A survey and investigation of the land will be necessary to determine the type and extent of contamination and the appropriate actions. These matters are covered in detail in chapter 3.

Table 11.2   The Duty of Care: a summary check-list

The following is a version of the summary check-list set out in the draft Code of Practice.

See section	Steps	Paragraph in the draft Code of Practice
11.4.12.2	(a) Is what you have waste? If yes	2.2–2.3
11.4.12.2	(b) Is it controlled waste? If yes	2.4
11.4.9.2	(c) While you have it, protect and store it properly	3.1–3.4

11.4.12.5	(d) Write a proper description of the waste, covering:
	  • Any problems it poses
	and, as necessary to others who might deal with it later, one or more of:
	  • the type of premises the waste comes from
	  • What the waste is called
	  • The process that produced the waste
	  • A full analysis
		2.9–2.10
		2.5–2.7
		
		2.12–2.13
		2.14
		2.15–2.16
		2.17–2.18

11.4.12.5	(e) Select someone else to take the waste; they must be one or more of the following and must prove that they are:
	  • A registered waste carrier
	  • Exempt from registration
	  • A waste manager licensed to accept the waste
	  • Exempt from waste licensing
	  • A waste collection authority
	  • A waste disposal authority (Scotland only)
		
		4.3–4.5
		4.6
		4.7–4.9
		4.10
		4.2
		4.12

11.4.12.5	(f) Pack the waste safely when transferring it	3.5–3.7
11.4.12.5	(g) Check the next person's credentials when transferring waste to them	4.4–4.6, 4.8–4.14
11.4.12.5	(h) Complete and sign a transfer note	Annex C, C3–C4
11.4.12.5	(i) Hand over the description and the transfer note when transferring the waste	2.9 and Annex C, C3–C4
11.4.12.5	(j) Keep a copy of the transfer note signed by the person it was given to, and a copy of the description, for 2 years	Annex C, C5
11.4.12.5	(k) When receiving waste, check that the person who hands it over is one of those listed in (e), obtain a transfer note and a description from them and keep the documents for 2 years	5.2–5.5 and 6.4
11.4.12.5	(l) Whether transferring or receiving waste, be alert for any evidence or suspicion that the waste you handle is being dealt with illegally at any stage; in case of doubt, question the person involved and if not satisfied, alert the waste regulation authority	6.5–6.11

- It will be necessary to obtain planning permission to carry out the works, and it is during this stage that trigger levels for the concentrations of various contaminants will be established. These will determine the scope of the restoration work prior to redevelopment of the site. The determination and source of trigger levels are discussed in detail in chapter 3 and reference data sources are given in Appendix I.

- Where waste is to be treated on the land, a Waste Management Licence will be required (S33 EPA) unless the operation falls within a prescribed exception. The reader is directed to section 11.4.11.2 for the definition of treated.
- Where waste is to be removed from the land it will be necessary to determine a suitable reception site and system for complying with the duty of care for disposal.

Figure 11.4 outlines the main steps to be taken in dealing with the disposal of a waste. The specific requirements of the latest Draft Code of Practice on the Duty of Care (issued in September 1991) dealing with the mechanics of disposal are set out in section 11.4.12.

11.4.11 *How will control be imposed?*

11.4.11.1 *Planning permission.* The Town and Country Planning Act 1990 requires that planning permission is obtained for any development and it is at this stage that controls as to working arrangements, clean-up levels and restoration conditions will be applied.

Of major importance at this stage is the choice of clean-up levels for the restoration. These may be imposed by the planning authority as a condition of the planning permission after consultation with internal departments, the NRA and normally the applicant. The United Kingdom does not have a mandatory set of clean-up levels, although guidance documents issued by the Department of the Environment exist and there are a number of other recognized documents in use from which values may be obtained (Appendix I). It is therefore often a matter of local interpretation as to the particular clean-up levels specified, how they are chosen and how they are applied. Once engrossed in a planning permission, change is difficult to achieve and the applicant should therefore endeavour to take a positive part in their agreement from the outset. It is likely in the future that planning authorities will make increasing use of formal agreements under S106 of the 1990 Act as an alternative to issuing planning approvals with conditions specifying clean-up levels.

11.4.11.2 *Waste management licence.* S33 of the Environmental Protection Act provides that no one can "treat, keep or dispose of controlled waste . . . in or on any land except under and in accordance with a waste management licence". A number of limited exceptions, subject to conditions, are set out in the Collection and Disposal of Waste Regulations 1988 Schedule 6 (see Appendix II hereto). In particular in paragraph (4) there is an important exception for "waste arising from works of construction or demolition, including waste arising from work preparatory thereto and waste arising from tunnelling or any other excavation or of oil, slag or clinker."

Section S36(2) of the Environmental Protection Act states that a waste management licence cannot be issued unless a planning permission is in force

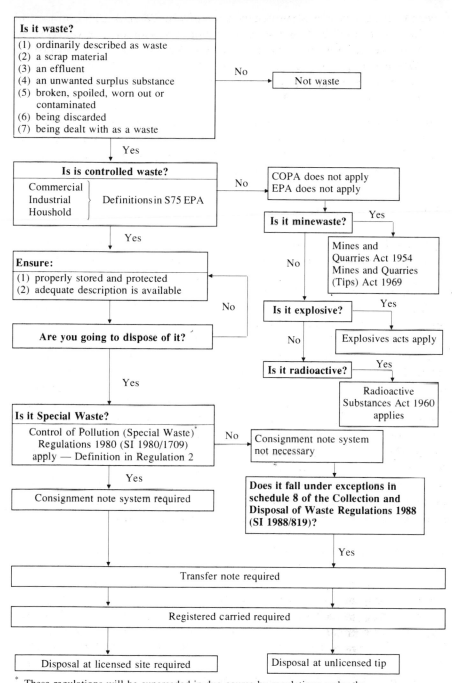

Is it waste?
(1) ordinarily described as waste
(2) a scrap material
(3) an effluent
(4) an unwanted surplus substance
(5) broken, spoiled, worn out or contaminated
(6) being discarded
(7) being dealt with as a waste

No → Not waste

Yes

Is is controlled waste?
Commercial
Industrial } Definitions in S75 EPA
Houshold

No → COPA does not apply
EPA does not apply

Is it minewaste? — Yes

No → Mines and Quarries Act 1954 Mines and Quarries (Tips) Act 1969

Ensure:
(1) properly stored and protected
(2) adequate description is available

Yes

No

Are you going to dispose of it?

Yes

Is it explosive? — Yes → Explosives acts apply

No

Is it radioactive? — Yes → Radioactive Substances Act 1960 applies

Is it Special Waste?
Control of Pollution (Special Waste)* Regulations 1980 (SI 1980/1709) apply — Definition in Regulation 2

No → Consignment note system not necessary

Yes

Consignment note system required

Does it fall under exceptions in schedule 8 of the Collection and Disposal of Waste Regulations 1988 (SI 1988/819)?

Yes

Transfer note required

Registered carried required

Disposal at licensed site required   Disposal at unlicensed tip

* These regulations will be superseded in due course by regulations under the Environmental Protection Act

Figure 11.4 Main steps for the disposal of waste.

for that purpose on the land, and S35(3) allows the WRA to attach conditions to the licence. Guidance on conditions is given in *Waste Management Paper No. 4* issued by Her Majesty's Inspectorate of Pollution. Each WRA will have its own standard form of licence and the scope for substitution is limited. It is important for the applicant to take an active part from the outset in the determination of the licence conditions that relate specifically to the proposed operation.

The 'disposal' of waste may leave little room for doubt, but 'treatment' is less straightforward. The term 'treated' is defined in S29(6) and S29(7) of the Environmental Protection Act:

> S29(6) The "disposal" of waste includes its disposal by way of deposit in or on land and, subject to subsection (7) below, waste is "treated" when it is subjected to any process, including making it re-usable or reclaiming substances from it and "recycle" (and cognate expressions) shall be construed accordingly.
>
> S29(7) Regulations made by the Secretary of State may prescribe activities as activities which constitute the treatment of waste for the purposes of this Part or any provision of this Part prescribed in the regulations.

At the time of writing no Regulations have been made by the Secretary of State under S29(7) of the Act. There is therefore some doubt as to the extent of the intended ambit of 'treatment' that could, on the wording of S29(6), encompass processes such as enhanced biodegradation and the like.

Although the term 'treated' did not appear in the corresponding section, S3, of the Control of Pollution Act 1974 the words "dealing in a prescribed manner" were present. One might therefore turn to regulations made under the Control of Pollution Act to provide an indication as to what might be pre-scribed under the EPA. The regulations known as The Collection and Disposal of Waste Regulations 1988 state in Regulation 8: "The manners of dealing with controlled waste set out in schedule 5 are prescribed for the purposes of Section 3(1)(b) (of the Control of Pollution Act)".

Schedule 5 of the regulations prescribed the following for the purposes of S3(i)(b):

(1) Bailing, compacting, incinerating, pulverizing, sorting of storing waste
(2) Processing or holding waste at a site designed or adapted for the recep-tion of waste with a view to its being disposed of elsewhere
(3) Shredding waste as a trade or business
(4) Treating waste by pyrolysis
(5) Producing fuel from waste
(6) Making compost from waste
(7) Processing or treating waste oil or waste solvent to permit its reuse
(8) Using untreated waste as fuel to produce electricity or heat

One might therefore anticipate that similar processes would be prescribed under the EPA. However, the change in wording from that used in the former Act indicates that the EPA intended a wider scope in order to permit control to be exercised over the whole range of reclamation processes. That wider interpretation is supported by the definition of 'treatment' adopted in the European Community's proposed Directive on the landfill of waste 91/C 190/01. Com(91) 102 syn 335 of 23 April 1991. This proposed directive uses the definition from Directive 75/442/EEC on waste: " 'Treatment' means the physical, chemical or biological process that changes the characteristics of the waste in order to reduce its volume or hazardous nature, facilitate its handling or enforce recovery." Until the Regulations are made under S59(7) this question is likely to remain unanswered; however, indications would appear to suggest that a wide definition will be used.

11.4.11.3 *End use.*   Remediation works are normally instigated as the preparatory stage in the overall development of a site and the developer will require that the restoration is suitable for the end use for which he has obtained planning permission. Even if the planning authority does not·impose specific clean-up conditions, the developer will need to satisfy himself, purchasers and occupiers that all appropriate steps have been taken to ensure a safe environment during and after restoration. The developer will also need to consider the marketability of the site after clean-up. It may be that the end use of the site necessitates that the land is cleaned up to a much higher standard than would be necessary simply to comply with the requirements of the regulatory authorities.

11.4.12 *What if it is decided to remove the contamination?*

11.4.12.1 *What controls exist?*   The treatment and transfer of waste is regulated by Part II of the EPA and, as has been seen (section 11.4.9.1), S(34) imposes a duty of care to ensure that the waste:

- Is not disposed of in an illegal manner
- Does not escape from control
- Is only transferred to an authorized person
- Is transferred with an adequate description

A Code of Practice will be issued by the Department of the Environment to assist in the discharge of the duty of care. The steps outlined in the following sections are based on the advice given in the second draft of the code issued in September 1991.
    Reference to the actual current code of practice must be made when dealing with controlled waste, since it has been necessary in this chapter to paraphrase the full advice given.

11.4.12.2 *Is it waste and is it controlled?*   S33 of the EPA restricts the treatment and disposal of what is termed 'controlled waste'. It is therefore necessary to determine whether or not a material is controlled waste.

Waste. The definition of waste is given in S75(2) of the EPA and the Code of Practice states that the question should be asked from the point of view of the person producing or discarding it. This view has been taken consistently by the courts in a series of cases, and leads to the result that an innocuous and harmless substance such as natural quarry overburden is a waste when viewed from the standpoint of the producer, despite the fact that it can, and in the case of *Long v Brooke (1980) Crim LR109*, subsequently was used as landscaping material.

S75(2) "Waste" includes:
(a) any substance which constitutes a scrap material or an effluent or other un-wanted surplus substance arising from the application of any process; and
(b) any substance or article which requires to be disposed of as being broken, worn out, contaminated or otherwise spoiled;
but does not include a substance which is an explosive within the meaning of the Explosives Act 1875.
(3) Anything which is discarded or otherwise dealt with as if it were waste shall be presumed to be waste unless the contrary is proved.

Thus, any material surplus to requirements is likely to be a waste and subject to control as controlled waste.

Controlled waste. Controlled waste is defined rather unhelpfully in S75(4) of the EPA as meaning simply: "household, industrial and commercial waste or any such waste". Household waste is defined in S75(5). Industrial waste is defined in S75(6). Commercial waste is defined in S75(7). The only important exceptions relate to:

(a) Mines and quarries
(b) Agricultural activities
(c) Radioactive substances

Assistance in the allocation of particular materials between these classes is given in the Control of Pollution Act in schedules 1, 3 and 4 of the Collection and Disposal of Waste Regulations 1988 (SI No. 819 1988), which are repro-duced in Appendix II. The same classification is likely to be used under the Environmental Protection Act.

Special waste. Special Waste S75(9) is a particular form of controlled waste of the type or types yet to be specified in regulations made under S62 EPA. When made, these regulations are likely to be similar to those currently in force under COPA 1974, namely The Control of Pollution (Special Waste) Regulations 1980 (amended by SI 1988 No. 1790), except that the Government has proposed that the definition of special waste is amended and defined by reference to:

(a) A list of wastes, which by their nature are considered likely to have characteristic properties that make them dangerous or present disposal difficulties

(b) A list of substances, which, if they are present in waste, may cause the waste to have those characteristic properties

(c) A list of twelve characteristic properties, which it is proposed to use in the definition of special waste: for example, explosive, oxidizing, toxic, infectious, ecotoxic, etc.

(d) A separate list of readily identifiable household wastes that have those characteristic properties

The definition is proposed to be formulated in a way that enables the source or nature of the waste, its components, and its characteristic properties to be identified by a reference number and coded.

Figure 11.4 outlines the steps to be taken in determining the type of waste. The major question is not whether the material is 'waste,' which it is almost certain to be, but whether it is controlled waste thereby requiring a licence to allow it to be 'kept, treated or disposed'.

11.4.12.3 *How is a disposal site chosen?*   As already mentioned, controlled waste may only be disposed of at a licensed site (S33 of the EPA), and anyone having control of waste, i.e. the landowner in this analysis, has a Statutory Duty of Care (S34 of the EPA) to ensure that the waste is disposed of at such a site.

The Code of Practice states that in selecting a suitable disposal site, a check should be made to ensure that a licence for the site exists and that it permits the deposit of the particular waste. The disposal site management should be able to provide such evidence, but if there is any doubt a secondary check should be made with the Waste Regulation Authority.

Table 11.3 gives a list of the waste regulation authorities in England and Wales. In order to carry out this check, information from the site investigation on the chemical composition of the waste will be required to enable correct description of the waste. If the checking of the licence requires the appointment of a specialist consultant, the producer or holder of the waste will need to bear in mind that he remains primarily liable for discharging the Duty of Care and cannot delegate this duty to an agent (see section 11.4.9).

Table 11.3   Waste regulation authorities: Who are they?

In the non-metropolitan areas, the County Councils are waste regulation authorities.

In the metropolitan areas, the position is more complicated. In the following list WRAs are shown in bold type.

MERSEYSIDE

Merseyside Waste Disposal Authority

Knowsley MBC
Liverpool CC
St. Helens MBC
Sefton MBC
Wirral MBC

Table 11.3 *(Cont.)*

GREATER MANCHESTER

Greater Manchester Waste Disposal Authority

Bolton MBC
Bury MBC
Manchester CC
Oldham MBC
Rochdale MBC
Salford CC
Stockport MBC
Tameside MBC
Trafford MBC
Wigan MBC

SOUTH YORKSHIRE

Barnsley MBC
Doncaster MBC
Rotherham MBC
Sheffield CC

Note: There is a joint committee operating under a 'rolling' coordinating agreement with rights of withdrawal on 12 months' notice. Rotherham MBC is the 'host' authority.

TYNE AND WEAR

Gateshead MBC
Newcastle upon Tyne CC
North Tyneside MBC
South Tyneside MBC
Sunderland MBC

Note: There is a joint committee operating under a 5 year coordinating agreement expiring in 1991. Gateshead MBC is the 'host' authority.

WEST MIDLANDS

Birmingham CC
Coventry CC
Dudley MBC
Sandwell MBC
Solihull MBC
Walsall MBC
Wolverhampton MBC

Note: There is a joint committee operating under a 5 year coordinating agreement with rights of withdrawal on 12 months' notice after 1991. Walsall MBC is the 'host' authority.

WEST YORKSHIRE

Bardford CC
Calderdale MBC
Kirklees MBC
Leeds CC
Wakefield CC

Note: There is a joint committee operating under a 7 year coordinating agreement with rights of withdrawal on 3 years' notice after 1993.

LONDON

West London Waste Authority

L.B. Brent
L.B. Ealing
L.B. Harrow
L.B. Hillingdon
L.B. Hounslow
L.B. Richmond upon Thames

North London Waste Authority

L.B. Barnet
L.B. Camden
L.B. Enfield
L.B. Hackney
L.B. Haringey
L.B. Islington
L.B. Waltham Forest

East London Waste Authority

L.B. Barking and Dagenham
L.B. Havering
L.B. Newham
L.B. Redbridge

Western Riverside Waste Authority

L.B. Hammersmith and Fulham
Royal L.B. Kensington and Chelsea
L.B. Lambeth
L.B. Wandsworth

Central London Group

City of London
City of Westminster
L.B. Tower Hamlets

Note: There is a joint committee operating under a coordinating agreement.

South London Group

L.B. Bromley
L.B. Croydon
Royal L.B. Kingston upon Thames
L.B. Merton
L.B. Sutton

Note: There is a joint committee operating under coordinating agreement.

South East London Group

L.B. Greenwich
L.B. Lewisham
L.B. Southwark

Note: There is a joint committee operating under a coordinating agreement.

L.B. Bexley

Note: There is an agreement with Kent County Council, which expires in 1991.

Note: (1) The proposed Environment Agency may well assume the responsibilities of the
County Councils and other WRAs listed above.
(2) Until Part II of the Environmental Protection Act 1990 is implemented, WRAs are
strictly speaking still waste disposal authorities.

11.4.12.4 *How does it get there?* S34(1)(c) of the EPA requires that waste is transferred only to an authorized person, with a description of the waste that will enable that person to comply with S33 (Disposal) and the Duty of Care. S34(3) states that authorized persons are:

(a) Any authority which is a waste collection authority for the purposes of this Part
(b) Any person who is the holder of a waste management licence under section 35 below or of a disposal licence under section 5 of the Control of Pollution Act 1974
(c) Any person to whom section 33(1) above does not apply by virtue of regulations under subsection (3) of that section
(d) Any person registered as a carrier of controlled waste under section 2 of the Control of Pollution (Amendment) Act 1989
(e) Any person who is not required to be so registered by virtue of regulations under section 1(3) of that Act and
(f) A waste disposal authority in Scotland

For the present purposes an authorized person will be:

(a) A waste collection authority
(b) The holder of a waste management or waste disposal licence
(c) A registered carrier of controlled waste

11.4.12.5 *Working with the registered carrier.* Carriers of controlled waste must be registered under the Control of Pollution (Amendment) Act 1989 and the Controlled Waste (Regulation of Carriers and Seizure of Vehicles) Regulations 1991.

Checking a carrier's authority. The Draft Code of Practice on the duty of care states that:

> Anyone intending to transfer waste to a carrier will need to check that the carrier is registered or is exempt from registration. A registered carrier's authority for transporting waste is either his certificate of registration or a copy of his certificate of registration *if it was provided by the waste regulation authority.* The certificate or copy certificate will show the date on which the carrier's registration expires. All copy certificates must be numbered and marked to show that they are copies and have been provided by the waste regulation authority. Photocopies are not valid and *do not* provide evidence of the carrier's registration.

If in doubt a check should be made with the issuing authority.

Two other requirements must be met: (i) to package the waste adequately; and (ii) to describe it.

Handling waste from source to disposal (packaging). The way in which waste is handled obviously depends on the nature of the waste material. For the typical contaminated land muckshift, this will require only careful excavation practice into closed or sheeted wagons; liquid wastes or solvents on the other hand will require barrels or tankers. The handling or packaging method employed must be such as safely to contain the waste from source to disposal.

The transfer note. The requirement that waste be transferred with an adequate description of it may be met by use of the model waste transfer note annexed to the Draft Code of Practice on the Duty of Care. In describing the waste, any special problems likely to be associated with its packaging, handling, transport, storage and disposal should be stated. Where possible, information on the following should also be stated.

(a) The type of premises or business from which the waste comes
(b) The name of the substance or substances
(c) The process that produced the waste
(d) A chemical and physical analysis

Consignment notes. Waste classified as special waste by Regulation 2(1) of the Control of Pollution (Special Waste) Regulations 1980 must, under the provisions of Part II of these Regulations, be transferred using the consignment note system prescribed therein.

The Draft Code of Practice states that the consignment note system can be used in place of the transfer notice for discharge of that element of the duty of care and, although not suggested in the Draft Code of Practice, since it provides a full audit chain for the waste, its use with all waste as a matter of course should be considered. Figure 11.5 provides a summary of the main steps to be taken when using a waste carrier.

11.4.12.6 *Can it now be forgotten?* The Duty of Care does not extend to pursuing the waste after if has been transferred to an appropriate person. However, an audit of the waste's final destination and the transport chain, and periodic site visits are recommended as a prudent means of protecting the waste producer's position by being able to demonstrate the steps taken to prevent subsequent illegal treatment of the waste.

Producers may wish to use the consignment note system for all waste, coupled with periodic supervision of the disposal chain as an administrative means of providing evidence of the steps they have taken.

11.4.13 *The various parties: their different perspectives*

11.4.13.1 *General.* The purpose of preceding sections has been to describe the legal framework within which the various parties involved in the reclamation of contaminated land have to work. The following paragraphs consider briefly the perspective of each of the parties in turn.

11.4.13.2 *The regulators.* The regulators, i.e. local government, Waste Regulation, and National Rivers Authority and Her Majesty's Inspectorate of

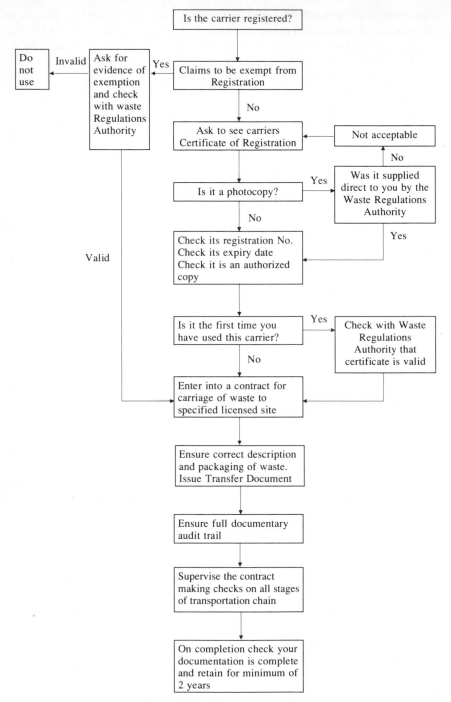

Figure 11.5 Using a waste carrier.

Pollution or, as appears to be on the horizon, the Environmental Protection Agency, are essentially the policemen of environmental legislation. It is their task to issue consents and licences, and to set threshold levels for the various applications; it may be anticipated that a more adversorial approach will develop, with prosecutions for relatively minor offences becoming commonplace. In carrying out their functions, the regulators must bear in mind the limitations on their powers and that rights of appeal are available against almost all decisions taken by them. Should attitudes to the need for licences and permissions become unduly restrictive, owners and developers who must foot the bill are likely to look to the appeal process in a similar fashion to that pertaining in the Town and Country Planning arena, i.e. take all unfavourable decisions to appeal as a matter of course. If this course of action is contemplated, applicants will need to assess very carefully the probability of success, the time delay associated with an appeal and the effect on their cash flow, against the cost of simply accepting the restrictions. It will be unfortunate should this scenario become the norm since to date most regulators have tried to work closely with developers to solve problems in the interest of economic development.

In the setting of planning conditions, in particular clean-up levels, the regulators will need to ensure that they are not 'guaranteeing' the acceptability of the land for its ultimate purpose. This responsibility is clearly that of the owner or developer and must remain fairly in his court.

11.4.13.3 *Owners and developers.* The prime concern of landowners and developers will be to ensure that the remediation work is carried out efficiently and legally. When working with a team of professional advisors and contractors, they will need to ensure that each party is in no doubt as to the precise functions he/she is to perform with regard to assisting the landowner or developer to discharge the Duty of Care. In addition, landowners and developers may wish to seek warranties from the professional team and the contractor. Those warranties would need to refer expressly to the legal provisions in this and the preceding section in order to put the extent and nature of the contractual duties beyond doubt. The warranties would need to be made available to subsequent owners and occupiers of the reclaimed site.

11.4.13.4 *The professional team.* The role of the professional team can be diverse. A major area of concern would be that they inadvertently accept the role of waste holder and thereby take upon themselves the Duty of Care. In order to avoid that position, consultants must ensure that they restrict their activities to advice, the construction of documents, and the supervision of contracts under which the contractor is clearly responsible for holding, handling, treating and disposing of the waste.

Extreme care will be necessary in drafting contracts for the disposal or treatment of waste to ensure adequate control over the contractor and compliance with the law. Particular attention will be required to ensure that the works contract includes sufficient and adequate checking procedures in order to discharge the employer's duty to prevent others from dealing with controlled waste unlawfully.

When designing or advising on the treatment of contaminated land, care must be taken to ensure that the advice is based on a proper knowledge of the site, its problems, and its end use such that appropriate methods may be used. Where limitations on the investigation of the site or the provision of information are imposed by the landowner or developer, steps must be taken to ensure that any warranties given by the professional team are suitably qualified and that adequate professional indemnity insurance is in place.

During supervision of the works, consultants will need to ensure their staff comply with checking procedures and avoid directing the contractor on how to execute the physical work.

11.4.13.5 *Contractors.* Contractors will be required to provide the practical means by which landowners and developers discharge their legal duties. They must therefore be wary of undertaking obligations that are not precisely defined and which are not capable of proper investigation. Since pressure from the regulatory authorities will become intense, and because landowners and developers will seek to protect themselves by guarantees from contractors, it will be important for contractors to make themselves fully conversant with the legal pitfalls and review their contracts with appropriate care.

When dealing with contaminated land, contractors must in turn ensure that they are provided with all the information about the waste necessary for them to discharge their Duty of Care. Where there is any doubt as to such provision, contractors would be advised to register their concern with the employer immediately and refrain from dealing with the waste until such time as the information is available. Alternatively, of course, they may take steps to obtain the necessary information themselves.

Particular care should be exercised in the selection of subcontractors and the form of contract used in respect of waste transfer, treatment or disposal so as to ensure coordination of duties and liabilities. Contractors will also need to increase their level of supervision on contracts to provide the necessary checks and audit chain required by the Code of Practice on the Duty of Care.

11.5 Future development

11.5.1 *The law*

Environmental law is developing so quickly that it is difficult to predict future changes with confidence. The following represent merely a selection:

  (a) *Environmental liability.* The draft directive on civil liability for damage caused by waste is currently 'stalled' in the EC legislative process and may be replaced in due course by a directive on general environmental liability. The strict (i.e. 'no fault') liability provisions in the draft directive on civil liability for damage caused by waste were eroded somewhat during that directive's passage and it may be that further erosion will take place if the directive is subsumed in a wider measure; but they are unlikely to be jettisoned entirely.

(b) *Landfill regulation.* The draft directive on the landfill of waste, if implemented in its present form, will introduce a regulatory regime that may well be more stringent than that established by Part II of the EPA. It will certainly make the UK practice of co-disposal more expensive and thus less attractive. Landfill operators will be required in effect to guarantee the long-term performance of their sites and will seek to pass on this liability to their customers.

(c) *Disclosure obligations.* At present landowners can take refuge in the principle of caveat emptor (let the buyer beware). The onus is currently on purchasers to find out what they can about the state and condition of a site. There have been moves to eliminate, or at least modify, the caveat emptor principle and, although the Law Commission has recommended that the principle remain, public and Government pressure will mount. Even if there is no formal change, the market place may well force landowners to be candid. Otherwise, purchasers may decide to go elsewhere.

(d) *Environment Protection Agency.* The Government has recently announced its desire to integrate the activities of the various regulatory bodies, including HMIP, the NRA and the WRA. The EC also plans to establish an Environmental Agency with a remit merely to gather information but with the possibility that it may later become involved in compliance work. Whatever the regulatory mechanisms, public opinion will ensure that the owners of contaminated sites will be faced with increasingly powerful and confident enforcement agencies.

(e) *Nuisance.* Despite the proliferation of environmental legislation, there will still be a role for the common law principles of nuisance and negligence. As already discussed there are various technical obstacles for plaintiffs to overcome (see section 11.3.3). For the present, the courts seem disinclined to relax the rules on public policy grounds: they fear a flood of claims that may seriously hamper industrial activity. Whether this approach will be sustainable in the medium- to long-term is rather doubtful.

11.5.2 *The approach to land reclamation*

Increasing regulation at all levels and the more stringent application of the 'polluter pays principle' are likely to have a two-fold effect on the approach to the treatment of contaminated land.

Contaminated land is likely to become considerably more expensive to treat by removing material to off-site landfill. On-site and in situ treatment will become comparatively more cost-effective and of wider application as technology improves to meet the market. Nevertheless, the cost of treatment of contaminated land will still exceed today's landfill prices. As a result, developers and purchasers of land will insist on receiving 'clean' land or will demand a heavy price discount to carry out the restoration themselves. It may be that in certain instances, the owner of contaminated land will be unable to dispose of that land due to the onerous associated legal responsibilities and clean-up costs. When restoration is undertaken, each party involved will demand a full site history and an audit trail. Regarding the de-contamination, it is also probable that the use of warranties will increase, and professionals and contractors will be asked to guarantee their work and insure against the consequences of default or failure. This search for security is likely to cause delays as the parties seek to ensure appropriate contractual protections, increase insurance premiums, encourage the production of over cautious designs and require rigidly controlled site works. All these developments will increase costs.

On the positive side, integrated pollution control coupled with a greater awareness of pollution mechanisms, their environmental effects and legal consequences, will tend to minimize the creation of contaminated land. When contamination does occur, the approach is likely to be one of immediate damage limitation and clean-up, rather than the 'leave it until tomorrow' attitude of previous generations. It may be anticipated that the role of professionals will be to advise on avoidance rather than treatment, and that of the contractors to install prevention measures instead of cleaning up the mess.

The result should be to ensure not only more effective clean-up programmes but also an ever diminishing supply of contaminated land.

12 Introduction to US waste management approach

M.K. MEENAN

12.1 Introduction

At times it seems difficult to identify the real objectives and achievements of the much maligned and criticized waste management programmes that the United States has devised to address hazards resulting from wastes, in what arguably is today, and will continue to be, a major risk to human health and the environment. Many contend that if the development and contents of the individual environmental programmes are examined, then many of these programmes fall short of their intended objectives and cost far more than the benefit society derives from them; furthermore, they might quite possibly be ill-conceived and too 'over-the-top'.

Prior to the 1970s in the United States, a decentralized form of environmental management or lack thereof existed within government and industry, with the result that our environment and health were seriously threatened. This threat, by the time it was characterized, was the largest on earth in comparison to any combination of industrialized nations and was continuing to grow at a pace unabated and inconceivable to other nations and the US public. Something had to be done sooner rather than later, and at a cost that was very difficult to identify because the solutions were unavailable at the time. In fact, since the problem had never really been identified properly, costs could not be determined with any reliability.

What was ultimately settled on was the adoption of a nationally legislated solution that would provide a minimum standard and uniform approach for managing environmental health. Unfortunately, while this system offers these advantages, it is also subject to the political influences that affect its success. Additionally, while US industry was slow to recognize the potential costs, and politicians were reluctant to oppose popular sentiment and curtail the extent of these regulations, government continued to construct ambitious and ultimately expensive environmental programmes.

Today in the United States, a strong regulatory framework of environmental programmes ensures that the public and the environment are protected from the threat of the mismanagement of all forms of wastes. These stringent requirements have goals and objectives that have inalterably changed the way in which US industry conducts its business both nationally and globally, and have spurred the growth of education, research, technologies, and service industries specializing in hazardous waste management. At present, much of the industry supporting the implementation of these requirements, including research and development of waste treatment technologies, is tailored to the legislatively defined US waste management programmes.

In spite of their search to find other more practical means of environmental management, many countries around the world today are adopting and/or considering some form of centralized and uniform practices similar to the US model, to address the waste management problems that are besetting the world as a whole (chapter 1). This chapter attempts to examine the legislative evolution of US waste management practices and to provide an overview of its direction today, focusing on the control, prevention, treatment and disposal of wastes in a world that is increasingly conscious of its limited resources and global responsibilities.

12.2 Historical development

The historical development of waste management in the United States began in the 1960s, at a time when the general public's attention was galvanized by a number of social issues. There have been many suggestions as to the reason for the development of the environmental consciousness of the 1960s, one of which, the publication of Rachel Carson's *Silent Spring* in 1963 [1], was particularly controversial for associating health risks with agri-chemical industrial practices. By today's standards this book was an oversimplification of the connections between exposures to various chemicals, and life-threatening diseases such as leukaemia, and in many scientific quarters it was dismissed as more fiction than fact [2]. However, the book served to heighten public health concerns, and identified to legislators and public alike an area that required federal regulatory involvement.

As a result, the US Congress dealt in general with the solid waste problem in statutes enacted in 1965 (Solid Waste Disposal Act) and in 1970 (Resources Recovery Act). In the meantime, other environmental legislations being enacted, such as the National Environmental Policy Act in 1969 and the Clean Air Act of 1970, were making it increasingly evident that consolidating the administration of these policies, which were formerly the responsibility of several federal agencies, into one government agency, would be more effective and therefore necessary. This brought about the creation of the US Environmental Protection Agency (EPA) whose powers have expanded to the extent that today it is one of the more influential and powerful agencies within the US Federal Government, with a 1991 annual budget of $6.4 billion. At present, a proposal for upgrading the EPA to cabinet level status, which would further enhance the EPA's role and influence within the national and world environmental scene, is under consideration.

Today the EPA and its several thousand employees administer federal environmental policies from offices in ten regions, encompassing all 50 states and land possessions of the United States. As mentioned, the EPA is responsible for administering the myriad of environmental legislations enacted by the US Congress. The enabling legislation and Congress intentions are interpreted into regulations by the EPA's 'technocrats', encompassing a formal judicial and informal public collaborative process leading to final rulemaking. It is often during the rulemaking process that major disagreements between the Agency, industry and environmentalists emerge. Through the years, all groups have certainly become aware of the importance of this function and are adept at employing lobby tactics to affect aspects of individual regulations whose results have long-term influential impacts on the enforcement and compliance of the regulations.

12.3 Federal legislation

To understand fully the US waste management process, and therefore its complications, it is useful to examine the myriad of environmental programmes that influence, through an overlapping and intertwining process, the regulation of solid waste. A very significant key to understanding this process involves examining the definitions of terms used in the regulations, because of their nature as laws of the land and US litigious propensities. Several of the important terms are defined later in this chapter. A brief overview of the waste regulations and significant associated environmental programmes will help to understand the present waste management system.

12.3.1 *National Environmental Policy Act*

A first step in the US environmental process began with the Congress enactment of the National Environmental Protection Act in 1969 (NEPA) [3]. NEPA establishes the policy and goals, and provides the means for carrying out the policy commonly referred to as the basic national charter for the protection of the environment. Quite simply, the basic policy and goals state:

- *Policy* The Federal Government shall "use all practical means and measures ... to create and maintain conditions under which man and nature can exist in productive harmony, and fulfill the social, economic, and other requirements of present and future generations of Americans." — NEPA Section 101(a).
- *Goals* Federal plans, functions, programmes and resources must be used to achieve six general goals specified in Section 101(b), including the assurance of "safe, healthful, productive, and aesthetically and culturally pleasuring surroundings" for all Americans.

The means to carry out this policy and achieve these goals is developed through NEPA's 'action enforcing provision' (Section 102). This provision directs all federal agencies, not only the EPA, to the fullest extent possible, to meet the following requirements:

(1) Utilize a systematic interdisciplinary approach in planning and decision making that may effect the environment — NEPA Section 102(2)(A).
(2) Identify and develop methods and procedures to ensure that presently unquantified amenities and values may be given appropriate consideration in decision making along with economic and technical considerations — NEPA Section 102(2)(B).
(3) Include an *Environmental Impact Statement* (EIS) in every recommendation or report on proposals for legislation, and other major Federal Actions significantly affecting the quality of the human environment — NEPA Section 102(2)(C).
(4) Study, develop and describe appropriate alternatives to recommended courses of action — NEPA Section 102(2)(E) [4].

This extremely influential legislation gives the federal government, and all relevant agencies, the ability to direct the development and mitigate the impacts of waste management for the good of human health and the environment. Federal agencies such as the US Fish and Wildlife Service, Army Corps of Engineers, Bureau of Land Management, etc., as consultees, possess the capacity through the EIS process to review and comment on any proposed waste management activity whose impacts significantly affect, either directly or indirectly, the quality of the human environment. Although the EIS process is not itself a permit or licensing system, it is in the case of almost every waste management development a requirement of a Federal permit, licence, loan, grant, lease, etc. State governments have their own influence on waste management projects through the enactment of NEPA type requirements (frequently called 'SEPAs' (State Environmental Protection Acts)), which require the environmental assessment of significant actions that are not addressed by federal requirements within state borders.

12.3.2 *First tier environmental statutes*

During the mid 1970s, following the establishment of NEPA, the passage of several major environmental bills was completed. This established the framework for future attention to the control and regulation of processes that were currently involved in the discharge to water, emission to air and disposal to land, of chemical pollutants that inalterably affect the resources of the United States. Initially the laws were intended to create national quality standards of maximum concentration levels for various pollutants, designed to protect public health and welfare. However, what became apparent through the years, after the initial legislation and adoption of the regulation, was the lack of success in achieving the emission/discharge standards. Each of the original environmental statutes (e.g. air, water, and waste) was followed by strengthening legislation in the form of amendments designed to increase the EPA's ability to monitor and enforce compliance with the standards. This then established the EPA as a major agency within the government, firmly closing the loop and setting the standards for environmental protection, and also having the capability to enter facilities and require information to ensure compliance, and impose penalties for those who failed to meet the requirements.

A brief chronology and introduction of 'first tier' US environmental legislation includes:

(a) *Air*   Major federal statutes: The Air Quality Act of 1967, the Clean Air Amendments (CAA) of 1970, the Clean Air Act Amendments of 1977 and the Clean Air Act of 1990. These Acts established the basic framework for federal and state programmes to regulate and control the emission of airborne pollutants. Established national emission standards roughly translate into source standards related to six criteria pollutants (e.g. sulphur oxide, total suspended particulates, carbon monoxide, ozone, nitrogen dioxide, and lead) and hazardous air pollutants (for example, asbestos, benzene, mercury, beryllium, radionuclides and vinyl chloride).

(b) *Water*   Major federal statutes: The Water Pollution Control Act of 1972, the Clean Water Act of 1976 and reauthorization of 1986, and the Safe Drinking Water Act (SDWA) of 1974 and reauthorization of 1986. These comprise basically a series of two major environmental legislations, one designed to regulate municipal and industrial discharges into public sewer systems and surface waters, and the other designed to establish and update continuously drinking water standards for public consumption. The former legislations regulated solid waste impacts, limited direct discharges and established pretreatment standards, while the later legislations have gained widespread acceptance as indicators of acceptable levels of contamination.

(c) *Waste*   Major federal statutes: The Toxic Substances Control Act (TSCA) of 1976, the Resource Conservation and Recovery Act (RCRA) of 1976 and 1980, the reauthorization Hazardous and Solid Waste Amendments (HSWA) of 1984, and the Comprehensive Environmental Response, Compensation and Liability Act (CERCLA) of 1980 and reauthorization Superfund Amendments and Reauthorization Act (SARA) of 1986. These legislations are designed to: (i) introduce mandatory testing prior to the commercial manufacture of any new chemical, and the disclosure of information about its toxicity; (ii) regulate the generation, transportation, storage, treatment and disposal of waste materials defined in the legislation as hazardous waste; and (iii) establish a 'no fault' liability of owners and operators, generators, and certain transporters of hazardous substances for releases of hazardous substances (addresses past activities resulting in releases of pollutants from waste sites). With the enactment of these legislations the US had closed the loop on pollution control and provided for the 'cradle-to-grave' administration of solid and chemical wastes.

In addition, several other 'secondary statutes' were designed and enacted mainly to fill loopholes in existing legislations or address immediate needs on a smaller scale than the major pieces of federal environmental control programmes.

- The Federal Insecticide, Fungicide, and Rodenticide Act (FIFRA)
- The Occupational Safety and Health Act (OSHA)
- The Coastal Zone Management Act
- Surface Mining Control and Reclamation Act
- The Noise Control Act
- The Consumer Product Safety Act
- Department of Transportation acts related to the handling, packing and transportation of hazardous materials.

12.3.3 Performance standards

The US EPA has increasingly encouraged individual states to seek and obtain authorization to operate waste management programmes in place of part or all of the federal programmes. The performance standards of these state programmes must at a minimum be as stringent as the federal requirements and administratively equivalent. The RCRA programme in particular has detailed guidance within the regulations which, when achieved, leads to state authorisation of the programme, with provisions for federal (EPA) oversight. Further still, many states have adopted equivalent superfund programmes to address contaminated sites unaffected by federal requirements.

For federally administered superfund site remediations, including source control and residual contamination management, clean-up criteria must comply with legally applicable or relevant and appropriate requirements (ARARs) from all other federal environmental laws, as well as the more stringent state environmental laws. This policy, loosely equivalent to that applied within the United Kingdom, is also the direction that the European Community is tending towards concerning its environmental remediation programme for some types of clean-ups. As costs increased and limitations in treatment technology capability were realized, the EPA has increasingly accepted a risk assessment approach to selecting clean-up criteria standards for soil contamination remediation. This approach, albeit flexible and case-by-case specific, is, its critics say, subjective; as a result, conservative criteria are often selected, and result in the specification of costly clean-ups. Risk assessment, as yet to be accepted in the United Kingdom, will, for the foreseeable future, be embraced in the United States as a means to determine, particularly for soil remediation, clean-up criteria that are achievable and protective of human health and the environment.

12.3.4 *CERCLA*

Prior to the passage of CERCLA, there were no statues in place that would address the problems created by releases to the environment from the historical disposal of toxic and hazardous wastes. By 1978, as a result of a series of nationally publicized incidents (e.g. Love Canal in New York; Stringfellow Acid Pits in California; Valley of the Drums in Kentucky; Seymour Recycling Facility in Indiana; and many others) in which the release of toxic chemical wastes from abandoned or uncontrolled sites had threatened the health and safety of the nearby public, regulatory agencies were found powerless under the existing statutes to regulate the chronic effects on the environment of these waste disposal sites. The problem for the regulatory agencies was compounded by the fact that in most cases the parties responsible for these sites were either unknown, no longer in existence or otherwise unavailable to address the situation. What was apparent to everyone involved, including the public, was the fact that the pollution cycle had not been closed with the enactment of statutes to date.

Pressed by the public to respond to environmental health and safety concerns associated with abandoned or uncontrolled, inactive hazardous waste disposal sites, the US Congress passed the Comprehensive Environmental Response, Compensation and Liability Act in December 1980, commonly referred to as Superfund. CERCLA was initially authorized for a 5 year period at a funding of $1.6 billion. Much maligned from the beginning, as it is still today, CERCLA's basic concept seeks to establish a "no fault liability of facility owners and operators, generators, and certain transporters of hazardous substances for release of those hazardous substances." [5] Enforcement through the courts, or self-financing and recovery (through lawsuits against potentially responsible parties) of costs to clean up sites, where releases have or may occur, are the means created within the Superfund statue to accomplish its goal to permanently remedy these industrial legacies. Broadly, requirements under Superfund fall into four categories:

- Reporting and recordkeeping (CERCLA Sections 102 and 103)
- Investigation and response (CERCLA Sections 104 and 105)
- Liability and financial responsibility (CERCLA Sections 106 through 109); and
- Environmental taxes and trust funds (Title II of CERCLA).

The first five years of CERCLA were characterized by litigation concerning the implementation and constitutionality of the statute. Although heavily criticized for the lack of progress in cleaning up sites, and the large amount of monies expended on administrative undertakings, the first five years did see agreement of approximately $600 million in private party clean-ups, initation of 580 removal actions at priority sites, completion of 470 remedial investigations/feasibility studies (RI/FSs) filing of 200 lawsuits by the government, and consideration by the EPA of 10 sites to be cleaned up [6]. As a result of the criticism, Congress, in its 1986 reauthorization, strengthened the original statutes and increased funding for the programme five-fold at $8.5 billion over the initial programme. This reauthorization, signed into law on 17 October 1986, is entitled the Superfund Amendments and Reauthorization Act (SARA).

The Reauthorization of CERCLA emphasized the view of the US Congress that the public should participate in the process of determining the level of concern and commitment to addressing the regulation and remediation of hazards associated with waste sites. As a result of this view, some of the more important provisions of SARA are related to identifying the responsibilities of owners/operators and other specified individuals to the public and enforcement agencies. Some of these important additions of SARA to the CERCLA programme include:

- The continuation of the Superfund programme with a significantly expanded budget aimed at accelerating enforcement actions.
- The development of new and independent regulatory programmes, such as the Emergency Planning and Community Right-to-Know Act, and deadlines attached to the more important Superfund activities.
- The development of an administrative record to record and document remedial action decisions and activities at Superfund sites.
- Greater citizen input and involvement in the selection of the remedial action alternative and clean-up activities.
- Significantly more state involvement, including consideration of state standards and provisions for state challenges of remedial actions [7].

12.4 Integrated waste management planning

By the mid-1980s, federal environmental policies had clearly closed the loop on the regulation of all facets of pollution, from the siting of waste-disposal facilities and control of toxic substances and process waters, to limiting discharges of contaminants to the air, water and soils through the permitting process. Although many of these laws are not directly related to the contaminated land problem, some, if not all of the primary environmental statutes indirectly affect the planning and licensing of remedial schemes during the process of cleaning up a contaminated land site.

First, it is important to differentiate between CERCLA and RCRA. The principal concern of CERCLA is the clean-up of toxic releases at uncontrolled or abandoned hazardous waste sites. In contrast, RCRA aims to regulate the management of active hazardous waste treatment, storage and disposal (TSD) facilities in order to avoid new Superfund sites in the future. However, in terms of the legal and scientific differences one of the most important considerations is probably in defining what constitutes 'hazardous waste'. Under these two statutory authorizations, a distinction exists between a hazardous waste and a hazardous substance. The former is regulated under RCRA, while the latter is regulated under the Superfund programme. The terms are defined as follows:

Hazardous waste. US EPA definition (40 CFR 260.10):

> a solid waste that may cause or significantly contribute to an increase in mortality or an increase in serious, irreversible, or incapacitating reversible illness; or pose a substantial present or potential hazard to human health or the environment when it is improperly treated, stored, transported, disposed of or otherwise managed; and, the characteristic can be measured by a standardized test or reasonably detected by generators of solid waste through their knowledge of their waste.

Under RCRA regulations, a waste is considered hazardous if it is reactive, ignitable, corrosive or toxic, or if the waste is listed as a hazardous waste. Listed wastes are incorporated into lists published by the EPA and organized into three basic categories (as of 1988 there were approximately 450 listed wastes): (i) source specific wastes, including wastes from specific industries; (ii) generic wastes, including wastes from common manufacturing and industrial processes; and (iii) commercial chemical products. All 'listed' wastes are presumed to be hazardous, regardless of their concentrations.

Hazardous substances. The CERCLA definition for a hazardous substance is any substance designated in Section 311(b)(2)(A) of the Federal Water Pollution Control Act, any hazardous waste having characteristics identified or listed in Section 3001 of the Solid Waste Disposal Act, "any toxic pollutant listed under Section 307(a) of the Federal Pollution Control Act, . . . any hazardous air pollutant listed under Section 112 of the Clean Air Act, and . . . any imminently hazardous chemical substance or mixture" with respect to action taken under Section 7 of the Toxic Substances Control Act. Accordingly, Superfund encompasses approximately 880 hazardous substances, priority pollutants, wastes, and wastestreams [8].

Although the definitions have been further complicated within EPA guidance notes, they are particularly useful to those concerned with the hazardous waste problem, particularly generators, transporters, storers or disposers of hazardous waste. Simply defined, a hazardous waste under RCRA is a hazardous substance under CERCLA, which, under current policies, means that once hazardous wastes at a Superfund site are removed, whether they are transported off-site or not, that action constitutes generation and the wastes must be disposed of in a RCRA permitted or authorized facility. RCRA commitments are often long-term, rigorous and costly, and therefore when dealing with a CERCLA site they should be minimized, which is the intent of the CERCLA/RCRA interface. While the RCRA programme is intended to discourage the disposal of hazardous waste by preventing its generation, CERCLA's ultimate goal is to encourage the permanent solution of hazardous waste legacies through treatment. This then is the basis for the integrated approach to waste management in the United States. Today, the key components of this include:

- Controlling wastes
- Pollution prevention

A description of each of these programmes and the mechanisms for promoting their agendas are discussed in the following sections.

12.4.1 *Controlling wastes*

Few would argue that past practices of hazardous waste mismanagement have resulted in detrimental effects to human health and the environment. These results have been reflected in polluted groundwaters, lakes and streams, and in areas now void of vegetation and wildlife. Improper disposal of hazardous waste has been linked to elevated levels of toxic contaminants in humans, aquatic species, and livestock. Illegal dumping of hazardous waste on road-sides or in open fields has resulted in explosions, fires, contamination of underlying groundwater, and generation of toxic vapours [9]. To eliminate these practices the EPA has designed a programme under RCRA that ensures the proper management of hazardous waste from 'cradle to grave'. This programme is designed such that the EPA and states can monitor and control hazardous waste at every point in the waste cycle. This approach has three key elements: (i) a tracking system; (ii) an identification and permitting system; and (iii) a system of restriction and controls on placement of hazardous waste.

The tracking system. The EPA places the responsibilities and burden of determination and documentation of a hazardous waste squarely on the generator. Under RCRA, the generator must ensure the ultimate fate of the hazardous waste. The hazardous waste generator must determine if waste is hazardous by definition and, if it is, obtain from the EPA an identification number for each site at which the hazardous waste is produced. Although the EPA estimates that 96% of US hazardous waste generators manage their RCRA hazardous waste on site, for the percentage that transport the waste off-site, the generator must package and label the waste properly and prepare a Uniform Hazardous Waste Manifest before the waste is shipped [10].

On the manifest, generators must certify that they are minimizing the amount and toxicity of their waste, and that the method of treatment, storage and/or disposal that they have chosen will minimize the risk to human health and the environment. This certification reflects the EPA's intention to promote reduction of the volume of hazardous waste produced and provide higher levels of protection. The manifest accompanies the waste wherever it travels until it reaches its ultimate destination, at which point the owner of the TSD facility returns the manifest to the generator confirming its arrival [11].

The permitting system.   Facilities that treat, store and dispose of hazardous wastes must obtain a permit to ensure that they meet standards established by the RCRA programme for proper waste management. These standards are designed to minimize the risks of exposure of the facilities operations to human health and the environment. For example, facilities must:

- Analyse and identify wastes prior to treatment, storage or disposal
- Prevent the entry of unauthorized personnel into the facility, by installing fences and surveillance systems and posting warning signs
- Periodically inspect the facility to determine if there are any problems
- Adequately and periodically train employees
- Prepare a contingency plan for emergencies and establish other emergency response procedures
- Comply with the manifest system and with various reporting and record-keeping requirements
- Comply with performance and technology requirements for various commonly used treatment, storage and disposal processes (e.g. tank systems, surface impoundments, waste piles, land treatment, landfills and incinerators)
- Design and implement closure and post-closure operation procedures for the processes
- Provide financial assurances for closure and post-closure, including liability insurance to cover third-party damages arising from accidents or waste mismanagement

The restrictions.   As part of the integrated approach to waste management activities, the EPA has been establishing standards to limit the emissions to air, water and land from hazardous waste treatment, storage and disposal facilities. Faced with the irrefutable evidence that improper land disposal practices had previously endangered public health and the environment, and that approximately 80% of hazardous waste disposal in the United States continued to be to land in the form of landfills, surface impoundments, waste piles, lagoons, and underground injection wells, the EPA (as a result of the 1984 RCRA amendments) developed regulations requiring the minimization of wastes by reduction, recycling, and treatment and banning of unsafe and untreated wastes from land disposal. These 'land ban' provisions of the 1984 RCRA amendments have encouraged the very considerable and rapid development of economic and effective means to treat hazardous wastes. Thus, as intended, the amendments have favoured the permanent treatment rather than disposal of certain types of hazardous wastes, with the result that industry has grown to support these goals. The goals are further fostered and encouraged through provisions requiring that the EPA undertake to sponsor research on new and innovative treatment technologies and establish treatment standards for banned wastes, which specify a level or method of treatment that substantially reduces the toxicity or mobility of the hazardous constituents so as to minimize long-term threats to human health and the environment. As a result of financial and technical support provided through the EPA's programmes, commercial, private and academic research into the development of innovative and alternative technologies around the world has greatly enhanced the methods available for the treatment of hazardous waste.

12.4.2 *Pollution prevention*

The systems to control hazardous wastes in the United States (e.g. tracking, permitting and restrictions) have altered the approaches that industries have traditionally undertaken to rid themselves of their waste problem. Because of the disposal costs and liability risks, generators of hazardous wastes currently look to treat and dispose of their wastes on-site, and ship only the more concentrated and toxic wastes off-site for treatment and disposal. As regulatory, statutory and economic factors combine to produce a limited amount of both on- and off-site disposal and treatment capacities, it is hoped that US industries will seek to develop techniques to prevent the generation of the waste. Recently enacted legislation, the Pollution Prevention Act of 1990, has established a framework for a comprehensive national pollution prevention strategy. This act requires the EPA to focus its approach on the prevention of pollution in all media (i.e. air, water and land), rather than on pollution control.

Many of the steps necessary to undertake the pollution prevention policies for the management of wastes are enforceable through the waste minimization requirements under RCRA. This requirement has encouraged hazardous waste generators to reduce their waste volumes in several ways: (i) source separation; (ii) recycling; (iii) manufacturing process changes; and (iv) substitution of products. Economically, as companies develop new processes, they will seek to decrease their costs of waste management through waste reduction; older processes that are unable to change will have to treat waste prior to disposal, to reduce the waste volume or eliminate hazardous constituents. In addition to being encouraged by the control of waste requirements, the development of these treatment processes will now have an additional incentive through the pollution prevention programme.

12.5 Conclusions

Although slow in development and still controversial, the US waste management programmes are a distinctly proactive approach to the long-term control and prevention of the hazardous waste problem. On recognition of the problem, the US practice has been to react through additional legislation and regulation in order to manage and ultimately reduce the risks of toxic and hazardous wastes towards human health and the environment.

Compliance with waste requirements has created a financial burden to both the US industry and the public. Alternatively, 'spin-off' research, education, technologies and service industries resulting from these programmes and their arguable benefits are becoming more widely acknowledged and adopted worldwide. On balance, the effectiveness of these waste management programmes have yet to be determined; however, in view of the diminishing land resources, their eventual success is more surely guaranteed.

13 Health and safety

S.A. SIMMONS and W.K. LEWIS

13.1 Introduction

Given the existence of a rigorous system of legal controls and strict criminal and civil liabilities, it is always surprising that management of health and safety is so often given scant attention in the planning and implementation of work on contaminated sites. While it is to some extent understandable that clients' and contractors' overriding priorities are to reclaim sites in as short a time-scale as possible and at least cost, this must not be at the expense of the health and safety of those involved in or affected by the reclamation works.

There is now growing importance being attached to ensuring health and safety during reclamation of contaminated land. The main enforcing authority, the Health and Safety Executive, is planning to issue new guidance on the standards that are required. Once this guidance is issued to inspectors and contractors engaged in reclamation, significantly higher standards will be expected. Failure to meet these basic standards will undoubtedly expose those involved to risks of prosecution. This chapter therefore sets out the current legal framework, and points to the measures that should be considered to ensure that reclamation is achieved without breaching fundamental legal requirements.

Notwithstanding the anticipated tightening of controls on health and safety during reclamation, the current framework of laws and regulations means that failure to ensure health, safety and welfare for those at work and others may have severe consequences including:

- Criminal prosecution possibly leading to imprisonment and/or an unlimited fine; responsibility for an offence usually attaches to the employer but may also attach to employees, directors, managers, the company secretary or other officers
- The issue of a Prohibition Notice (section 22 of the Health and Safety at Work etc. Act 1974) requiring immediate cessation of an activity
- The issue of an Improvement Notice (section 21 of the Act) requiring corrective action within a specified time-scale
- Delay or suspension of work on site

- Unfavourable publicity and increased scrutiny by other enforcing authorities such as the local authority, National Rivers Authority, etc.
- Contract penalties or even loss of contracts
- Civil actions leading to an injunction or compensation for injury or damage
- Increased or unanticipated costs

These adverse consequences and the precautions that need to be taken to avoid them are not unique to work on contaminated land as they may apply to any construction project where hazardous substances are used or encountered. However, there are special problems inherent in reclamation work, which means that the precautions that are taken need to be even more rigorously applied.

The main hazards that may be encountered in work on contaminated sites can be identified as follows:

- Toxic substances including gases, dusts, liquids, solids
- Corrosive or irritant substances
- Carcinogenic, mutagenic or teratogenic substances
- Flammable or explosive atmospheres, liquids or solids
- Asphyxiating atmospheres
- Pathogenic substances including bacteria, viruses, etc.
- Radioactive or otherwise harmful substances
- Physical safety hazards such as buried tanks, high temperatures associated with combustion of fill materials, unstable fill materials, etc.

Many of these types of hazard are by no means unique and may be found elsewhere in the construction industry. What makes contaminated sites different is that there is often little available information about the nature and extent of hazards, there may be multiple hazards and very complex mixtures of substances, and conditions may be highly variable across sites and with depth. Taken together, these factors mean that if reclamation is to proceed safely, rigorous precautions must be followed. This chapter deals with the basic problems that may be encountered, the main legal requirements and the practical measures that can be applied to ensure effective management of health, safety and welfare. Throughout this chapter attention is focused on those aspects of work on contaminated sites that may present special risks not generally encountered in work on relatively uncontaminated sites.

13.2 Health and safety problems: overview

13.2.1 Hazard and risk

It is important in any assessment of the potential consequences for health and safety associated with a particular activity or operation, that the concepts of hazard and risk are fully appreciated.

A material that is hazardous is one that has the capacity to cause harm. Risk on the other hand is the probability that harm will result in the specific circumstance in which a hazardous substance is encountered or used. For example, blue asbestos is a hazardous substance that is capable of causing asbestosis, lung cancer and mesothelioma. The main exposure route giving rise to concern with respect to health would be inhalation of airborne fibres. It is possible that an operation in which blue asbestos is present in a form unlikely to give rise to airborne fibres could be conducted with low risk of exposure by inhalation, for example if the asbestos was sealed to prevent damage, abrasion or dust. Operations in circumstances where airborne fibres are likely to be present in significant concentrations would, however, present a substantially greater risk to health, for example during dry stripping of old blue asbestos insulation. In these two cases, the hazard is the same, it is only the degree of risk that is different. This distinction is often forgotten or poorly understood. It is quite common to see examples of situations where workers are required to take precautions that are totally out of proportion to the risk of exposure, simply because a material encountered on site is toxic. For example, it is not uncommon to see workers engaged in strenuous work on a slightly contaminated site in mid-summer wearing respirators, despite the risk of inhalation being low as the site surface has been sprayed to minimize dust generation. At the other extreme are the cases where site workers or the general public are placed at risk because the precautions taken are inadequate or take no account of the main route of exposure. A good example of this poor understanding of exposure risk would be the site worker who is given a simple dust mask to protect against inhalation of toxic solvent vapours, and protective gloves (but no eye protection) whilst dismantling a tank and pipework still containing highly caustic residues. The dust mask would give no appreciable protection against solvent vapours and a small splash of caustic liquor in the eye could do serious damage and may result in loss of sight.

13.2.2 *Types of hazard*

While it is often the case that damage to health is seen as an important matter of concern, this is not the only type of health and safety problem affecting reclamation of contaminated land. It is possible to categorize the main health and safety problems into three groups:

(a) Firstly, there are the risks to health for workers and others (including local residents and site visitors). Hazardous substances can be present as contaminants in soil and fill materials; as components of buildings or structures still present on the site; as wastes deposited in the ground, in drums, tanks or pipes; as materials or substances used or produced during the reclamation works; or as contaminants in surface or ground-water [1, 2]. Examples include:

- Toxic chemicals such as cyanides and phenols, or metals such as lead, cadmium and arsenic, which are frequently present as contamination of soils and fills
- Toxic gases such as hydrogen sulphide and hydrogen cyanide in soil atmospheres and voids; these gases can be produced by chemical reactions or by the actions of soil microorganisms
- Corrosive liquids or solids such as strong acids or alkalis
- Carcinogenic substances in soil, water, or building materials such as asbestos, tars, soots or benzene
- Pathogenic agents in soil or water, e.g. *Leptospirosis* or *Anthrax*
- Asphyxiating gases such as carbon monoxide, carbon dioxide or oxygen deficient atmospheres; these gases may be present as a result of chemical reactions, combustion or microbiological activity
- Toxic gases produced by circumstances where vapours or gases are subject to heating or combustion, such as welding in areas contaminated by chlorinated solvents; this may give rise to highly toxic gases, including phosgene and hydrogen chloride
- Dusts, which may be a nuisance, or may contain toxic or otherwise harmful substances such as silica or man-made mineral fibre

These hazardous substances may present a risk to health in circumstances where they are in contact with or may penetrate the skin, or where they can be inhaled or ingested. Examples of the types of health hazard associated with former uses and the main exposure routes are given in Table 13.1.

(b) Secondly, there may be conditions that represent physical risks to safety. These include the presence of flammable or explosive materials and substances; presence of explosive or flammable atmospheres; high temperatures associated with underground fires [3, 4]; presence of sub-surface voids; or radioactive substances such as radium. There may be many other types of physical risk, such as work in trenches, or restricted and confined spaces; however, these may also be found on relatively uncontaminated sites and so will not be dealt with in detail.

(c) Finally, there are several conditions which, although not immediately hazardous to health, may cause nuisance or offence, or may impede work and contribute to increased risk of accidents. Examples include substances having offensive odours at low concentrations, for example some amines, sulphides or mercaptans; or ground soaked in some types of mineral oils, which may be unstable or slippery.

Table 13.1 Examples of hazardous substances, exposure risks and typical sites where these may be encountered

Types of hazard	Exposure risk	Industries
Toxic metals such as lead, cadmium, mercury, arsenic, nickel	Inhalation of dusts Ingestion	Iron and steel foundries Smelters Electroplating Galvanizing Engineering Shipbuilding Scrap yards Chemicals Waste disposal Minerals Glass Sewage works Gasworks
Toxic, irritant or corrosive substances such as cyanide, chromium salts, etc.	Inhalation of dusts Skin contact Ingestion	Electroplating Scrap yards Tanneries Gas works
Asbestos, man-made mineral fibres	Inhalation of dusts	Petrochemicals Energy Metal works Transport Docks Waste disposal Factories
Corrosive acids or alkalis	Skin contact	Petrochemicals Chemicals Electroplating Gasworks
Toxic solvents, phenols, and other organic substances	Inhalation of vapours Ingestion Skin contact	Chemicals Petrochemicals Gasworks Transport Waste disposal Tanneries Energy Rubber works

Note: Some hazards are frequently encountered on a wide range of sites. These include sulphates, polychlorinated biphenyls, oils, asbestos, metals.

The types of hazard that may be present depend to a large extent on the nature of the former land uses. Previous chapters have examined the types of contamination that may be expected in different circumstances. However, there is not always a clear association between former use and the nature of the hazards left behind. In some cases, sites may have been subject to a variety of uses, each of which has left a legacy of different hazards. In other cases, unexpected hazards may be present due to the unrecorded tipping of wastes brought onto sites, or from the migration of contaminants from adjoining sites.

In Section 13.1, the different types of hazardous properties associated with different substances or conditions were introduced briefly. These will now be reviewed in more detail.

Toxic substances. A large number of substances are moderately to highly toxic, including many that are in day to day use around the home (e.g. cleaning agents, pesticides, leaded paint). Toxic substances can be present on contaminated land in many different forms:

- Gases such as hydrogen sulphide and hydrogen cyanide liberated as a result of chemical or biological reactions with sulphates or cyanides in the soil. These may accumulate in voids, tanks or confined spaces representing a significant risk to health for workers who are exposed to them, mainly through inhalation.
- Liquids such as petroleum spirits, aromatic hydrocarbons including benzene and xylene, or groundwater containing toxic substances such as arsenic or phenol. Risks to health may be caused by skin contact allowing some substances to be absorbed directly into the body (e.g. phenol), or by ingestion either directly by drinking or indirectly by entry into the mouth as a result of eating contaminated food or through hand-to-mouth contact. Some liquids may also evaporate to form vapours that can then be inhaled (e.g. benzene, xylene, toluene, petroleum). Liquids in open vats, tanks or lagoons are particularly dangerous as people may inadvertently fall in.
- Solids in a variety of forms such as crystals, powders or lumps of material. Some toxic substances may be present as contamination in soil. Examples include toxic metals, cyanides and naphtha. Risks to health may be caused by skin contact and absorption (e.g. organophosphorous compounds), inhalation of dusts (which is a major route of exposure for many toxic solids such as lead), and ingestion (particularly through smoking, eating and drinking, or hand-to-mouth contact).

Control of exposure to toxic substances may be complicated by the fact that many are highly persistent and tend to become widely dispersed across sites during reclamation, by vehicle movements, wind blown dusts, surface and groundwater, and uncontrolled movement of soil/subsoil. Furthermore, with some toxic substances, there may be a tendency for accumulation in critical organs in the body. Daily exposures to quite small intakes into the body may eventually lead to the accumulation sufficient to cause damage to health (e.g. mercury or cadmium). Generation of fine dusts that can be inhaled is particularly hazardous, as many toxic substances may directly affect the lungs or may be absorbed from the lungs into the body.

Corrosive or irritant substances. Strong acids or alkalis directly destroy body tissue by corrosion. These may be present as liquors or tars, in drums, sumps or pipework. They attack whatever tissue they come into contact with, such as the skin, eyes, or lining of the nose, mouth, throat and lungs. Fine mists may be inhaled or may corrode the eyes. Some corrosive substances such as ammonia can be present in gaseous form; when inhaled these can burn the throat and respiratory passages. More dilute acids and alkalis and other substances such as chromium, some organic solvents and cutting oils irritate the skin or eyes to produce redness, blistering or dermatitis. Some workers may become sensitized to a particular substance, such as nickel salts, with an allergic reaction being initiated even at very slight and transient exposure. In general, skin or eye contact are the main corrosive risks to health in reclamation activities.

Carcinogenic substances.   Substances capable of causing cancer are called carcinogens. Examples of carcinogenic substances commonly encountered on contaminated sites include:

- Asbestos (lung cancer)
- Coal tar (skin cancer)
- Coal soots (skin cancer)
- Pitch (skin cancer)
- Certain mineral oils and used oils (skin cancer)
- Arsenic (skin and lung cancer)
- Benzene (leukaemia)

It is a widely held opinion that for most carcinogens, there is no 'safe' level of exposure below which there is no risk that cancer will develop. Consequently, it is not possible to set standards that give absolute protection for workers or the public exposed to carcinogens. In order to minimize risks to health of workers and local residents, exposure to potentially carcinogenic substances should be kept as low as possible.

The exposure routes vary from substance to substance. Benzene for example can be inhaled and absorbed into the body through the lungs or can be absorbed directly through the skin. Asbestos fibres are inhaled and deposited in the lungs. Coal tar and other skin carcinogens cause cancer at the points of most frequent exposure (typically the hands and forearms, neck and face, scrotum, etc.).

Asphyxiant substances.   Asphyxiation can be caused through exposure to oxygen deficient atmospheres or atmospheres in which asphyxiant gases are present. Oxygen deficient atmospheres can occur in situations where oxygen has been depleted through chemical and biological reactions, or by thermal combustion. Such situations are most likely to occur on waste landfill sites where landfill gas is known to be low in oxygen, and in areas affected by underground fires, e.g. colliery spoil heaps. Asphyxiation due to accumulations of asphyxiant gases such as carbon dioxide, carbon monoxide or inert gases can occur in many situations, such as underground fires, on landfill sites, and where lime is in contact with acidic substances. The main risks of asphyxiation occur in situations where air is stagnant for long periods or where there is inadequate ventilation such as trenches, basements, tanks, sewers, etc. [5].

Pathogenic substances.   Pathogenic substances are those that are capable of producing infection. They may be bacteria, viruses, or microorganisms. These may be present in soil, water, or the atmosphere (on dust or droplets). Examples of pathogenic risks include *anthrax* spores present at sites of former abattoirs, Weil's disease in areas where urine from animals such as rats may be present, and *Legionella* bacteria in cooling ponds or air conditioning units. Pathogens may produce infection by entry through cuts or abrasion of the skin, or by inhalation or ingestion.

Flammable/explosive substances.   Contaminated sites may contain solid materials of high calorific value which may be capable of combustion, e.g. coal wastes, domestic wastes and some types of ash. This type of hazard is less likely to present a significant risk to health and safety than situations where highly flammable liquids or gases are present, as these may ignite and burn rapidly or explode. Examples include solvents, petroleum vapours, methane, hydrogen sulphide or hydrogen, all of which may ignite and explode in confined spaces.

Buried explosives may also represent a risk to safety at sites formerly occupied by explosives factories and magazines, or areas that were used for war-time activities or were subject to bombing. Specialist advice may be required to identify the location of explosives prior to excavation or construction work.

Radiation.   Sites contaminated with radioactive materials will probably require specialist advice and contractors to ensure that risks to health are minimized. Examples of situations where radioactive contamination may be present include gas mantle factories, laboratories and scrap yards. Sites where radioactive contamination is present may be subject to the Ionising Radiation Regulations 1985 and the Radioactive Substances Act 1960 [6, 7]. There are also special regulations covering the carriage of radioactive substances by road.

Dusts.   The hazards associated with asbestos dust have already been highlighted. A wide variety of other substances, if inhaled in fine dust form, can present risks to health. Examples include:

● Silica: if inhaled this can lead to the development of silicosis; silica may be found naturally as a component of rocks and sands or as a contaminant on sites of former foundries, glass works, potteries, sand blasting works, etc.

- Man-made mineral fibres: these can cause irritation to the skin, eyes and upper respiratory airways; in addition, some types may possibly be carcinogenic (namely rock wool, glass wool, slag wool and ceramic fibres)
- Coal dust and other rock dusts: these may lead to pneumoconiosis (meaning dusty lung)

13.2.3 *Operations with particular risks*

While a great many factors contribute to risk, it is possible to identify particular types of operation or circumstances that carry special risks to health and safety on contaminated sites. The most frequently encountered high risk activity is the need to enter trenches, tanks, vats or other confined spaces where there is poor ventilation and also a risk of skin contact with hazardous substances in the soil, groundwater, liquors or sludges. In such circumstances, multiple hazards may be present. For example, in ground heavily contaminated by aromatic and some chlorinated solvents, an unprotected welder working with an oxyacetylene flame in a below-ground trench with groundwater seeping in could be exposed to the following health and safety risks:

- Inhalation of high concentrations of toxic and possibly narcotic solvent vapours
- Skin absorption of toxic solvents through the hands, leaking wellington boots and wet clothing
- Skin irritation from contact with organic solvents
- High risk of explosion or fire from accumulations of flammable organic vapours
- Asphyxiation risk from carbon dioxide from the cutting torch
- Possibility of toxic gases such as phosgene and hydrogen chloride being formed through heat breakdown of the chlorinated solvent vapours present in the trench
- Splashes of solvents into the eye causing irritation or more severe consequences

In such circumstances, stringent precautions would need to be taken to minimize risks to the worker. These would include:

- Wearing protective boots, gloves and clothing
- Having at least two people supervising the operation from the surface
- Safety harness and lifeline
- Provision of fire extinguishers and warning/emergency siren or whistle
- Entry using supplied air or self-contained breathing apparatus during initial entry to test the atmosphere for oxygen content and asphyxiant gases
- Test of the flammability of the atmosphere using an appropriate explosimeter

- Test of the atmosphere for solvent vapour using a portable meter or indicator tube
- Provision of audible alarm monitoring for oxygen and carbon dioxide/ monoxide
- Depending on solvent vapour concentrations, operation may be continued using breathing apparatus or an appropriate respirator
- External forced air ventilation could be provided during cutting
- Checks with indicator tubes for phosgene and hydrogen chloride
- On completion, the operator exits the trench and then proceeds to a decontamination unit where protective equipment is removed for cleaning, and the worker showers [5]

A second potentially high risk situation is where drums, underground tanks or vats containing substances are uncovered. In such situations it is important that immediate action is taken to assess the nature of the hazards involved and determine the level of risk. Strict procedures are required on sites where this type of problem may be encountered. These procedures would include immediate removal of all non-essential personnel; consultation with an experienced site environmental health and safety officer or chemist; sampling and analysis where appropriate to determine the nature of the substance involved; emergency action to contain any leakage or spillage; wearing of appropriate breathing apparatus if there is a risk of gases or vapours; design of an appropriate method of work to treat the problem.

A third type of situation that represents a significant risk is where inadequate information has been obtained on the nature of the hazards at a site and site workers are therefore exposed to substances that may have no warning properties such as odour or visibility. Finely divided asbestos fibres mixed into a soil matrix may not be visible to the naked eye but may nevertheless present a significant risk if inhaled during dusty operations. Toxic metals may be present in many forms, some of which may be considerably more toxic than others. For example, metallic lead is less easily absorbed into the body than certain lead salts such as lead chloride. Organic lead compounds such as tetra ethyl lead can be rapidly absorbed and are highly toxic to the brain and nervous system. A site investigation report may have identified lead contamination only as the total concentration of lead in soil, but with no indication of the chemical composition. An appreciation of the nature of the contamination is essential if risks are to be assessed and appropriate control measures identified.

A final situation in which potentially significant risks can occur is where the control over site access and operations is inadequate to prevent unprotected persons coming into contact with hazardous materials. For example, commonly encountered situations are where children can enter a site and be totally unaware of the hazards that may be present. This is a particular problem in many city areas, where the ingenuity of vandals may outwit even the most sophisticated security measures. Another common problem is where there are a number of contractors operating on site, some of whom are unaware of the hazards and appropriate precautions. It is quite common to find scrap recovery operations involving flame cutting of structural steelwork taking place in areas where there is heavy contamination by potentially flammable or explosive substances. Another example would be where only workers involved directly in work with toxic substances are given appropriate protection, despite the fact that operations such as contaminated soil stripping and haulage across the site cause the generation of dust, which may affect all workers on site, and even nearby residents. Dust is a significant problem in most reclamation work and requires stringent controls to limit dust generation and spread of contamination. Contaminated materials falling from lorry wheels and loads onto nearby roads may pose similar risks to health for the general public.

13.3 The legal framework

UK health and safety legislation consists of a framework of principal Statutes or Acts that are linked to a great number of subordinate Regulations and Orders. The main provisions that are relevant to work on contaminated sites are summarized in this section. There are a great number of Acts, Regulations or Orders that may apply to construction work in general and the reader should consult with the appropriate enforcing authority, trade bodies, and official guidance and reference works such as the *Construction Site Safety Manual* [8] for further information.

13.3.1 *The Health and Safety at Work etc. Act 1974*

Section 2 of the Act places employers under the general duty to ensure, so far as is reasonably practicable, the health, safety and welfare of their employees. In particular, employers have obligations to:

(a) Provide and maintain plant and systems of work that are safe and without risks to health
(b) Ensure that the use, transport, handling and storage of articles and substances are safe and without risks to health
(c) Provide such information, instruction, training and supervision as is necessary to ensure the health and safety at work of employees
(d) Ensure that any workshop under the employer's control is safe and healthy and that proper means of access and egress are maintained, particularly with respect to standards of housekeeping, cleanliness, disposal of rubbish and stacking of goods in the proper place
(e) Keep the working environment safe and healthy so far as is reasonably practicable, to ensure that no health risks exist, and that there are adequate facilities and arrangements for employees' welfare at work

Further duties include preparing and keeping up to date a safety policy where there are five or more employees and bringing this to the notice of all employees (S2(3)); consulting safety representatives (S2(6)); if requested, establishing a safety committee (S2(7)).

Section 3 places a duty on employers and self-employed persons to conduct their activities in such a way as to ensure that persons other than their employees are not exposed to risks to their health and safety. On contaminated sites this would include other contractors working on the site, visitors, clients and the general public.

13.3.2 *The Control of Substances Hazardous to Health Regulations 1988 (COSHH)*

COSHH applies to all workplaces where substances that may damage human health are present. On contaminated sites this would include substances present as contaminants, substances used in reclamation and substances produced during reclamation operations (i.e. dusts). These substances could cause harm by being inhaled, ingested, or by contact with the skin or eyes. Biological hazards such as bacteria, viruses or microorganisms are also covered. The only substances that are excluded from COSHH are asbestos, lead and ionizing radiations, as these are subject to separate legislation.

Regulation 6 requires all those involved in work on contaminated sites, including surveyors, contractors, and transport contractors, to assess the potential risks to health entailed in the work, and the precautions required to protect workers or the public. No work should be carried out unless a suitable and sufficient assessment has been undertaken. Assessments should be repeated if they are no longer valid, or if circumstances change, i.e. due to the discovery of new contamination or changes in the methods of work etc. Employers should ensure that exposure of employees and others to hazardous substances is prevented or, where this is not reasonably practicable, controlled with or without personal protective equipment.

Most contractors will have already established procedures for undertaking assessments for substances used in construction, e.g. solvents, welding fumes, etc. However, the requirements of COSHH also extend to substances encountered during reclamation works. It should be comparatively straightforward to use these procedures for substances present on sites.

To enable an adequate assessment to be made, the person in charge of a site should obtain and provide sufficient information on the nature and patterns of contamination to allow others to assess risks. Those in charge of sites should be satisfied that all contractors involved have undertaken satisfactory assessments and implemented appropriate controls. COSHH also requires that control measures are properly maintained and tested, and that non-disposable protective equipment is examined and tested at suitable intervals. Records of examinations should be kept for at least 5 years.

Employee exposure monitoring should be conducted where it is requisite for ensuring the maintenance of proper control systems or protecting the health of employees. Records should be kept for 30 years for personal exposures and 5 years in any other case (e.g. boundary monitoring).

Regulation 11 requires health surveillance of employees, for scheduled processes involving certain substances or where it is appropriate to protecting health. Examples of situations that may require health surveillance would include reclamation of sites heavily contaminated by mercury, cadmium, arsenic, beryllium, phosphorus or organophosphorous compounds. Sites contaminated by coal tar, pitch or mineral or used oils may require skin inspections to detect signs of dermatitis, oil acne or cancer.

Health surveillance would be inappropriate in most cases, as exposure will be for only limited durations. However, where workers are engaged for long periods or where it could be expected that they would be exposed to unavoidable and significant levels of contamination, health surveillance may be advisable. In these situations, specialist advice should be sought from the Employment Medical Advisory Service or from an occupational physician. The Approved Code of Practice gives more details of the requirements of COSHH [9, 10].

13.3.3 The Factories Act 1961

The main provisions of the Factories Act are not directly relevant to contaminated sites. However, certain provisions are of relevance, including Sections 30 and 31. These deal with dangerous fumes and lack of oxygen, and precautions with respect to explosive or inflammable dust, gas, vapour or other substances. Section 30 requires that work in confined spaces such as chambers, vats, pipes or flues should not be undertaken unless:

(a) There is an adequate manhole or means of egress.
(b) The space is certified by a responsible person as being safe for entry without breathing apparatus. A space shall be certified until effective steps have been taken to prevent ingress of dangerous fumes, all sludge and other deposits have been removed, and there is adequate ventilation.
(c) If a space is not certified as being safe, no person shall enter or remain in it unless authorized to do so; they must then be equipped with breathing apparatus, and, where practicable, wear a belt with rope attached. A colleague must always be stationed outside the space on watch.
(d) A supply of approved breathing apparatus, belts, ropes, reviving apparatus and oxygen is kept available and in good working order. These shall be examined at least once a month.
(e) Sufficient training has been given to persons to use the apparatus employed.

Where any process or operation gives rise to dust that can explode on ignition, under Section 31 the plant should be enclosed, dust should not be allowed to accumulate and sources of ignition should be excluded. Any plant, tank or vessel that has or still contains explosive or flammable substances should not be welded, brazed, soldered, cut or heated to open until the substance and any fumes from it have been removed or rendered non-flammable.

13.3.4 *The Construction (General Provisions) Regulations 1961*

These Regulations apply to all construction sites. The specific provisions relevant to contaminated land include:

  (a) Safety supervisors must be appointed in writing by contractors employing more than 20 workers
  (b) Explosives should not be handled or used unless by a competent person
  (c) Adequate ventilation of working places in excavations and confined spaces must be provided and atmospheres checked where appropriate
  (d) Demolition work must be supervised by a competent and experienced person, and precautions should be taken to prevent fire, gas explosion, flooding and accidental collapse

There are also provisions with respect to protection of trenches and excavations from collapse and flooding.

13.3.5 *The Construction (Health and Welfare) Regulations 1966*

These regulations also apply to all construction sites. Specific provisions affecting contaminated sites include requirements to provide facilities for eating; a supply of clean drinking water; and facilities for washing, cleaning and drying, including hot and cold, or warm water.

13.3.6 *The Control of Lead at Work Regulations 1980*

Work causing exposure to lead is controlled by the Lead at Work Regulations and the Approved Code of Practice entitled Control of Lead at Work [11, 12]. Examples of circumstances where these regulations would apply would include welding or cutting materials coated in lead paint or lead coatings, or handling dusts containing high concentrations of lead. The requirements include the duty to assess risks prior to commencement of work and to provide adequate information, instruction and training to those likely to be affected. Controls must be introduced to provide adequate protection from inhalation and ingestion of lead. Adequate washing facilities must be provided, and eating, drinking or smoking must be prohibited in places likely to be contaminated by lead. Employers and employees also have a duty to prevent the spread of contamination. On a contaminated site this would include dust suppression, wheel washing and covering of lorries to prevent spillage of wastes taken off-site. Where appropriate, air monitoring, medical surveillance and biological monitoring of exposed employees shall be undertaken.

13.3.7 *The Control of Asbestos at Work Regulations 1989*

These regulations impose duties on employers with respect to employees and other persons who may be affected by operations involving any asbestos or asbestos-containing materials, including contaminated soils. Prior to undertaking any work with asbestos, the employer should assess the risks involved. This requires an identification of the type of asbestos or, if unidentifiable, that it be treated as either amosite or crocidolite. Measures should be taken to reduce exposure to the lowest reasonably practicable level. Unless an employer is licensed under the Asbestos (Licensing) Regulations 1983, or the extent of exposure will not exceed the action level (of a cumulative exposure over a 12-week period for crocidolite or amosite of 48 fibre-hours/ml of air, or, for other types of asbestos, 120 fibre-hours/ml of air as specified in Guidance Note EH40/91), no work shall commence for the first time until 28 days after the Health and Safety Executive have been informed in writing. Other provisions concerning control measures and protection etc. are similar to those of COSHH and the Control of Lead at Work Regulations.

13.3.8 *The Reporting of Injuries, Diseases and Dangerous Occurrences Regulations 1985 (RIDDOR)*

RIDDOR covers all accidents occurring in any employment and lays down the reporting and recording requirements. Instances of deaths at work, certain types of injuries, diseases, and dangerous occurrences must be reported to the Health and Safety Executive (principally the factory inspector). Examples of the types of occurrence that must be reported and may be relevant to work on contaminated sites include:

- Loss of consciousness or injury from lack of oxygen or absorption of any substance, or exposure to any pathogen or infected material
- Any injury requiring immediate admission to hospital for more than 24 h
- Poisoning by such substances as arsenic, benzene, beryllium, cadmium, lead, manganese, mercury or phosphorus or their compounds
- Skin diseases resulting from contact with chromic acid or chromium compounds; mineral oil, tar, pitch, or arsenic; ionizing radiations
- Lung diseases including occupational asthma etc.
- Infections such as leptospirosis, hepatitis, anthrax or other pathogens
- Gas accidents

13.4 Managing health and safety

13.4.1 *Management structure*

In the preceding sections, the main types of hazard and the legal obligations of the employer have been examined. It is clear that reclamation activity on contaminated sites is subject to a great many hazards, and sometimes significant risks that are subject to strict statutory duties. If any employer engaged in work on a contaminated site or responsible for investigating, supervising or managing such projects is to avoid potentially serious risks to health and safety, a clear system of management controls will be essential. In this section, the management systems of controls that would be required to ensure compliance with the main health and safety laws is explained. However, all of these provisions will not necessarily be required in every case. Small projects on relatively uncontaminated sites will require little more than is routinely done for normal construction projects. On larger schemes or where there are significant levels of contamination, many of the following provisions will be required. The decision as to what measures will be advisable on each site can only be made after an adequate examination of the hazards and risks, and a review of the health and safety management arrangements.

Prior to the commencement of any site investigation, demolition or remediation works, a management structure must be put in place to ensure that through all stages of the reclamation, due control is exerted over health and safety issues. This must extend to identifying information requirements, obtaining sufficient information about hazards, obtaining expert advice where appropriate and making available sufficient information to all those involved or likely to be affected by the project, including contractors, clients, the public and relevant emergency services. An effective management structure will also have additional benefits in helping to facilitate better quality assurance and cost control.

Health and safety policy. The first key requirement is a Health and Safety Policy, which sets out:

- A declaration of intent — what is intended
- Organizational arrangements for health and safety management — how it will be done
- Responsibilities for implementing the policy — who will do it
- Procedures for distribution to all employees and contractors, and revision arrangements
- Resources available and required to implement the policy

The policy should be on prominent display at the site and should be brought to the notice of all employees and visitors.

This policy will be a dynamic document, which must be amended as circumstances change and more information becomes available. It is likely that a number of contractors will be involved at different stages in the project, and it is essential that each contractor is aware of the requirements set out in the main health and safety policy. Furthermore, all contractors employing over five people should be required to provide a copy of their specific health and safety policy before they are awarded contracts. The adequacy of their policies should be assessed as part of the tendering process.

Responsibilities. Overall responsibility for health and safety should be assigned to a suitably qualified and experienced manager. Most construction companies will already have clearly assigned responsibilities for health and safety. Where the client/contractor relationship follows the design and build model, the contractors may also take on the overall responsibility for health and safety across the site. In circumstances where a resident engineer is appointed to act on behalf of the client to supervise a number of contractors, the division of responsibilities should be clear and unambiguous and be accepted by all parties concerned.

It is increasingly common practice on larger sites for a full-time health and safety officer to be appointed to police all health and safety matters, provide advice and guidance, train and provide information, and ensure that the health and safety policy is implemented and updated. This role may be fulfilled by full-time appointed staff, by retained consultants, or by short-term appointed staff. The role of the health and safety officer may require knowledge of chemistry, air sampling and safety issues. It is usual that such officers report directly to the resident engineer. It is essential that the health and safety officer and the responsible manager are aware of their responsibilities for all work conducted on site. Contractors engaged by the client have their own duties under law; however the overall site manager may still retain liability in certain circumstances. The individual responsibilities for all matters concerning health and safety must be clearly assigned. For example, the responsibility for ensuring that restrictions on eating, smoking and naked flames are followed must rest with the supervisors responsible.

Contractors should also have clearly designated responsibilities set out in contract documents. They are also subject to the requirements of the Health and Safety at Work etc. Act, and other acts or regulations such as COSHH. Each contractor may appoint his own responsible manager and health and safety officer, and appropriate communication mechanisms will be essential to ensure that information is shared. Safety representatives should be encouraged to take part in safety activities. Safety committees, where they exist, should also be involved in safety management and liaison.

13.4.2 *Information, training and instruction, and supervision*

Once the appropriate procedures for investigation and reclamation have been defined, information, instruction and training needs should be identified so that all employees and others connected or affected by the project are aware of the hazards and the steps that they should take to minimize risks. It is important to ensure that sufficient information is made available to reassure workers and the public that hazards have been fully identified and safe systems of work that will minimize risks to health and safety are established. The risks involved must be explained rationally, with care taken to avoid frightening employees or the public whilst not allowing complacency to develop. The emphasis should be on explaining that, providing the safe systems of work are followed, risks will be minimized. Situations likely to increase risk or contribute to dangerous situations should be explained. In most circumstances it will be appropriate to provide written information about the hazards, circumstances likely to represent greatest risk, precautions to be taken, signs and symptoms of exposure or ill-health associated with the substances involved, indicators such as characteristic odours, visual indicators of contamination, etc. The required action in the event of ill-health, unusual odour, visible contamination, etc. should be written down and given to employees and visitors. In some situations, it may be appropriate to distribute colour photographs of the main hazardous materials so that employees can recognize these when they are encountered. A typical set of advisory instructions given to employees involved in a reclamation of a coke works site is given in Table 13.2.

Table 13.2 Site safety instructions for staff and contractors: reclamation of Newtown coke works

The site is an old cokeworks and the ground is contaminated by harmful and flammable chemicals. The levels of contamination are generally low, except in certain limited areas of the site. However, as there may be patches of contamination anywhere on site, it must be assumed that all areas are contaminated.

Reclamation can be achieved in safety, providing you follow all of the instructions that are given in this leaflet and those given by the health and safety officer and other responsible managers.

The toxic substances that are present can be harmful if you:

- eat (ingest) contaminated materials or soil
- breath (inhale) contaminated dusts, fumes or gases
- get splashes or materials on your skin, eyes and particularly in scratches or wounds

The following safety instructions should be followed at all times whilst you are working in the marked contaminated area:

(1) Protective clothing should be worn at all times. Your supervisor will define the type of clothing required for each job.

(2) You must wash thoroughly before eating, drinking, smoking or going to the toilet. At the end of your shift, you must pass through the hygiene unit and obey all instructions and notices. You should shower all over with soap and water before putting on your own clothes.

(3) No protective clothing is to be taken home or worn outside of the contaminated area. Your respirator must be stored in the marked lockers between shifts. On no account must respirators be shared.

(4) You must not smoke on any area of the site except for the mess room. Smoking may cause you to ingest harmful chemicals and may ignite flammable liquids or gases leading to explosion or fire.

(5) You must not eat or drink on any area of the site, except for the mess room because of the risk that you will ingest harmful chemicals.

Table 13.2   *(Cont.)*

(6) If you get any tarry or oily materials on your skin, you must wash them off immediately using the cleaning agent in the hygiene unit followed by soap and water. Contaminated clothes must be changed.

(7) If you have the slightest cut or wound it must be cleaned at once by your First Aider and then covered by a waterproof dressing. If your skin develops any red or sore patches, these must be reported to your supervisor or the health and safety officer at once.

(8) No non-working clothes must be worn on-site, even under your overalls. Your own clothes should be left in the lockers provided in the hygiene unit. Clean overalls will be provided as required.

(9) Dirty overalls must be placed in the bags provided in the hygiene unit.

(10) You must report any of the following immediately to your supervisor or the First Aider:
- headache
- stomach ache
- red or sore skin
- giddiness
- loss of appetite
- nausea or diarrhoea
- constipation
- coughs or chest pains
- painful joints
- or any other symptoms which you are concerned about

(11) No one is allowed to enter a trench, excavation or other confined space unless they are given a current permit to work from the site health and safety officer.

(12) If you notice any of the following, leave the area immediately and report it to your supervisor or the health and safety officer:
- patches of unusually coloured or textured materials
- any smells, particularly bad eggs
- any sign of visible vapours
- any signs of fire
- any containers, drums, tanks, pipes are uncovered

(13) If you are uncertain about these instructions or any aspect of your work, please discuss it straight away with your supervisor or health and safety officer or representative.

23 September 1991

Signed

Resident Engineer

Hazard data sheets for substances likely to be encountered during reclamation may be required by the COSHH Regulations. These provide general information on the substances, specific risks and actions to be taken to minimize risk in normal and emergency situations. An example of a hazard data sheet for phenol present at a contaminated site is given in Table 13.3.

Additional information by way of prominent notices and signs should be provided to indicate any prohibitions, dangerous areas, protective equipment requirements or specific hazards. All signs should conform with the Safety Signs Regulations 1980.

Supervision and regular inspections will be essential to ensure that systems of work and any prohibitions are adhered to, and that standards are not allowed to lapse. Safety audits may be a useful way of ensuring that health and safety management systems are in place and working effectively. Proprietary audit systems have been developed recently for use in the construction industry, and an example of a section covering reclamation of contaminated land is given in Table 13.4.

Table 13.3 Hazard data sheet

SUBSTANCE NAME: PHENOL

UN No:	1671	CAS No: 108–95–2	Formula: C_6H_5OH		
OEs 8 h:	5 8 hppm	19 8 hmg/m^3	10 min 10 ppm	38 10 min mg/m	

General risk: toxic,
Other effects: suspected or confirmed human carcinogen
Health survelliance: State: S; Vap.Den: 3.2;   Vap.Press: 0.3 mmHg;   Expl.limit 1.7–8.6%
BP   182°C;   FlP   78°C;   ignition temp: 605°C

Reactivity:
Reacts with strong oxidizers; reacts with calcium
Hypochlorite; attacks Al, Zn and lead

Notes: On heating, emits toxic fumes; explosive mixtures formed > 78 °C; very rapid absorption through skin

RISK PHRASES
24/25   Toxic in contact with skin and if swallowed
34   Causes burns

SAFETY PHRASES:
2   Keep out of reach of children
28   After contact with skin, wash immediately with plenty of water
44   If you feel unwell, seek medical advice

HAZARD DATA
Haz. Chem:
2X   Use water fogs (fine water spray in absence of fog equipment)
       No danger of violent reaction or explosion
       Full protective clothing
       Contain

Fire extinguishant: CO_2, Halons, Powder, Alcohol-res., foam, water fog/spray

HEALTH PHRASES
1   May be absorbed via inhalation, ingestion and the skin
12   Corrosive to the eyes, skin and respiratory tract
39   Can affect the central nervous system, liver and kidney
30   Avoid all skin contact
50   Prevent dispersion of dust
60   Inhalation may cause lung oedema; serious cases may be fatal

FIRST-AID PHRASES
M1   Call doctor, or send/take to doctor/hospital; show medical
       staff subst data sheet or ensure info accompanies patient.
IH2   Inhalation: fresh air, rest and half upright position;
       especially in cases of irritating or corrosive substance
E1   Eyes: Rinse continuously with water for at least 10 min;
       transport to a doctor or hospital
S7   Skin: Place contaminated clothing in sealable thick plastic bag;
       wash skin with water and soap; rescuers must wear PPE

Table 13.3   *(Cont.)*

_____

CONTAINMENT PHRASES
9   Sweep up spilled substance
15   Carefully collect remainder
21/22   Extra PPE: Self-contained breathing apparatus and
                   complete protective clothing
26   Contain: avoid spillage entering drains or water courses

.PERSONAL PROTECTIVE EQUIPMENT PHRASES
4   Breathing protection: Chemical cartridge respirator,
                   self-contained breathing apparatus
8   Impervious protective clothing and gloves

_____

Training of employees in safe systems of work, use of any control measures, use of personal protective equipment, and in emergency procedures will be required before starting work. Refresher training will be required at intervals. Details of all training given should be recorded.

13.4.3 *Site investigation*

The requirements of site investigations have been dealt with in chapter 3. This is an essential aspect of management of health and safety as this stage of reclamation should provide sufficient details of the types of hazard that may be present to allow the selection of a reclamation design that will involve minimum risks of exposure to hazardous substances, and suitable conditions for workers and the general public [13]. Site investigation data should be sufficiently detailed to allow the broad patterns of contamination both across the site and with depth to be identified. The survey should also provide information on any physical risks that may be present, such as buried tanks, pipelines, unstable ground, flammable gas, shallow mineworkings, etc. The site investigation phase itself is a high risk activity, since the very object of the investigation is to identify the presence of contamination. Inevitably the site investigation team must be able to respond to a wide variety of situations that may put themselves and others at risk. While specialists in site investigation may be employed in assessing contamination and are likely to be sufficiently knowledgeable and experienced to handle most situations, other types of investigation may also be conducted by persons not as familiar with the health and safety problems. For example, civil and structural engineers may be required to supervise the excavation of trial pits and take samples to determine the structural stability and characteristics of the site. Any possible hazards and the appropriate precautions should be explained prior to commencement of such investigations.

13.4.4 *Assessment of risk*

Once the results of the preliminary site investigation are available and an initial reclamation programme has been developed, it will be possible to

Table 13.4 Example of a health and safety audit checklist

HEALTH AND SAFETY ON SITE MANAGEMENT OF THE PLACE OF WORK II	B 2

B2.9   Work on Contaminated Sites

Answer
Yes or No

B2.9a   Is any work carried out on a previously
contaminated site?

NO   If NO, score 0 points and write points total below. Enter carried
Maximum Section Score in cell B2 in Maximum Scores column.
Enter 0 in cell B2.9, and then GO TO next section.

YES   If YES, add 24 points to carried Maximum Section Score and
continue this page.

B2.9b   Have you undertaken a through examination
of the previous uses of the site?

B2.9c   Have soil samples and samples from existing
buildings and plant been taken and analysed?

B2.9d   Has an agreed Method Statement been
produced to eliminate or reduce contact with
contaminants?

B2.9

B2.9e   Have you made arrangements to monitor
regularly the environmental conditions on the
site and in the vicinity of the site boundaries?

B2.9f   Have you made arrangements for medical
examinations of workers who are likely to be
exposed to contaminants?

B2.9g   Have you made arrangements to monitor
regularly the health of all personnel on site?

B2.9h   Have you given all personnel on site
information about the hazards involved and
on what action to take?

For guidance on best practice in respect of
Contaminated Sites, see Construction Safety
Manual Section 31

SCORE   Count 3 points for each YES answer. Write points total
below and enter your score in cell B2.9   **46**

TOTAL		ENTER

CARRIED MAX SCORE _____ + _____ = ____
(from previous page)   (0 or 24)
ENTER FINAL MAX SECTION SCORE IN MAX SCORES COLUMN CELL (B2)

Source: Building Advisory Service and Health And Safety Technology And Management Ltd
(1990), Construction–CHASE Manual, BAS London.

identify those aspects of the reclamation project that involve specific hazards. The project management team should review the planned operations and the hazards that are likely to be encountered, in order to prepare an outline strategy for health and safety management. The first important aspect will be to assess the risks of exposure, and to health and safety.

The first task in assessing risks is to consider how the presence of hazardous substances or conditions may affect different members of the reclamation work-force and the general public. Based on the nature and properties of the substances concerned, the main exposure routes and harmful effects can be identified. For substances presenting risks though inhalation, there may be Occupational Exposure Limits (OELs) as set out in the Guidance Note EH40/92, which should not be exceeded [14]. In these cases, the likely exposures of workers engaged in different operations can be assessed directly against the OELs for the substances concerned. In some situations, particularly where there are mixtures of substances or where exposure may occur by other routes (principally skin absorption or ingestion), lower exposure limits may be advisable. Expert guidance may be needed in such situations. Risks associated with other forms of exposure, such as skin contact or ingestion, should also be taken into account.

Once the main exposure routes and the types of health hazard have been identified, the next stage in the assessment of risks should be to examine the proposed operations and consider what precautions may be needed to ensure that exposure is kept as low as reasonably practicable and, in the cases of substances that are listed in Table 1 of EH40, below the maximum exposure limit (MEL). Substances for which MELs have been set include benzene, arsenic and its compounds, cadmium and its compounds, hydrogen cyanide, lead and its compounds, and styrene.

In deciding what precautions are required, the nature of the work and the susceptibility of different groups of individuals should be taken into account. For example, an excavator driver in a cab with a filtered air supply is likely to be better protected than a worker in the open. On the other hand, if the site is adjacent to a hospital or school, it is likely that the infirm or young children will be more susceptible to some types of harmful substances than healthy site workers.

In some circumstances, exposure monitoring may be required to provide information upon which the assessment of risk can be based or validated. It may be appropriate to stage small-scale trial operations in which the workers' exposure can be monitored using appropriate monitoring equipment. The health and safety executive may be able to offer assistance in selection of appropriate monitoring and analytical techniques.

13.4.5 *Risk control*

Where an assessment has identified potential risks, some form of control should be implemented to prevent or control the risk. In many cases, the responsibility of selecting control measures may be placed upon contractors.

Clients may provide documentary evidence of the extent and nature of hazards likely to be encountered and require that prospective contractors submit Methods of Work Statements, which specify the controls that will be used. There are advantages in seeking information on how health and safety matters will be handled early on in the process of selecting contractors, since this allows the client to assess how well the potential contractor has understood the complexity of the site and the requirements for safe working.

In practice, it is probable that a final working plan based on the detailed discussions with the preferred contractors and the in-house reclamation team will be produced. The working plan should cover all aspects of health and safety management, including the controls, any monitoring and inspection, auditing and evaluation. The following may need to be considered:

Method of Work Statements. Appropriate measures should be introduced to control risks through the adoption of safe working methods. It is important that all methods of work by all contractors are reviewed to ensure compatibility and consistency across the site and throughout the reclamation phase. The Method of Work Statements should summarize the risks involved and document procedures for work in contaminated areas, incident notification, emergency procedures, etc. The advice of a health and safety professional is often desirable in order to ensure that controls are appropriate. Many problems can be solved by selecting methods of work that minimize risk. For example, keeping site surfaces damp will minimize dust release.

Personal protective equipment. Personal protective equipment (PPE) includes overalls, gloves, safety footwear or wellingtons, goggles and hearing defenders. Prior to selection, information from manufacturers should be obtained in order to find out the suitability and range of working conditions. All issues and withdrawals of PPE should be recorded. Recipients should be trained in the use and inspection of PPE. The following factors should be taken into account:

● Overalls will require regular cleaning and must not be removed from site by workers in any circumstance. Disposable overalls may be better in some circumstances to avoid the requirement for specialized laundering. PVC overalls are better in wet weather, or work in or near water, but are heavy and may present problems due to excessive perspiration. Overalls should not have external pockets as these may trap contaminants. Close fitting ankles and cuffs are more effective for preventing skin contamination. Integral hoods may also be required to prevent contamination of the hair and head. Once used, contaminated overalls should be removed in a decontamination unit and placed in a sealed and clearly marked double-skin plastic bag. Bags containing disposable overalls should be treated as contaminated waste. Laundries should always be advised of the nature of the contamination prior to laundering.

- Safety footwear or wellingtons may be required on some sites. Work in contaminated water always requires specially selected wellingtons. Some contaminants such as solvents may rapidly penetrate or degrade certain kinds of rubber. Wellingtons should be regularly inspected to ensure that there are no leaks.
- Gloves will be required in many cases to prevent skin contact, abrasion and damage. Gloves also help to ensure that toxic contaminants are not trapped under finger nails and in cracked skin, which could present risks through hand-to-mouth contact, e.g. biting finger nails. Selection of gloves will be important. There are a wide variety of materials and designs, each of which is better at withstanding some conditions than others. Some contaminants, such as benzene, may pass rapidly through most materials, therefore glove life may be short.

In some circumstances, respiratory protective equipment (RPE) will be required. RPE should be suitable for the purpose and of a type approved by the HSE [15]. Arrangements for the cleaning, storage and changing of filters should be set in place and all recipients should be trained in their use and testing. Respirators should never be stored overnight in cabs or unprotected areas, nor should they be taken home. Used filters should be treated as contaminated waste. Disposable respiratory protection should never be used for longer that their design limit. One-shift disposable respiratory protection should be disposed of at the end of each shift. Breathing apparatus should be stored in an appropriate location and regularly inspected and tested. Suitable training should be given to all those likely to use breathing apparatus.

Environmental monitoring. In order to ensure safe working environments, in situations where there is likely to be atmospheric contamination, it is important that environmental and personal exposure monitoring is conducted at regular intervals. Air monitoring at working positions, within hygiene units, and in clean parts of the site can be used to check on the effectiveness of controls. Site boundary monitoring is valuable in helping to provide confirmatory evidence that the public is not placed at risk. Local authorities are more frequently asking for boundary monitoring to ensure the protection of public health. Water courses should also be checked to ensure that water pollution is not taking place. Visual inspection for dust should be conducted frequently and should always lead to appropriate action being taken to control any dust emissions. Testing of atmospheres for flammable, explosive, or toxic gas should be undertaken when appropriate. Tests of confined spaces for asphyxiant gases or lack of oxygen should also be carried out prior to entry [5].

Health surveillance. It may, in some cases, be appropriate, or a requirement, that health surveillance is undertaken for all workers whose exposure is likely to be significant. Specialist advice should be sought on the type of surveillance that is appropriate, the records that will be required and any other relevant factors. Advice may be obtained from the Employment Medical Advisory Service or an occupational physician.

Enforcement procedures.   There should be clear allocation of responsibilities to ensure that safe working practices are always followed, and that controls and personal protective equipment are always used where required, and inspected, maintained, etc. Each person working on the site should be aware of the health and safety issues and their own individual responsibilities. Written instructions may be appropriate. There should be regular inspections of the site and clear disciplinary procedures for any breach of procedure.

Record keeping.   With increasing demands to demonstrate that all that is reasonably practicable has been done to ensure health and safety, and with formal requirements governing the maintenance of records, it is important that appropriate records are kept and made available for inspection by clients, enforcing authorities, etc. Examples of the types of information that may be required include:

- Daily work activities
- Protective equipment issues and withdrawals
- Inspections, tests results
- Health/medical surveillance records (some may be subject to confidentiality restrictions)
- Personal monitoring results (some may be subject to confidentiality restrictions)
- Environmental monitoring results
- Training records
- Maintenance of any control measures etc.

Emergency plans.   Whilst the Method of Work Statement will indicate the actions to be taken during the normal work on site, it is vital that there is a central plan covering the entire site. This should identify all foreseeable emergencies, their likely consequences, and the actions that should be taken should they occur. Examples may include accidental spillages, fire, toxic substance release, collapse or unconsciousness of personnel, or an accident. The emergency plan should be available at all times and personnel should be trained in the procedure to be followed. Where appropriate, copies should be given to the local emergency services and the local hospital, who should also be advised of the substances that are involved so that they can make contingency preparations. For example, work with phenol may require special medical supplies in the case of poisoning and the local hospital would require advance notice so that supplies can be arranged. Emergency telephone numbers should be displayed in a prominent position on site.

Site facilities. All contaminated sites should have, at a minimum, the following facilities:

- Toilets
- Showers
- Facilities for eating, drinking, smoking, etc.
- First aid room and equipment [16]

No smoking should be allowed whilst working on the site, and signs indicating no smoking should be on prominent display. The reclamation team should have a number of members qualified in first aid.

A hygiene facility will be required on some sites. Transportable units are available to hire or purchase. The minimum requirement should be a first stage storage area for ordinary clothing, personal effects, etc. and a second stage with high standard washing facilities, including at least two showers. Warm water should be available. A third stage should consist of a changing area for contaminated clothing, facilities for storage/disposal of contaminated clothing, storage of PPE/RPE. A boot wash should be provided at the entrances to the contaminated area and the hygiene unit.

Designated dirty areas of the site should be fenced with security fencing and suitable warning signs posted. The only access to or from the dirty area should be through the hygiene facility or through a control gate where a wheel wash is installed for vehicles. Records of all personnel passing into or out of the contaminated area should be kept in a log held at the control gate.

Housekeeping. Rigorous standards of housekeeping will be required on all contaminated sites. Examples of the issues that may need to be considered include:

- Regular sweeping of access roads
- Sheeting of all wastes transported off-site
- Clearly marked areas for waste handling/storage
- Dust suppression during dry periods on all active site areas
- Cleaning of cabs of all vehicles
- Wheel washing for all vehicles leaving the site
- Decontamination of vehicles, particularly tracked vehicles, prior to transport off-site
- Control of surface water run-off to prevent spread of contamination
- Appropriate treatment of wheel wash effluent (consultation with the local authority and/or the National Rivers Authority is advisable prior to installation).

13.5 Case studies

The first case study demonstrates that work on contaminated sites will inevitably involve discovery of previously unidentified hazards. The lesson from this case is that the management system should be sufficiently flexible to enable the appropriate responses to be made following frequent reassessment of the hazards and risks.

The site was a former ironworks, now surfaced with slags, inert rubble and ashes. The site investigation had identified that wastes were variable in depth from 1 to 5 m. The rubble was scattered across the site and was underlain by slags with high levels of metal contamination. Bands of ash were present in the slags and had high calorific values, which posed a potential combustion risk. The site investigation had also detected pockets of flammable methane gas in some areas of the site, as well as areas of elevated carbon dioxide. The local authority had set conditions governing the reclamation, which took account of the need to minimize dust generation through surface wetting etc. Workers on the site were given appropriate respiratory protection to reduce inhalation of metals. Washing facilities were provided.

Shortly after commencement of excavation operations, a worker collapsed. He was removed from the area and later fully recovered from his experience. An immediate investigation was launched to determine the cause of the loss of conciousness. After initial enquiries it became clear that the reason for the problem was not an accumulation of carbon dioxide gas, but was in fact due to oxygen starvation. The initial site investigation had not included measurements of oxygen levels and this hazard had therefore gone undetected. Once this problem was identified, Safe Systems of Work were introduced to ensure that atmospheres were tested for oxygen as well as flammable gas and carbon dioxide.

The second case study illustrates the merits of establishing an appropriate management structure. The project was the redevelopment of a derelict waterside site.

As part of the site investigation phase of the reclamation programme, the site history desk study revealed some information about the site's past. The ground investigation identified the presence of both superficial and deeply buried wastes, which were contaminated to varying degrees with a wide range of substances. Some areas of the site had only slight surface contamination with ashy residues, others had buried structures that had been backfilled with oily wastes and rubble. Significant areas of the site had been subject to indiscriminate tipping of contaminated materials including boiler ash, barrels of oil, asbestos, etc. The depths of contamination extended in places to 5 m below surface level.

Having identified the highly variable nature of the contamination, the environmental consultants who undertook the initial site investigation were retained by the site purchasers to provide advice during the pre-tender stage of drawing up contracts and specification for requirements of removal of contamination. Detailed Method of Work Statements were prepared governing all major stages of reclamation. In addition, there was a clearly specified chain of responsibility and communication to ensure that all contractors obeyed the

rules governing health and safety. The responsibility for instructing the immediate cessation of work rested with the full time on-site chemist who was
supported by a laboratory.

Once the reclamation commenced, operations were controlled so that no
general soil movements or excavation took place until the worst areas of
contamination had been treated or removed. No works were allowed until all
asbestos had been removed for disposal.

The Method of Work Statements also documented: (i) the procedures to be
adopted by all vehicles on site, including vehicle movement routes, cleaning
requirements, waste disposal requirements and record keeping; (ii) environmental monitoring, including continuous boundary monitoring; (iii) personal
exposure monitoring; (iv) sampling and analysis of soils and materials; and
(v) post-excavation sampling and analysis to ensure complete removal of
contamination.

Specific arrangements for health and safety included:

- Concise information leaflets for all personnel on site
- Regular updating of the work force by on-site training
- Provision of a full-time qualified nurse and first aid facilities
- Provision of a hygiene unit
- Provision of a mess room outside of the dirty area
- Provision of adequate and sufficient protective clothing, respirators and
  breathing apparatus
- Continuous monitoring of the atmosphere for the presence of asbestos
  fibres during the excavation of asbestos materials
- Close liaison with the local Environmental Health Department, HSE and
  the public by the publication of air monitoring results on a daily basis
  and the random sampling of air in gardens and streets

At the outset of the project, the Method of Work Statement was made available
for public inspection and comment; this allowed local residents to monitor the
effectiveness of the contractor's performance against the original programme
and methods. Around 50 000 m^3 of contaminated materials and 300 m^3 of
asbestos were removed from the site without any health and safety or environmental incidents to the workforce or the local population.

Appendix I: Soil and water quality criteria

The purpose of this appendix is to provide a summary of relevant soil quality and environmental reference values and standards. The following points should be noted:

(1) Information on the more commonly required soil quality and environmental reference values and standards is summarized for convenient reference.
(2) Before use is made of any values, the original publications should be consulted, since their legal standings, appropriate contexts and applicability differ. References to the original publications are listed in chapter 3.
(3) Reference values and standards are listed for
  - Soil qualities
  - Groundwater qualities (on Netherlands A-B-C system)
  - Environmental aquatic reference values and standards for water use
  - Proposed leachability criteria for landfill materials
  - Trade effluent regulations
  - Landfill gas concentration reference levels

Soil and groundwater quality standards

ICRCL 59/83 (chapter 3 [2]): Guidance on the assessment and redevelopment of contaminated land

Use codes

(1) Domestic gardens and allotments
(2) Parks, playing fields, open space
(3) Landscaped areas
(4) Any use where plants are grown (applies to contaminants that are phytotoxic, but not normally hazards to health)
(5) Buildings
(6) Hard cover

Conditions

(1) Tables are invalid if reproduced without the conditions and footnotes.
(2) All values are for concentrations determined on 'spot' samples based on adequate site investigation carried out prior to development. They do not apply to analysis of averaged, bulked or composited samples, nor to sites that have already been developed.
(3) Many of these values are preliminary and will require regular updating. For contaminants associated with former coal carbonization sites, the values should

not be applied without reference to the current edition of the report, *Problems Arising from the Redevelopment of Gas Works and Similar Sites* (chapter 3 [12]).

(4) If all samples values are below the threshold concentrations then the site may be regarded as uncontaminated as far as the hazards from these contaminants are concerned, and development may proceed. Above these concentrations, remedial action may be needed, especially if the contamination is still continuing. Above the action concentration, remedial action will be required or the form of development changed.

Footnotes

NS Not specified.

NL No limit set as the contaminant does not pose a particular hazard for this use.

(1) Soluble hexavalant chromium extracted by 0.1 M HCl at 37°C; solution adjusted to pH 1.0 if alkaline substances present.

(2) Determined by standard ADAS method (soluble in hot water).

(3) Total concentration (extraction by $HNO_3/HClO_4$).

(4) Total phytotoxic effects of copper, nickel and zinc may be additive. The trigger values given here are those applicable to the worst-case phytotoxic effects that may occur at these concentrations in acid, sandy soils. In neutral or alkaline soils phytotoxic effects are unlikely at these concentrations.

The soil pH value is assessed to be about 6.5 and should be maintained at this value. If the pH falls, the toxic effects and uptake of these elements will be increased.

Grass is more resistant to phytotoxic effects than most other plants and its growth may be adversely affected at these concentrations.

(5) Many of these values are preliminary and will require regular updating. They should not be applied without reference to the current edition of the report *Problems Arising from the Redevelopment of Gas Works and Similar Sites* (chapter 3 [12]).

(6) Used here as a marker for coal tar, for analytical reasons. See *Problems Arising from the Redevelopment of Gas Works and Similar Sites*, Annex A1.

(7) See *Problem Arising from the Redevelopment of Gas Works and Similar Sites* for details of analytical methods.

(8) See also BRE Digest 250: Concrete in sulphate-bearing soils and groundwater.

Contaminants	Use code	Reference value trigger concentrations (mg/kg air-dried soil)	
		Threshold	Action
Group A: Selected inorganic contaminants that may pose hazards to health			
Arsenic	1	10	NS
	2	40	NS
Cadmium	1	3	NS
	2	15	NS
Chromium total	1	600	NS
	2	1000	NS
Chromium (hexavalant) (1)	1, 2	25	NS
Lead	1	500	NS
	2	2000	NS
Mercury	1	1	NS
	2	20	NS
Selenium	1	3	NS
	2	6	NS
Group B: Contaminants that are phytotoxic, but not normally hazards to health			
Boron (water soluble) (2)	4	3	NS
Copper (3)(4)	4	130	NS
Nickel (3)(4)	4	70	NS
Zinc (3)(4)	4	300	NS
Contaminants associated with former coal carbonization sites			
Polyaromatic hydrocarbons (5)(6)(7)	1	50	500
	3, 5, 6	1000	10 000
Phenols (5)	1	5	200
	3, 5, 6	5	1000
Free cyanide (5)	1, 3	25	500
	5, 6	25	500
Complex cyanides (5)	1	250	1000
	3	250	5000
	5,6	250	NL
Thiocyanate (5)(7)	All	50	NL
Sulphate (5)	1, 3	2000	10 000
	5	2000 (8)	50 000 (8)
Sulphide	All	250	1000
Sulphur	All	500	20 000
Acidity (pH less than)	1, 3	pH 5	pH 3
	5, 6	NL	NL

ICRCL 70/90 (Chapter 3[3]): Notes on the restoration and aftercare of metalliferous mining sites for pastures and grazing

Table III Guidelines for toxic element trigger concentrations in minespoil-contaminated 'soils' (total element concentration in mg/kg air dry weight)

Element	Threshold trigger concentrations [a]	Maximum (action trigger) concentrations (values not to be exceeded for use as specified)	
		For grazing livestock [b]	For crop growth [c] (risk of phytotoxicity)
Zinc	1000	3000[d]	1000
Cadmium	3	30[d]	50
Copper	250	500	250
Lead	300	1000	–
Fluoride	500	1000	–
Arsenic	50	500	1000

[a] These concentrations are acceptable only where the soil contamination is derived from mine spoil. In other situations the elements may be present in forms that are more available to plants and animals when lower trigger concentrations will be appropriate.

[b] For calves, sheep and horses, assuming that plant uptake is normal, the stock are continuously exposed to these concentrations and that it is proposed to manage the sward in such a way that only a relatively low level of soil contamination of herbage will occur. In such case, soil may comprise up to about 5% of dry matter intake. Under less favourable conditions soil ingestion may be much higher. See Section 7, grazing access. (The corresponding EDTA-extractable phytotoxic limits for Zn and Cu are 130 and 70 mg/dm^3 soil, respectively. Soil material should be considered a phytotoxic risk if either the total or EDTA-extractable limits are exceeded.)

[c] For clover and the more productive, sown grass species, assuming a soil pH of at least 6.0. Metal tolerant cultivars are available, but these are intended for amenity/recreation after-uses and advice should be sought before they are used in agricultural seed mixtures.

[d] The possibility of subclinical antagonistic effects on copper metabolism cannot be ruled out if concentrations of zinc and cadmium in soils exceed 2000 and 15 mg/kg, respectively.

The guidelines given in Table III should not be regarded as standards to which concentrations of contaminants in the soil could be raised if this were avoidable. On heavily contaminated sites or where there is a lack of suitable soil-forming materials, reclamation of the land for pasture and grazing is not acceptable and more appropriate end uses should be considered.

Greater London Council Definitions of Contaminated Soils: suggested range of values (mg/kg on air-dried soils, except for pH) (chapter 3 [31])

Parameter	Typical values for uncontam- inated soils	Slight contam- ination	Contaminated	Heavy contam- ination	Unusually heavy contam- ination, more than
pH (acid)	6–7	5–6	4–5	2–4	< 2
pH (alk.)	7–8	8–9	9–10	10–12	12
Antimony	0–30	30–50	50–100	100–500	500
Arsenic	0–30	30–50	50–100	100–500	500
Cadmium	0–1	1–3	3–10	10–50	50
Chromium	0–100	100–200	200–500	500–2500	2500
Copper (avail.)	0–100	100–200	200–500	500–2500	2500
Lead	0–500	500–1000	1000–2000	2000–1.0%	1.0%
Lead (avail.)	0–200	200–500	500–1000	1000–5000	5000
Mercury	0–1	1–3	3–10	10–50	50
Nickel (avail.)	0–20	20–50	50–200	200–1000	1000
Zinc (avail.)	0–250	250–500	500–1000	1000–5000	5000
Zinc equiv.	0–250	250–500	500–2000	2000–1.0%	1.0%
Boron (avail.)	0–2	2–5	5–50	50–250	250
Selenium	0–1	1–3	3–10	10–50	50
Barium	0–500	500–1000	1000–2000	2000–1.0%	1.0%
Beryllium	0–5	5–10	10–20	20–50	50
Manganese	0–500	500–1000	1000–2000	2000–1.0%	1.0%
Vanadium	0–100	100–200	200–500	500–2500	2500
Magnesium	0–500	500–1000	1000–2000	2000–1.0%	1.0%
Sulphate	0–2000	2000–5000	5000–1.0%	1.0–5.0%	5.0%
Sulphur (free)	0–100	100–500	500–1000	1000–5000	5000
Sulphide	0–10	10–20	20–100	100–500	500
Cyanide (free)	0–1	1–5	5–50	50–100	100
Cyanide total	0–5	5–25	25–250	250–500	500
Ferricyanide	0–100	100–500	500–1000	1000–5000	5000
Thiocyanate	0–10	10–50	50–100	100–500	2500
Coal Tar	0–500	500–1000	1000–2000	2000–1.0%	1.0%
Phenol	0–1	2–5	5–50	50–250	250
Toluene extract	0–5000	5000–1.0%	1.0–5.0%	5.0–25.0%	25.0%
Cyclohexane extract	0–2000	2000–5000	5000–2.0%	2.0–10.0%	10.0%

Additional guidelines by the Greater London Council. A combustibility hazard may exist if the calorific value of a soil sample is greater than 7 MJ/kg. Surface deposits containing asbestos should be removed. Guidance for heavy metals should be taken from the values proposed by the ICRCL.

Dutch test table (chapter 3 [32])

The treatment of polluted soil depends on the nature and the concentrations of polluted substances present in it. In connection with this, a test framework is used in The Netherlands, built up of three values which must be distinguished. These values, consisting of different, ascending levels of concentration. A, B and C are differentiated according to the nature of the pollution:

- Level A acts as a reference value. This level may be regarded as an indicative level above which there is demonstrable pollution and below which there is no demonstrable pollution.
- Level B is an assessment value. Pollutants above the B level should be investigated more throughly. The question asked is: to what extent are the nature, location, and concentration of the pollutant(s) of such a nature that it is possible to speak of a risk of exposure to man or the environment ?
- Level C is to be regarded as the assessment value above which the pollutant(s) should generally be treated.

Present in:		Soil (mg/kg dry matter)			Groundwater µg/l		
component/concentration		A	B	C	A	B	C
I. Metals							
Cr		*	250	800	*	50	200
Co		20	50	300	20	50	200
Ni		*	100	500	*	50	200
Cu		*	100	500	*	50	200
Zn		*	500	3000	*	200	800
As		*	30	50	*	30	100
Mo		10	40	200	5	20	100
Cd		*	5	20	*	2.5	10
Sn		20	50	300	10	30	150
Ba		200	400	2000	50	100	500
Hg		*	2	10	*	0.5	2
Pb		*	150	600	*	50	200
II. Inorganic compounds							
NH_4	(as N)	–	–	–	*	1000	3000
F	(total)	*	400	2000	*	1200	4000
CN	(total, free)	1	10	100	5	30	100
CN	(total, combined)	5	50	500	10	50	200
S	(total, sulphide)	2	20	200	10	100	300
Br	(total)	20	50	300	*	500	2000
PO_4	(as P)	–	–	–	*	200	700
III. Aromatic compounds							
Benzene		0.05	0.5	5	0.2	1	5
Ethylbenzene		0.05	5	50	0.2	20	60
Toluene		0.05	3	30	0.2	15	50
Xylenes		0.05	5	50	0.2	20	60

Present in: component/concentration	Soil (mg/kg dry matter)			Groundwater μg/l		
	A	B	C	A	B	C
Phenols	0.05	1	10	0.2	15	50
Aromatics (total)	–	7	70	–	30	100
IV. Polycylic hydrocarbons						
Napthalene	*	5	50	0.2	7	30
Anthracene	*	10	100	0.005	2	30
Phenanthrene	*	10	100	0.005	2	10
Fluoranthene	*	10	100	0.005	1	5
Chrycene	*	5	50	0.005	0.2	2
Benzo(a)anthracene	*	5	50	0.005	0.5	2
Benzo(a)pyrene	*	1	10	0.005	0.2	1
Benzo(k)fluoranthene	*	5	50	0.005	0.2	2
Indeno(1,2,3cd) phyrene	*	5	50	0.005	0.5	2
Benzo(ghy)perylene	*	10	100	0.005	1	5
Total polycyclic	1	20	200	0.2	10	40
V. Chlorinated hydrocarbons						
Aliphatic (indiv.)	*	5	50	0.01	10	50
Aliphatic (total)	–	7	70	–	15	70
Chlorobenzenes (indiv.)	*	1	10	0.01	0.5	2
Chlorobenzenes (total)	–	2	20	–	1	5
Chlorophenols (indiv.)	*	0.5	5	0.01	0.3	1.5
Chlorophenols (total)	–	1	10	–	0.5	2
Chlor. polycyclic (tot)	*	1	10	–	0.2	1
PCBs (total)	*	1	10	0.01	0.2	1
EOCl (total)	0.1	8	80	1	15	70
VI. Pesticides						
Chlor. organics (indiv.)	*	0.5	5	1/0.1	0.2	1
Chlor. organics (total)	–	1	10	–	0.5	2
Non-chlor. (indiv.)	*	1	10	1/0.1	0.5	2
Non-chlor. (total)	–	2	20	–	1	5
VII. Other pollutants						
Tetrahydrothurane	0.1	4	40	0.5	20	60
Pyridine	0.1	2	20	0.5	10	30
Tetrahydrothiophene	0.1	5	50	0.5	20	60
Cyclohexanes	0.1	5	60	0.5	15	50
Styrene	0.1	5	50	0.5	20	60
Phalates (total)	0.1	50	500	0.5	10	50
Total polycyclic hydrocarbons oxidized	1	200	2000	0.2	100	400
Mineral oil	*	1000	5000	50	200	600

* Reference value (level A) for soil quality. (For I and II, please see Tables 1 and 2)

Table 1   Reference values (level A) for heavy metals, arsenic and fluorine

Component	Soil (mg/kg dry matter)		Groundwater (µg/l)
	Method of calculation	Standard soil (H=10/L=25)	
Cr (chromium)	50 + 2L	100	1
Ni (nickel)	10 + L	35	15
Cu (copper)	15 + 0.6(L + H)	36	15
Zn (zinc)	50 + 1.5(2L + H)	140	150
As (arsenic)	15 + 0.4(L + H)	29	10
Cd (cadmium)	0.4 + 0.007(L + H)	0.8	1.5
Hg (mercury)	0.2 + 0.0017(2L + H)	0.3	0.05
Pb (lead)	50 + L + H	85	15
F (fluorine)	175 + 13L	500	–

Notes on Table 1:   Reference values (level A) for heavy metals, arsenic and fluorine in all types of soil can be calculated by means of the formula given for each element. This formula expresses the reference value in terms of the clay content (L) and/or the organic materials content (H). The clay content is taken to be the weight of mineral particles smaller than 2 µm as a percentage of the total dry weight of the soil. The organic materials content is taken to be the loss in weight due to burning as a percentage of the total dry weight of the soil. As examples, the references values (level A) are given for an assumed standard soil containing 25% clay (L) and 10% organic material (H). For groundwater in the saturated zone, the reference are considered independently of the type of soil.

Table 2   Reference values (level A) for other inorganic compounds

Component	Groundwater	Remarks
Nitrate	5.6 mg N/l	Lower values may be
Phosphate	0.4 mg P/l sandy soils	specified for the
(Total phosphate)	3.0 mg P/l clay and peaty soils	protection of soils low in nutrients
Sulphate	150 mg/l	In maritime areas higher
Bromides	0.3 mg/l	values occur naturally
Chlorides	100 mg/l	(saline and brackish
Fluorides	0.5 mg/l	groundwater)
Ammonium compounds	2 mg/N/l sandy soils 10 mg N/l clay and peaty soils	

Environmental reference values

DoE Circular 7/89 (chapter 3 [39]): Environmental quality objectives

List I substances
Mercury
   0.3 mg/kg (wet weight) in a representative sample of fish flesh
   1 µg/l (annual mean) total mercury in inland surface waters
   0.5 µg/l (annual mean) dissolved mercury in estuary waters
   0.3 µg/l (annual mean) dissolved mercury in marine waters
Cadmium
   5 µg/l (annual mean) total cadmium in inland surface waters
   5 µg/l (annual mean) dissolved cadmium in estuary waters
   2.5 µg/l (annual mean) dissolved cadmium in marine waters
Hexachlorocyclohexane (HCH)
   0.1 µg/l total HCH (annual mean) in inland surface waters
   0.02 µg/l total HCH (annual mean) in estuary and marine waters
Carbon tetrachloride
   12 µg/l (annual mean) in all waters
DDT
   0.01 µg/l (annual mean) for the isomer *para-para*-DDT in all waters
   0.025 µg/l (annual mean) for total DDT in all waters
Pentachlorophenol (PCP)
   2 µg/l (annual mean in all waters)
The drins' (aldrin, dieldrin, endrin and isodrin)
   0.03 µg/l (annual mean) total drins for all waters, with a maximum of 0.005 µg/l for
      endrin .
   From 1 January 1994
   0.01 µg/l (annual mean) aldrin for all waters
   0,01 µg/l (annual mean) dieldrin for all waters
   0.005 µg/l (annual mean) endrin for all waters
   0.005 µg/l (annual mean) isodrin for all waters
Hexachlorobenzene (HCB)
   0.03 µg/l (annual mean) in all waters
Hexachlorobutadiene (HCBD)
   0.1 µg/l (annual mean) in all waters
Chloroform
   12 µg/l (annual mean) in all waters

List II substances (a)

Quality objective (e)		Lead (f) TR208*	Chromium TR207	Zinc TR209	Copper (g) TR210	Nickel TR211	Arsenic TR212
Fresh water							
Direct	A1(b)	50PT	50PT	3000PT	20PT	50PT	50PT
abstraction to potable supply	A2(b)	75MT	75MT	5000PT	50PT	50PT	50PT
	Total hardness (as mg/l CaCO₃)						
Protection of	0–50	4AD	5AD	8AT(30P)	1AD(5P)	50AD	50AD
sensitive	50–100	10AD	10AD	50AT(200P)	6AD(22P)	100AD	50AD
aquatic life	100–150	10AD	20AD	75AT(300P)	10AD(40P)	150AD	50AD
(e.g. salmonid	150–200	20AD	20AD	75AT(300P)	10AD(40P)	150AD	50AD
fish (c)	200–250	20AD	50AD	75AT(300P)	10AD(40P)	200AD	50AD
	250+	20AD	50AD	125AT(500P)	28AD(112P)	200AD	50AD
Protection of	0–50	50AD	150AD	75AT(300P)	1AD(5P)	50AD	50AD
other aquatic	50–100	125AD	175AD	175AT(700P)	6AD(22P)	100AD	50AD
life (e.g.	100–150	125AD	200AD	250AT(1000P)	10AD(40P)	150AD	50AD
cyprinid fish)	150–200	250AD	200AD	250AT(1000P)	10AD(40P)	150AD	50AD
	200–250	250AD	250AD	250AT(1000P)	10AD(40P)	200AD	50AD
	250+	250AD	250AD	500AT(2000P)	28AD(112P)	200AD	50AD
Salt water							
Protection of salt water life		25AD	15AD	40AD	5AD	30AD	25AD

* WRC Report reference number .
All values are given as μg/l. A, Annual average; P, 95% of samples (d); M, maximum allowable concentration; D, dissolved; T, total.

List II substances (a)

Quality objective (e)		Boron (h) TR256*	Iron (h)(i) TR258	pH TR259	Vanadium TR253	
Fresh water						
Direct abstraction to	A1(b)	1000PT	300PD	6.5–8.5P		
potable supply	A2(b)	1000PT	2000PD	5.5–9.0P		
					Total hardness (as mg/l CaCO₃)	
Protection of sensitive		2000AT	1000AD	6.0–9.0P	0–200	20AT
aquatic life (e.g. salmonid fish) (c)					200+	60AT
Protection of other		2000AT	1000AD	6.0–9.0P	0–200	20AT
aquatic life (e.g. cyprinid fish)					200+	60AT
Salt water						
Protection of salt water life		7000AT	1000AD	6.0–8.5P(k)	100AT	

* WRC Report reference number.
All values given as μg/l, except pH where 95% of samples must lie within the range shown.
A, annual average; P, 95% of samples (d); D, dissolved; T, total.

List II substances (a)

Quality objective (e)		Triorganotin Compounds TR255 [*]	
		Tributyltin	Triphenyltin
Fresh water			
Direct abstraction to potable supply	A1(b)	0.02MT	0.09MT
	A2(b)	0.02MT	0.09MT
Protection of sensitive aquatic life (e.g. salmonid fish) (c)		0.02MT	0.02MT
Protection of other aquatic life (e.g. cyprinid fish)		0.02MT	0.02MT
Salt water			
Protection of salt water life		0.002MT(1)	0.008MT(1)

[*] WRC Report reference number.
All values given as µg/l. P, 95% of samples (d); M, maximum allowable concentration; T, total.

List II substances (a)

Quality objective (e)		Mothproofing agents TR261[*]				
		PCSDs	Cyfluthrin	Sulcofuron	Fulcofuron	Permethrin
Fresh water						
Direct abstraction	A1(b)		0.001PT			0.01PT
to potable supply	A2(b)		0.001PT			0.01PT
Protection of sensitive aquatic life (e.g.salmonid fish) (c)		0.05PT	0.001PT	25PT	1.0PT	0.01PT
Protection of other aquatic life (e.g. cyprinid fish)		0.05PT	0.001PT	25PT	1.0PT	0.01PT
Salt water						
Protection of salt water life		0.05PT	0.001PT	25PT(j)	1.0PT(j)	0.01PT(j)

[*] WRC Report reference number.
All values given as µg/l. P, 95% of samples (d); T, total.

Notes

(a) These quality standards are set for the purpose of controlling discharges of dangerous substances under Directive 76/464/EEC. A number of other EC directives set down standards for some of the substances listed here in respect of particular uses of water. For the purposes of implementing these directives, certain provisions may apply (for example in relation to definitions, sampling frequency or possible derogations) which are not set out here, and authorities should consult those directives separately (see also paragraph 45 of this circular).

(b) These categories correspond to those defined in Directive 75/440/EEC. Waters in category A1 should be suitable for abstraction for drinking after simple physical treatment and disinfection; waters in category A2 for drinking after normal physical treatment, chemical treatment and disinfection.

(c) In some cases more stringent values may be appropriate locally to protect particularly sensitive flora or fauna (see appropriate WRC Report).

(d) Notwithstanding the advice in paragraph 45 of this circular, where the values shown for certain substances are applied as 90 percentiles under other EC directives, authorities may apply them here also on the basis that 90% of samples should comply. Otherwise 95% of samples should fall within the quality standard shown.

(e) Other standards may be applicable for other particular water uses, notably irrigation of crops and livestock watering. In some cases these are identified in the appropriate WRC report and in published advice from ADAS. In other cases authorities should consult ADAS as necessary.

(f) Where breeding populations of rainbow trout are present the quality standard for lead should be 50% of that recommended for sensitive aquatic life. The standards given also assume that the lead present is almost entirely inorganic; if a significant proportion of organic lead is present, more stringent standards may be necessary.

(g) Higher concentrations of copper may be acceptable where the presence of organic matter may lead to complexation.

(h) Certain crops are particularly sensitive to these substances and may require especially stringent standards for irrigation (see appropriate WRC report).

(i) The toxicity of iron increases at pHs below 7, and authorities may need to set more stringent quality standards, especially where pH is below 6.5.

(j) These standards may need to be reviewed when more adequate data become available.

(k) A more restricted range of 7.0–8.5 should be applied for the protection of shellfish.

(l) Further analytical development is likely to be needed before these standards could be verifiable in receiving waters. They can, however, be used in calculating acceptable concentrations in effluents.

Standards for various water uses

EC Directive 75/440 EEC (Chapter 3[42]): Characteristics of surface water intended for the abstraction of drinking water

	Parameters		A1 G	A1 I	A2 G	A2 I	A3 G	A3 I
1	pH		6.5–8.5		5.5–9		5.5–9	
2	Coloration (after simple filtration)	mg/l Pt scale	10	20 (O)	50	.100 (O)	50	200 (O)
3	Total suspended solids	mg/l SS	25					
4	Temperature	°C	22	25 (O)	22	25 (O)	22	25 (O)
5	Conductivity	$\mu s/cm^{-1}$ at 20°C	1000		1000		1000	
6	Odour	(dilution factor at 25°C)	3		10		20	
7*	Nitrates	mg/l NO_3	25	50 (O)		50 (O)		50 (O)
8¹	Fluorides	mg/l F	0.7–1	1.5	0.7–1.7		0.7–1.7	
9	Total extractable organic chlorine	mg/l Cl						
10*	Dissolved iron	mg/l Fe	0.1	0.3	1	2	1	
11*	Manganese	mg/l Mn	0.05		0.1		1	
12*	Copper	mg/l Cu	0.002	0.05 (O)	0.05		1	
13	Zinc	mg/l Zn	0.5	3	1	5	1	5
14	Boron	mg/l B	1		1		1	
15	Beryllium	mg/l Be						
16	Cobalt	mg/l Co						
17	Nickel	mg/l Ni						
18	Vanadium	mg/l V						
19	Arsenic	mg/l As	0.01	0.05		0.05	0.05	0.1
20	Cadmium	mg/l Cd	0.001	0.005	0.001	0.005	0.001	0.005
21	Total chromium	mg/l Cr		0.05		0.05		0.05
22	Lead	mg/l Pb		0.05		0.05		0.05
23	Selenium	mg/l Se		0.01		0.01		0.01
24	Mercury	mg/l Hg	0.0005	0.001	0.0005	0.001	0.0005	0.001
25	Barium	mg/l Ba		0.1		1		1
26	Cyanide	mg/l Cn		0.05		0.05		0.05

(Table cont.)

	Parameters		A1 G	A1 I	A2 G	A2 I	A3 G	A3 I
27	Sulphates	mg/l SO$_4$	150	250	150	250 (O)	150	250 (O)
28	Chlorides	mg/l Cl	200		200		200	
29	Surfactants (reacting with methyl blue)	mg/l (laurylsulphate)	0.2		0.2		0.5	
30*,2	Phosphates	mg/l P$_2$O$_3$	0.4		0.7		0.7	
31	Phenols (phenol index) paranitraniline 4 aminoantipyrine	mg/l C$_6$H$_5$OH		0.001	0.001	0.005	0.01	0.1
32	Dissolved or emulsified hydrocarbons (after extraction by petroleum ether)	mg/l		0.05		0.2	0.5	
33	Polycyclic aromatic hydrocarbons	mg/l		0.0002		0.0002		0.001
34	Total pesticides (parathion, BHC, dieldrin)	mg/l		0.001		0.0025		0.005
35*	Chemical oxygen demand (COD)	mg/l O$_2$					30	
36*	Dissolved oxygen saturation rate	% O$_2$	> 70		> 50		> 30	
37*	Biochemical oxygen demand (BOD) (at 20°C without nitrification)	mg/l O$_2$	< 3		< 5		< 7	
38	Nitrogen by Kjeldahl method (except NO$_3$)	mg/l N	1		2		3	
39	Ammonia	mg/l NH$_4$	0.05		1	1.5	2	4 (O)
40	Substances extractable with chloroform	mg/l SEC	0.1		0.2		0.5	
41	Total organic carbon	mg/l C						
42	Residual organic carbon after flocculation and membrane filtration (5 μm) TOC	mg/l C						
43	Total coliforms 37°C	/100 ml	50		5000		50 000	
44	Faecal coliforms	/100 ml	20		2000		20 000	
45	Faecal streptococci	/100 ml	20		1000		10 000	
46	Salmonella		Not present in 5000 ml		Not present in 1000 ml			

I, mandatory; G, guide; O, exceptional climatic or geographical conditions; *, see Article 8(d).

1 The values given are upper limits set in relation to the mean annual temperature (high and low).

2 This parameter has been included to satisfy the ecological requirements of certain types of environment.

Article 8(d). In the case of surface water in a shallow lake or virtually stagnant surface water, for parameters marked with an asterisk in the table in Annex II, this derogation being applicable only to lake with a depth not exceeding 20 m, with an exchange of water slower than 1 year, and without a discharge of waste water into the water body.

EC Directive 76/160 EEC (chapter 3 [43]): Quality requirements for bathing water

	Parameters	G	I	Minimum sampling frequency	Method of analysis and inspection
	Microbiological:				
1	Total coliforms /100 ml	500	10000	Fortnightly (1)	Fermentation in multiple tubes. Subculturing of the positive tubes on a
2	Faecal coliforms /100 ml	100	2000	Fortnightly (1)	confirmation medium. Count according to MPN (most probable number) or membrane filtration and culture on an appropriate medium such as Tergitol lactose agar, endo agar, 0.4% Teepol broth, subculturing and identification of the suspect colonies In the case of 1 and 2, the incubation temperature is variable according to whether total or faecal coliforms are being investigated
3	Faecal streptococci /100 ml	100	–	(2)	Litsky method. Count according to MPN (most probable number) or filtration on membrane. Culture on an appropriate medium
4	*Salmonella* /1 l	–	0	(2)	Concentration by membrane filtration. Inoculation on a standard medium. Enrichment, subculturing on isolating agar, identification
5	Entero viruses (PFU/10 l)	–	0	(2)	Concentrating by filtration, flocculation or centrifuging and confirmation

(Table cont.)

	Parameters	G	I	Minimum sampling frequency	Method of analysis and inspection
	Physico-chemical				
6	pH	–	6–9 (0)	(2)	Electrometry with calibration at pH 7 and 9
7	Colour	–	No abnormal change in colour (0)	Fortnightly (1)	Visual inspection or photometry with standards on the Pt.Co scale
		–	–	(2)	
8	Mineral oils (mg/l)	–	No film visible on the surface of the water and no odour	Fortnightly (1)	Visual and olfactory inspection extraction using an adequate volume and weighing the dry residue
		≤ 0.3	–	(2)	
9	Surface-active substances reacting with methylene blue (mg/litre, laurylsulfate)	–	No lasting foam	Fortnightly (1)	Visual inspection or absortion spectrophotometry with methylene blue
		≤ 0.3	–	(2)	
10	Phenols (phenol indices) (mg/l C_4H_5OH)	–	No specific odour	Fortnightly (1)	Verification of the absence of specific odour due to phenol or absorption spectrophotometry 4-aminoantipyrines (4 AAP) method
		≤ 0.005	≤ 0.05	(2)	
11	Transparency (m)	2	1 (0)	Fortnightly (1)	Secchi's disc
12	Dissolved oxygen (% saturation O_2)	80–120	–	(2)	Winkler's method or electrometric method (oxygen meter)
13	Tarry residues and floating materials such as wood, plastic articles, bottles, containers of glass, plastic, rubber or any other substance; waste or splinters	Absence		Fortnightly (1)	Visual inspection
14	Ammonia (mg/l NH_4)			(3)	Absorption spectrophotometry, Nessler method, or indophenol blue method
15	Nitrogen Kjeldahl (mg/l N)			(3)	Kjeldahl method

	Parameters	G	I	Minimum sampling frequency	Method of analysis and inspection
	Other substances regarded as indications of pollution				
16	Pesticides (parathion, HCH, dieldrin) mg/l			(2)	Extraction with appropriate solvents chromatographic determination
17	Heavy metals such as: Arsenic (mg/l As) Cadmium (mg/l Cd) Chrome VI (mg/l Cr^{VI}) Lead (mg/l Pb) Mercury (mg/l Hg)			(2)	Atomic absorption possibly preceded by extraction
18	Cyanides (mg/l Cn)			(2)	Absorption spectrophometry using a specific reagent
19	Nitrates (mg/l NO_3) Phosphates (mg/l PO_4)			(2)	Absorption spectrophotometry using a specific reagent

G, guide; I, mandatory

(0) Provision exists for exceeding the limits in the event of exceptional geographical or meteorological conditions.

(1) When a sampling taken in previous years produced results which are appreciably better than those in this Annex and when no new factor likely to lower the quality of the water has appeared, the competent authorities may reduce the sampling frequency by a factor of 2.

(2) Concentration to be checked by the competent authorities when an inspection in the bathing area shows that the substance may be present or that the quality of the water has deteriorated.

(3) These parameters must be checked by the competent authorities when there is a tendency towards the eutrophication of the water.

EC 78/659 EEC (chapter 3 [44]): Quality of fresh water to support fish life

Annex I List of parameters

Parameter	Salmonid waters		Cyprinid waters		Method of analysis or inspection	Minimum sampling and measuring frequency	Observations
	G	I	G	I			
1. Temperature (°C)	1. Temperature measured downstream of a point of thermal discharge (at the edge of the mixing zone) must not exceed the unaffected temperature by more than				Thermometry	Weekly, both upstream and downstream of the point of thermal discharge	Over-sudden variations in temperature shall be avoided
		1.5°C		3°C			
	Derogations limited in geographical scope may be decided by Member States in particular conditions if the competent authority can prove that there are no harmful consequences for the balanced development of the fish population						
	2. Thermal discharges must not cause the temperature downstream of the point of thermal discharge (at the edge of the mixing zone) to exceed the following:						
	21.5 (0)		28 (0)				
	10 (0)		10 (0)				
	The 10°C temperature limit applies only to breeding periods of species which need cold water for reproduction and only to waters which may contain such species						
	Temperature limits may, however, be exceeded for 2% of the time						

2. Dissolved oxygen (mg/l O₂)	50% ≥ 9 100% ≥ 7	50% ≥ > 9 When the oxygen concentration falls below 6 mg/l, Member States shall implement the provisions of Article 7 (3). The competent authority must prove that this situation will have no harmful consequences for the balanced development of the fish population	50% ≥ 8 100% ≥ 5	50% ≥ 7 When the oxygen concentration falls below 4 mg/l, Member States shall implement the provisions of Article 7 (3). The competent authority must prove that this situation will have no harmful consequences for the balanced development of the fish population	Winkler's method or specific electrodes (electrochemical method)	Monthly, minimum one sample representative of low oxygen conditions of the day of sampling However, where major daily variations are suspected, a minimum of two samples in one day shall be taken
3. pH	6–9 (0) [1]		6–9 (0) [1]		Electrometry calibration by means of two solutions with known pH values, preferably on either side of, and close to the pH being measured	Monthly

Parameter	Salmonid waters		Cyprinid waters		Method of analysis or inspection	Minimum sampling and measuring frequency	Observations
	G	I	G	I			
4. Suspended solids (mg/l)	≤ 25 (0)		≤ 25 (0)		Filtration through a 0.45 μm filtering membrane, or centrifugation (five minutes minimum, average acceleration of 2800 to 3200 g) drying at 105°C and weighing		The values shown are average concentrations and do not apply to suspended solids with harmful chemical properties Floods are liable to cause particularly high concentrations
5. BOD$_5$ (mg/l O$_2$)	≤ 3		≤ 6		Determination of O$_2$ by the Winkler method before and after 5 days incubation in complete darkness at 20 ± 1°C (nitrification should not be inhibited)		

6. Total phosphorus (mg/l P)		Molecular absorption spectrophotometry

In the case of lakes of average depth between 18 and 300 m, the following formula could be applied:

$$L \leq 10 \frac{\bar{Z}}{T_w} (1 + \sqrt{T_w})$$

where L is the loading expressed as mg P per m^2 of lake surface in 1 year, $\bar{Z}$ is the mean depth of lake (m) and, T_w is the theoretical renewal time of lake water (years)

In other cases limit values of 0.2 mg/l for salmonid and of 0.4 mg/l for cyprinid waters, expressed as PO_4, may be regarded as indicative in order to reduce eutrophication

7. Nitrites (mg/l NO_2)	$\leqslant 0.01$	$\leqslant 0.03$	Molecular absorption spectrophotometry

Parameter	Salmonid waters		Cyprinid waters		Method of analysis or inspection	Minimum sampling and measuring frequency	Observations
	G	I	G	I			
8. Phenolic compounds (mg/l C_6H_5OH)		[2]		[2]	By taste		An examination by taste shall be made only where the presence of phenolic compounds is presumed
9. Petroleum hydrocarbons		[3]		[3]	Visual By taste	Monthly	A visual examination shall be made regularly once a month, with an examination by taste only where the presence of hydrocarbons is presumed
10. Non-ionized ammonia (mg/l NH_3)	≤ 0.005	≤ 0.025	≤ 0.005	≤ 0.025	Molecular absorption spectrophotometry using indophenol blue or Nessler's method associated with pH and temperature determination	Monthly	Values for non-ionized ammonia may be exceeded in the form of minor peaks in the daytime

In order to diminish the risk of toxicity due to the non-ionized ammonia, of oxygen consumption due to nitrification and of eutrophication, the concentrations of total ammonium should not exceed the following:

Parameter	Salmonid waters		Cyprinid waters		Method of analysis or inspection	Minimum sampling and measuring frequency	Observations
	G	I	G	I			
11. Total ammonium (mg/l NH_4)	≤ 0.04	≤ 1 [4]	≤ 0.2	≤ 1 [4]			

No. Parameter	G	I	Method of analysis or inspection	Minimum sampling and measuring frequency	Observations
12. Total residual chlorine (mg/l HOCl)		≤ 0.005	DPD-method (diethyl-*p*-phenylenediamene)	Monthly	The I-values correspond to pH = 6; higher concentrations of total chlorine can be accepted if the pH is higher
13. Total zinc (mg/l Zn)	≤ 0.3	≤ 1.0	Atomic absorption spectrometry	Monthly	The I-values correspond to a water hardness of 100 mg/l CaCO₃; for hardness levels between 10 and 500 mg/l corresponding limit values can be found in Annex II
14. Dissolved copper (mg/l Cu)	≤ 0.04	≤ 0.04	Atomic absorption spectrometry		The G-values correspond to a water hardness of 100 mg/l CaCO₃; for hardness levels between 10 and 300 mg/l corresponding limit values can be found in Annex II

¹ Artificial pH variations with respect to the unaffected values shall not exceed ± 0.5 of a pH unit within the limits falling between 6.0 and 9.0 provided that these variations do not increase the harmfulness of other substances present in the water.

² Phenolic compounds must not be present in such concentrations that they adversely affect fish flavour.

³ Petroleum products must not be present in water in such quantities that they: form a visible film on the surface of the water or form coatings on the beds of water-courses and lakes; impart a detectable 'hydrocarbon' taste to fish; produce harmful effects in fish.

⁴ In particular geographical or climatic conditions, and particularly in cases of low water temperature and of reduced nitrification or where competent authority can prove that there are no harmful consequences for the balanced development of the fish population, Member States may fix values higher than 1 mg/l.

General observation: It should be noted that the parametric values listed in this Annex assume that the other parameters, whether mentioned in this Annex or not, are favourable. This implies, in particular, that the concentrations of other harmful substances are very low. Where two or more harmful substances are present in mixture, joint effects (additive, synergic or antagonistic effects) may be significant. G, guide; I, mandatory; (O), derogations are possible in accordance with Article 11.

Annex II: Particulars regarding total zinc and dissolved copper

Zinc concentrations [1] (mg/l Zn) for different water hardness values between 10 and 500 mg/l $CaCO_3$

	Water hardness (mg/l $CaCo_3$)			
	10	50	100	500
Salmonid waters (mg/l Zn)	0.03	0.2	0.3	0.5
Cyprinid waters (mg/l Zn)	0.3	0.7	1.0	2.0

[1] See Annex I, No. 13, Observations column.

Dissolved copper [1] concentrations (mg/l Cu) for different water hardness values between 10 and 300 mg/l $CaCO_3$

	Water hardness (mg/l $CaCo_3$)			
	10	50	100	300
mg/l Cu	0.005 (2)	0.022	0.04	0.112

[1] See Annex I, No. 14, observations column.

[2] The presence of fish in waters containing higher concentrations of copper may indicate a predominance of dissolved organo-cupric complexes.

SI 1989/2286 (chapter 3 [45]): The Surface Waters (Dangerous Substances) (Classification) Regulations

Schedule 1: Regulation 3(1)

Classification of inland waters (DSI)

Substance	Concentration (μg/l) (annual mean)
Aldrin, dieldrin, endrin and isodrin	(i) 0.03 for the four substances in total (ii) 0.005 for endrin
Cadmium and its compounds	5 (total cadmium, both soluble and insoluble forms)
Carbon tetrachloride	12
Chloroform	12
DDT (all isomers)	0.025
para-para-DDT	0.01
Hexachlorobenzene	0.03
Hexachlorobutadiene	0.1
Hexachlorocyclohexane (all isomers)	0.1
Mercury and its compounds	1 (total mercury, both soluble and insoluble forms)
Pentachlorophenol and its compounds	2

Schedule 2: Regulation 3(2)

Classification of coastal waters and relevant territorial waters (DS2)

Substance	Concentration (μg/l) (annual mean)
Aldrin, dieldrin, endrin and isodrin	(i) 0.03 for the four substances in total (ii) 0.005 for endrin
Cadmium and its compounds	2.5 (dissolved cadmium)
Carbon tetrachloride	12
Chloroform	12
DDT (all isomers)	0.025
para-para-DDT	0.01
Hexachlorobenzene	0.03
Hexachlorobutadiene	0.1
Hexachlorocyclohexane (all isomers)	0.02
Mercury and its compounds	0.3 (dissolved mercury)
Pentachlorophenol and its compounds	2

SI 1989/1148 (chapter 3 [46]): The Surface Waters (Classification) Regulations

Schedule: Regulation 3

Criteria for the classification of waters (the limits set out below are maxima)

No. in Annex II to 75/440/EEC	Parameters		DW1	DW2	DW3
2	Coloration (after simple filtration)	mg/l Pt scale	20	100	200
4	Temperature	°C	25	25	25
7	Nitrates	mg/l NO_3	50	50	50
8(1)	Fluorides	mg/l F	1.5		
10	Dissolved iron	mg/l Fe	0.3	2	
12	Copper	mg/l Cu	0.05		
13	Zinc	mg/l Zn	3	5	5
19	Arsenic	mg/l As	0.05	0.05	0.1
20	Cadmium	mg/l Cd	0.005	0.005	0.005
21	Total chromium	mg/l Cr	0.05	0.05	0.05
22	Lead	mg/l Pb	0.05	0.05	0.05
23	Selenium	mg/l Se	0.01	0.01	0.01
24	Mercury	mg/l Hg	0.001	0.001	0.001
25	Barium	mg/l Ba	0.1	1	1
26	Cyanide	mg/l Cn	0.05	0.05	0.05
27	Sulphates	mg/l SO_4	250	250	250

No. in Annex II to 75/440/EEC	Parameters		DW1	DW2	DW3
31	Phenols (phenol index) paranitraniline 4-aminoantipyrine	mg/l C_6H_5OH	0.001	0.005	0.1
32	Dissolved or emulsified hydrocarbons (after extraction by petroleum ether)	mg/l	0.05	0.2	1
33	Polycyclic aromatic hydrocarbons	mg/l	0.0002	0.0002	0.001
34	Total pesticides (parathion, BHC, dieldrin)	mg/l	0.001	0.0025	0.005
39	Ammonia	mg/l NH_4		1.5	4

Note: (1) The value given is an upper limit set in relation to the mean annual temperature (high and low).

Explanatory note (this note is not part of the Regulations). These Regulations prescribe the system of classifying the quality of inland waters (as defined in Section 103(1)(c) of the Water Act 1989) according to their suitability for abstraction by water undertakers for supply (after treatment) as drinking water.

The classifications DW1, DW2 and DW3 reflect the mandatory values assigned by Annex II to Council Directive 75/440/EEC (OJ No. L 194, 25.7.75, p. 26) (concerning the quality required of surface water intended for the abstraction of drinking water) to the parameters listed in the Schedule to the Regulations.

The classifications are relevant for the purposes of setting water quality objectives for rivers, lakes and other inland waters under Section 105 of the Act and for ascertaining the treatment to which the water is to be subjected before it is supplied for public use, in accordance with Part VI of the Water Supply (Water Quality) Regulations 1989 (S.I. 1989/1147).

Standards proposed for classifying wastes

C190/33 EC (chapter 3 [30]): Proposal for a Council Directive on the Landfill of Waste Control Criteria for Assessing Leachate Potential.

Treatment of the samples. The original structure of the sample used should be maintained as far as possible; large parts should be crushed. The proposed analytical method is DIN 38414-S4 (October 1984 issue) with the following additions and/or simplifications:

A wide-necked glass bottle (10 cm diameter) should be used
Shake, rotating bottle by 180° once/min for 24 h
Centrifuge, 250 µl filter syringes with 0.45 µm filters should be used for sampling

Assignment values. This table fixes the ranges by which wastes will be characterized for the purpose of landfilling according to the composition of their eluates:

> Wastes whose eluate concentration is in the range fixed for hazardous wastes will be considered as such with respect to landfilling; for eluate concentrations higher than the maximum values fixed, hazardous wastes will have to be treated prior to landfill, unless compatible for joint disposal with municipal waste, or, if treatment is not possible, destinated to a mono-landfill
>
> Wastes whose eluate concentration is not above the maximum values fixed for inert wastes will be considered as such
>
> Wastes whose eluate concentration falls in the range between inert wastes and the minimum value for hazardous wastes will be considered non-hazardous

		Hazardous waste range	Inert waste
1.01	pH value	4–13	4–13
1.02	TOC	40–200 mg/l	< 200 mg/l
1.03	Arsenic	0.2–1.0 mg/l	< 0.1 mg/l
1.04	Lead	0.4–2.0 mg/l	
1.05	Cadmium	0.1–0.5 mg/l	
1.06	Chromium	0.1–0.5 mg/l	
1.07	Copper	2–10 mg/l	The total of these metals: < 5 mh/l[1]
1.08	Nickel	0.4–2.0 mg/l	
1.09	Mercury	0.02–0.1 mg/l	
1.10	Zinc	2–10 mg/l	
1.11	Phenols	20–100 mg/l	< 10 mg/l
1.12	Fluoride	10–50 mg/l	< 5 mg/l
1.13	Ammonium	0.2–1.0 mg/l	< 50 mg/l
1.14	Chloride	1.2–6.0 mg/l	< 0.5 g/l
1.15	Cyanide [2]	0.2–1.0 mg/l	< 0.1 g/l
1.16	Sulphate [3]	0.2–1.0 g/l	< 1.0 g/l
1.17	Nitrite	6–30 mg/l	< 3 mg/l
1.18	AOX [4]	0.6–3.0 mg/l	< 0.3 mg/l
1.19	Solvents [5]	0.02–0.10 mgCl/l	< 10 µg Cl/l
1.20	Pesticides [5]	1–5 µg Cl/l	< 0.5 µg Cl/l
1.21	Lipophilic substances	0.4–2.0 mg/l	< 1 mg/l

[1] And no single value above the minimum fixed for hazardous water.

[2] Readily released.

[3] If possible < 500 mg/l.

[4] Adsorbed organically bound halogens.

[5] Chlorinated.

Notes

(1) For characterization purposes the components to be analysed in the eluates shall be chosen in function of the qualitative composition of the waste.

(2) In addition to these eluate criteria, a determination of asbestos on a representative sample of the crude inert waste shall be performed, according to the

annexes of the Council Directive 87/217/EEC on the prevention and reduction of environmental pollution by asbestos.

Analytical methods. The following ISO or DIN methods are proposed as reference methods. Any equivalent method after a certification procedure based on the use of a certified reference material will be accepted. In case of discrepancy of the results the proposed methods will be used as reference.

1.01	pH	ISO-DP 10 523 or DIN 38404-C5-84
1.02	TOC in eluate	DIN 38409-H3-85
1.03	Arsenic	ISO 6595-1982 or DIN 38405-E6-81
1.04	Lead	ISO 8288-1985 or DIN 38406-E6-81
1.05	Cadmium	ISO 8288-1985 or DIN 38406-E19-80
1.06	Chromium(VI)	ISO-DIS 9174-88 or DIN 38405-D24-87
1.07	Copper	ISO 8288-1985 or DIN 38406-E21-80
1.08	Nickel	ISO 8288-1985 or DIN 38406-E21-80
1.09	Mercury	ISO 5666-1/3-88 or DIN 38406-E12-80
1.10	Zinc	ISO 8288-1985 or DIN 3840-E8-85
1.11	Phenols	ISO 6439-1990 or DIN 38409-H16-84
1.12	Fluoride	ISO-DP 359-1 or DIN 38406-D4-85
1.13	Ammonium	ISO 7150-1983 or DIN 38406-E5-83
1.14	Chloride	ISO-DIS 9297 or DIN 38405-D1-85
1.15	Cyanide	DIN 38405-D14-88
1.16	Sulphate	ISO-DIS 9280-1 or DIN 38405-D5-85
1.17	Nitrite	ISO 6777-1983 or DIN 38405-D10-81
1.18	AOX	ISO-DIS 9562 or DIN 38409-H14-85
1.19	Chlorinated solvents [1]	ISO-DP 10 301 or GC head-space
1.20	Chlorinated pesticides [2]	GC (capillary column)
1.21	Extractible lip. substances [3]	cf. param. 27, Directive 80/778/EEC

[1] Needs 2 ml of eluate.
[2] After extraction of 1 l of eluate.
[3] Needs 250 ml of eluate; chloroform extract, results in 'dry residue' (mg/l).

SI 1989/1156 (chapter 3 [47]): Trade Effluents (Prescribed Processes and Substances) Regulations 1989

Section 74 of the 1989 Act (control of exercise of trade effluent functions in certain cases) shall apply to trade effluent in which any of the substances listed in Schedule 1 to these Regulations is present in a concentration greater than the background concentration.

Schedule 1: Regulation 3

Prescribed substances

Mercury and its compounds
Cadmium and its compounds
γ-Hexachlorocyclohexane

(Table cont)

Prescribed substances
DDT
Pentachlorophenol
Hexachlorobenzene
Hexachlorobutadiene
Aldrin
Dieldrin
Endrin
Carbon tetrachloride
Polychlorinated biphenyls
Dichlorvos
1,2-Dichloroethane
Trichlorobenzene
Atrazine
Simazine
Tributyltin compounds
Triphenyltin compounds
Trifluralin
Fenitrothion
Azinphos-methyl
Malathion
Endosulfan

Salient guidance on landfill gas concentrations

Waste Management Paper No. 27 (chapter 3 [48]): references to gas concentrations

Clause	
1.17 (a) and 7.9	Monitoring of tip to continue until flammable gas $< 1\%$, $CO_2 < 1.5\%$ for 24 months
3.3	Flammability range: methane 5–15%; hydrogen 4–74%
3.4	Asphyxiation: no one to enter confined space where oxygen $< 18\%$ by vol. Carbon dioxide: short-term exposure limit 1.5% by vol (over 10 min); occupational exposure standard (over 8 h) 0.5% by vol.
3.5	Toxicity: hydrogen sulphide: short-term exposure limit 15 ppm (over 10 min.); occupational exposure standard 10 ppm (over 8 h)
6.19	Natural concentrations of carbon dioxide up to 2 m deep may occur up to 7% by vol.
7.21	Trigger values in buildings: evacuation requirement when flammable gas $> 1\%$ by vol. and/or $> 20\%$ of LEL; carbon dioxide $> 1.5\%$ by vol.
C.16.2	Ventilation of rooms required when: methane $> 0.25\%$ by vol. and/or $> 5\%$ of LEL; carbon dioxide $> 0.5\%$ by vol.

Appendix II: Waste disposal regulations

This appendix comprises Schedules 1, 3, 4 and 6 of the Collection and Disposal of Waste Regulations 1988 SI 1988 No. 819.

Schedule 1: Regulation 3

Waste to be treated as household waste

(1) Waste from premises which are not a private dwelling for the purposes of the General Rate Act 1967(a) by virtue of paragraph 2 of Schedule 13 to that Act (rooms let singly for residential purposes)

(2) Waste from a garage or store used wholly in connection with a private dwelling

(3) Waste from premises occupied by a religious community and used wholly for the purposes of human habitation

(4) Waste from premises exempted from liability to be rated by virtue of Section 39 of the General Rate Act 1967 (relief from rates for places of religious worship)

(5) Waste from premises given relief from the payment of rates by virtue of Section 40(1)(a) of the General Rate Act 1967 (relief from rates for charitable organizations)

(6) Waste from a camp site

(7) Waste from a residential hostel

(8) Waste from a prison or other penal institution

(9) Waste from a hall or other premises used wholly or mainly for public meetings

(10) Waste from a royal palace

Schedule 3: Regulation 6

Waste to be treated as industrial waste

(1) Waste from premises used for maintaining vehicles, vessels or aircraft not being part of, or whose use is incidental to, a private dwelling

(2) Waste from a laboratory

(3) (1) Waste from a workshop or similar premises not being a factory within the meaning of Section 175 of the Factories Act 1961(c) because the people working there are not employees or because the work there is not carried on by way of trade or for purposes of gain

(2) In this paragraph, 'workshop' does not include premises at which the principal activities are computer operations or the copying of documents by photographic or lithographic means

(4) Waste from premises occupied by a scientific research association approved by the Secretary of State under Section 362 of the Income and Corporation Taxes Act 1970(d)

(5) Waste from dredging operations

(6) Waste arising from works of construction or demolition, including waste arising from work preparatory thereto

(7) Waste arising from tunnelling or from any other excavation

(8) Sewage or sewage sludge deposited on land other than:
   (a) Sewage or sewage sludge deposited within the curtilage of a sewage treatment works as an integral part of the operation of those works or
   (b) Sewage sludge deposited directly onto land for agricultural purposes

(9) Clinical waste other than that from a private dwelling or residential home

(10) Waste arising from any aircraft, vehicle or vessel which is not occupied as a private dwelling

(11) Waste which has previously formed part of any aircraft, vehicle or vessel and which is not household waste

(12) Waste removed from land on which it has previously been deposited and any soil with which such waste has been in contact

(13) Leachate from a deposit of waste

(14) Poisonous or noxious waste arising from any of the following processes undertaken on premises used for the purposes of a trade or business:
   (a) Mixing or selling paints
   (b) Sign writing
   (c) Laundering or dry cleaning
   (d) Developing photographic film or making photographic prints
   (e) Selling petrol, diesel fuel, paraffin, kerosene, heating oil or similar substances or
   (f) Selling pesticides, herbicides or fungicides

(15) Waste from premises used for the purposes of breeding, boarding, stabling or exhibiting animals

(16) Waste imported into Great Britain for disposal, treatment or re-export

(17) Waste oil or waste solvent other than from a private dwelling or residential home

(18) Waste which is scrap metal within the meaning of Section 9 of the Scrap Metal Dealers Act 1964(a) other than that from a private dwelling or residential home.

Schedule 4: Regulation 7

Waste to be treated as commercial waste

(1) Waste from an office or showroom

(2) (1) Waste from premises providing facilities for passengers at an airport, hoverport, seaport, railway station or bus station

(2) 'Hoverport' has the same meaning as in Section 4(1) of the Hovercraft Act 1968(b)

(3) Waste from a mixed hereditament within the meaning of Section 48(5) of the General Rate Act 1967

(4) Waste from premises occupied by a club, society or any association of persons (whether incorporated or not) in which activities are conducted for the benefit of the members

(5) Waste from premises (not being premises from which waste is by virtue of the Act or of any other provision of these Regulations to be treated as household waste or industrial waste) occupied by:

    (a) A court

    (b) A government department

    (c) A local authority

    (d) A body corporate or an individual appointed by or under any enactment to discharge any public functions, other than a body corporate established by or under any enactment for the purpose of carrying on under national ownership any industry or part of an industry or any undertaking or

    (e) A body incorporated by a Royal Charter

(6) Waste from a tent pitched on land other than a camp site

(7) Waste from a market or fair

Schedule 6: Regulation 9

Cases in which a disposal licence is not required

The deposit of waste on land

(1) The deposit of effluent or other waste matter in accordance with a consent given under Section 34

(2) The deposit of waste in accordance with a licence issued under Part II of the Food and Environment Protection Act 1985(a)

(3) The deposit of waste specified by an order under Section 7 of the Food and Environment Protection Act 1985 as an operation which does not need a licence under Part II of that Act

(4) (1) Subject to sub-paragraph (3), the deposit on land of wastes of the descriptions set out in paragraphs 6 and 7 of Schedule 3, or of ash, slag or clinker, provided that the deposit is made for the purposes of construction currently being undertaken on the land on which the waste is deposited

    (2) Subject to sub-paragraph (3), the deposit on land, for a period not exceeding three months, of wastes of the descriptions mentioned in sub-paragraph (1), provided that the deposit is made for the purposes of future construction on the land on which the waste is deposited

    (3) The deposit is made by, or with the consent of, the lawful occupier of the land

(5) The deposit of excavated material arising from peatworking

(6) (1) Subject to sub-paragraph (2), the deposit of excavated material arising from a borehole or an excavation which is:

      (a) Made in order to ascertain the presence, extent or quality of a deposit of a mineral other than petroleum, with a view to exploiting that mineral or

      (b) Made by the British Coal Corporation in the course of prospecting for coal workable by opencast methods or

      (c) Made by a water authority in accordance with that authority's functions, other than waste from dredging operations

    (2) The borehole or excavation is a development permitted by an order under Section 24 of the Town and Country Planning Act 1971(a), and the deposit complies with such conditions or limitations as may be specified in that order

(7) The deposit of spent railway ballast on land which is operational land (within the meaning of Section 222 of the Town and Country Planning Act 1971) of a railway undertaking

(8) (1) Subject to sub-paragraphs (2) and (3), the deposit of waste from dredging operations of any inland water within the meaning of Section 135 of the Water Resources Act 1963(b)

    (2) The deposit is made along the banks of the inland water from which the waste is dredged and is made as operations proceed

    (3) The waste is not deposited in a lagoon or container

(9) (1) Subject to sub-paragraph (2), the deposit of waste vegetable matter or waste soil in any park, sports ground, public garden or other recreation ground or any churchyard or cemetery

    (2) The deposit is made within the boundaries of the land on which the waste is produced and is made by, or with the consent of, the lawful occupier of that land

(10) (1) The deposit, in a secure lagoon or container, on land used for agriculture, of sewage sludge intended to be deposited directly onto land for agricultural purposes

    (2) A secure lagoon or container is one designed or adapted so that, as far as is practicable, waste cannot escape from it, and members of the public cannot have access to the waste contained within it

(11) The deposit of sewage from a sanitary convenience forming part of a passenger carrying rail vehicle

(12) The deposit by burial of sewage from a removable receptacle forming part of a sanitary convenience serving persons on premises other than a private dwelling.

(13) (1) Subject to sub-paragraphs (2) and (3), the deposit:

      (a) Of sewage sludge on land for the purpose of fertilizing or otherwise beneficially conditioning that land or

      (b) Of any waste, on land used for agricultural purposes, for the purpose of fertilizing or otherwise beneficially conditioning that land

    (2) The waste is deposited directly onto the land and not in a lagoon or container

    (3) The person depositing the waste shall furnish particulars to the disposal authority in whose area the deposit is to be made as follows:

      (a) Where there is to be a single deposit of waste, he shall furnish the following particulars in advance of making the deposit:

      (i) His name, telephone number and address

      (ii) A description of the waste, including the process from which it arises

      (iii) An estimate of the quantity of the waste and

      (iv) The location and intended date of the deposit

    (b) Where there are to be regular or frequent deposits of wastes of a similar composition he shall furnish the following particulars every six months:

      (i) His name, telephone number and address

      (ii) A description of the waste, including the process from which it arises

      (iii) An estimate of the total quantity of waste he intends to deposit during the next six months and

      (iv) The locations and frequency of the deposits

    and he may deposit wastes of a different description from that notified provided that he furnishes amended particulars in advance of making the deposit

(14) The deposit of waste on the premises on which it is produced, pending its disposal elsewhere

(15) (1) The deposit, on the premises on which it is produced, of special waste of the following descriptions, pending its disposal elsewhere:

    (a) Liquid waste of a total volume of not more than 23 000 l deposited in a secure container or containers and either

    (b) Non-liquid waste of a total volume of not more than 80 m^3 deposited in a secure container or containers or

    (c) Non-liquid waste of a total volume of not more than 50 m^3 deposited in a secure place or places

  (2) A secure container or place is one designed or adapted so that, as far as is practicable, waste cannot escape from it, and members of the public cannot have access to the waste contained within it

(16) (1) Subject to sub-paragraphs (2), (3) and (4), the deposit, outside the premises on which it is produced, of non-liquid waste of a total volume of not more than 50 m^3 in a container or containers so designed or adapted that, as far as is practicable, waste cannot escape from it or from them

  (2) The deposit is made for a period not exceeding 28 days

  (3) The deposit is made by, or with the consent of, the owner of the container

  (4) The deposit is not at a site designed or adapted for the reception of waste with a view to its being disposed of elsewhere

(17) The deposit of waste paper or rags pending disposal elsewhere, provided that the deposit is made by, or with the consent of, the lawful occupier of the land

The use of plant or equipment for the purpose of disposing of waste

(18) The disposal of waste as an integral part of the industrial process that produces it

The use of plant or equipment for the purpose of dealing with waste in a manner prescribed by Regulation 8

(19) Baling, compacting, pulverizing or sorting waste on the premises on which it is produced

(20) Baling, sorting or shredding waste paper or rags
(21) Storing waste of the descriptions set out in paragraphs 14 and 15 on the premises on which it is produced
(22) Incinerating waste, which is not special waste, on the premises on which it is produced by means of plant with a disposal capacity of not more than 200 kg/h
(23) Using waste oil as fuel to produce heat

References

Chapter 1

1. Evelyn, J. (1661) *Fumifugium: or the inconvenience of the air and smoke of London*.
2. Trevelyan, G.M. (1952) *Illustrated English Social History*, Vol. 4, Longmans, Green, London.
3. Carson, R. (1963) *Silent Spring*. H. Hamilton.
4. Rorsch, A. (1985) Foreword, in: *Proceedings of the International TNO Conference on Contaminated Soils*, eds. E.W. Assink and W.J. Van Der Brink, Martinus Nijhoff, Dordrecht.
5. Beckett, M.J. and Simms, D.L. (1984) The development of contaminated land, in: *Conference on Hazardous Waste Disposal and the Re-use of Contaminated Land*, SCI, London.
6. Failey, R.A. and Bell, R.M. (1991) Behaviour of pollutants in soils, in: *Land Reclamation — An End to Dereliction?*, Elsevier Applied Science, London, pp. 215–227.
7. Thornton, I. (1985) Metal contamination of soils in UK urban gardens, implications to health, in: *First international TNO Conference on Contaminated Soils*, Martinus Nijhoff, Dordrecht, pp. 203–210.
8. van Wijnen, J.H. (1985) Health risk assessment, population survey and contaminated soil, in: *First International TNO Conference on Contaminated Soils*, Martinus Nijhoff, Dordrecht, pp. 181–190.
9. Denner, J. (1992) Contaminated land: development of policy in the UK, in: *Contaminated Land — Policy, Regulation and Technology*, IBC Technical Services, London.
10. Soczo, E. and Meeder, T. (1992) Clean-up of contaminated sites in Europe and the USA — a comparison, in: *Eureco '92* (European Urban Regeneration Conference), Birmingham, UK.
11. de Bruijn, P. (1992) Biotreatment in soil remediation, in: *Contaminated Land — Policy, Regulation and Technology*, IBC Technical Services, London.
12. Edelman, T. and de Bruin, M. (1985) Background values of 32 elements in Dutch topsoils, in: *First International TNO Conference on Contaminated Soils*, Martinus Nijhoff, Dordrecht, pp. 89–99.
13. Leonard, M. and Privett, K. (1991) Environmental assessments of reclaimed land in the USA, in: *Land Reclamation — An End to Dereliction?*, Elsevier Applied Science, London, pp. 235–240.
14. Harris, M.R. (1987) Recognition of the problem, in: *Reclaiming Contaminated Land*, ed. T. Cairney, Blackie, Glasgow.
15. Swanson, A.E. (1992) Legal considerations and liabilities, in: *Contaminated Land — Policy, Regulation and Technology*, IBC Technical Services, London.
16. Sanning, D.E. (1992) The NATO/CCMS Pilot Study Program, in: *Contaminated Land — Policy, Regulation and Technology*, IBC Technical Services, London.

Chapter 2

1. Royal Commission on Environmental Pollution (1984) *10th Report: Tackling Pollution — Experience and Prospects*, Cmnd 9194, HMSO, London.
2. House of Commons Select Committee on the Environment (1990) *First Report on Contaminated Land*, 170-I, in three volumes, HMSO, London.
3. Department of the Environment (1990) *Contaminated Land, the Government's Response to the First Report from the House of Commons Select Committee on the Environment*, Cm. 1161, HMSO, London.
4. *Environmental Protection Act (1990)* HMSO, London.

5. Department of the Environment and the Welsh Office (1991) *Public Registers of Land which may be Contaminated — A Consultation Paper.*
6. Department of the Environment (1989) *A Review of Derelict Land Policy*, HMSO, London.
7. British Standards Institution (1988) Draft for Development, *DD175: Code of Practice for the Identification of Chemically Contaminated Land and its Investigation*, BSI, London.

Chapter 3

1. British Standards Institution (1988) Draft for Development, *DD175: 1988 Code of Practice for the Identification of Potentially Contaminated Land and its Investigation*, BSI, London.
2. Interdepartmental Committee on the Redevelopment of Contaminated Land (ICRCL) (1987) Guidance Note 59/83, *Guidance on the Assessment and Redevelopment of Contaminated Land*, Department of the Environment.
3. ICRCL (1990) Guidance Note 70/90, *Notes on the Restoration and Aftercare of Metalliferous Mining Sites for Pastures and Grazing*, Department of the Environment.
4. *Environmental Protection Act (1990)* HMSO, London.
5. Hobson, D.M. (1991) Planning for reclamation, in: *Land Reclamation. An End to Dereliction*, ed. M.C.R. Davies, Elsevier Applied Science, London, pp. 75–81.
6. Ground Board of the Institution of Civil Engineers (1991) *Inadequate Site Investigation*, Thomas Telford, London.
7. Department of the Environment (1991) *Public Registers of Land which may be Contaminated. A Consultation Paper.*
8. ICRCL (1978) Guidance Note 17/78, *Notes on the Redevelopment of Landfill Sites*, Department of the Environment.
9. ICRCL (1979) Guidance Note 18/79, *Notes on the Redevelopment of Gaswork Sites*, Department of the Environment.
10. ICRCL (1979) Guidance Note 23/79, *Notes on the Redevelopment of Sewage Works and Forms*, Department of the Environment.
11. ICRCL (1980) Guidance Note 42/80, *Notes on the Redevelopment of Scrap Yards and Similar Sites*, Department of the Environment.
12. Environmental Resources Ltd (1987) *Problems Arising from the Redevelopment of Gas Works and Similar Sites*, 2nd edition, Department of the Environment.
13. Bell, R.M., Gildon, A. and Parry, G.D.R. (1983) Sampling strategy and data interpretation for site investigation of contaminated land, in: *Reclamation of Former Iron and Steelworks Sites*, G.P. Doubleday, Durham.
14. Smith, M.A. and Ellis, A.C. (1986) An investigation into methods used to assess gas works sites for reclamation, *Reclamation and Revegetation Research*, **4**, 183–209.
15. Thornton, I. (1980) Background levels of heavy metals in soils and plants, in: *Reclamation of Contaminated Land* C5/1–C5/12, Society of Chemical Industry.
16. Berrow, M.C. and Burridge, J.C. (1980) Inorganic pollution and agriculture, in: *Ministry of Agriculture Fisheries and Food Reference Book 326*, HMSO, London, pp. 159–183.
17. British Standards Institution (1990) BS1377, *Methods of Test for Soils for Civil Engineering Purposes*, BSI, London.
18. British Standards Institution (1983) BS6068, *Water Quality — Part 2. Physical, Chemical and Biochemical Methods*, BSI, London.
19. British Standards Institution (1977–1984) BS1016, *Methods for the Analysis and Testing of Coal and Coke*, BSI, London.
20. Health and Safety Executive *Methods for the Determination of Hazardous Substances*, MDHS Series.
21. Agriculture, Development and Advisory Service (ADAS) (1986) *The Analysis of Agricultural Materials. A Manual of the Analytical Methods used by ADAS*, HMSO, London.
22. *Methods for the Examination of Waters and Associated Materials*, HMSO, London.
23. United States Environmental Protection Agency (1986) *Test Methods for Evaluating Solid Waste: Physical/Chemical Methods*, SW-846, 3rd edition, Office of Solid Waste and Emergency Response, US EPA.

24. American Public Health Association, American Water Works Association and Water Pollution Control Federation (1989) *Standard Methods for the Examination of Water and Waste Water*, 17th edition, APHA.
25. Cairney, T., Clucas, R.C. and Hobson, D.M. (1990) Evaluating subterranean fire risks on reclaimed sites, in: *Reclamation Treatment and Utilisation of Coal Mining Wastes*, Balkerina, Rotterdam.
26. Lord, D.W. (1987) Appropriate site investigations in: *Reclaiming Contaminated Land*, ed. T. Cairney, Blackie, Glasgow.
27. Blackheath Publishing Ltd (1991) *Construction Plant and Equipment Annual 1991*, Morgan Grampian.
28. Russel, A. and Gee, R. (1990) Use of the dynamic probe on polluted and marginal sites, in: *Proceedings of International Conference on Construction on Polluted and Marginal Land*, Glasgow, ed. Forde.
29. British Standards Institution (1981) BS5930, Code of Practice for Site Investigations, BSI, London.
30. European Communities (1991) Proposal for a Council Directive on the landfill of waste, 91/C190/01, *Official Journal of the European Communities* C190/33, EC.
31. Kelly, R.T. (1980) Site investigation and material problems, in: *Reclamation of Contaminated Land*, B2/1–B2/13, Society of Chemical Industry.
32. NVPG Dutch Association of Soil Treatment Companies (1990) *Soil Purification*.
33. House of Commons Select Committee on the Environment (1990) *First Report on Contaminated Land*, HMSO, London.
34. *Control of Pollution (Special Waste) Regulations (1980)* SI 1980 No.1709, HMSO, London.
35. Sax, N.I. and Lewis, R.J. (1989) *Dangerous Properties of Industrial Materials*, Van Nostrand Reinhold, New York.
36. Barry, D.L. (1983) CIRIA Report 98, *Material Durability in Aggressive Ground*, CIRIA.
37. European Communities (1976) Council Directive on pollution caused by certain dangerous substances discharged into the aquatic environment of the Community, 76/464/EEC, *Official Journal of the European Communities*, L129/23, EC.
38. European Communities (1979) Council Directive on the protection of groundwater against pollution caused by dangerous substances, 80/68/EEC, *Official Journal of the European Communities* L78/32, EC.
39. Department of the Environment (1982) Circular 4/82, *EC Directive on Protection of Groundwater Against Pollution caused by Certain Dangerous Substances* (80/68/EEC), HMSO, London.
40. Department of the Environment (1990) Circular 7/89, *Water and the Environment*, DOE.
41. Department of the Environment (1990) Circular 20/90, *EC Directive on Protection of Groundwater Against Pollution Caused by Certain Dangerous Substances (80/68 EEC): Classification of Listed Substances*, DOE.
42. European Communities (1975) Council Directive concerning the quality required of surface water intended for the abstraction of drinking water in the Member States, 75/440/EEC, *Official Journal of the European Communities* L194/26, EC.
43. European Communities (1975) Council Directive concerning the quality of bathing water, 76/160/EEC, *Official Journal of the European Communities* L31/1, EC.
44. European Communities (1978) Council Directive on the quality of fresh waters needing protection or improvement in order to support fish life, 78/659/EEC, *Official Journal of the European Communities* L222/1, EC.
45. *The Surface Waters (Dangerous Substances) (Classification) Regulations (1989)* SI 1989/2286, HMSO, London.
46. *The Surface Waters (Classification) Regulations (1989)* SI 1989/1148, HMSO, London
47. *The Trade Effluent (Prescribed Processes and Substances) Regulations (1989)* SI 1156, HMSO, London.
48. Her Majesty's Inspectorate of Pollution (1991) *Waste Management Paper No. 27, The Control of Landfill Gas*, 2nd edition, HMSO, London.

Chapter 4

1. ICRCL (1983) Guidance Note 59/83, 2nd edition, HMSO, London.

2. Anon (1989) Cleaning up sites with on-site process plants, *Environmental Science and Technology*, **23**, 912–916.
3. Anon (1991) Contaminated land treatment technology, *Croner's Environmental Management*, **3**, 325–338.
4. House of Commons Select Committee on the Environment (1990) *First Report on Contaminated Land*, HMSO, London.
5. Burnett-Hall, R. (1990) Legal aspects, in: *Contaminated Land, Policy, Regulation and Technology*, IBC Technical Services, London.
6. Petts, J. (1990) Contaminated land — is the UK cleaning up or covering up? in: *Contaminated Land, Policy, Regulation and Technology*, IBC Technical Services, London.
7. European Communities (1991) Proposal for a Council Directive on the landfill of waste, *Official Journal of the European Communities*, C190/91/C190/01, 33.
8. National Rivers Authority (1991) *Policy and Practice for the Protection of Groundwater*. Draft for Consultation.
9. *Texaco News* (1991) *Hi-tech clean up at Aberdeen terminal*. Issue 142.
10. Hinsenveld, M. (1990) Alternative physico-chemical and thermal cleansing technologies for contaminated soil, in: *Contaminated Land, Policy, Regulation and Technology*, IBC Technical Services, London.
11. Haiges, L. *et al.* (1989) Evaluation of underground fuel spill clean-up technologies, in: *Haztech International Conference*, San Francisco.
12. Anon (1985) "Getting to grips with waste solidification" *ENDS Report* (120), pp. 11–13.
13. Hubbard, S.J. (1990) Practical examples of the USEPA site program Superfund Innovative Technology Evaluation Program, in: *Contaminated Land, Policy, Regulation and Technology*, IBC Technical Services, London.
14. Boelsing, F. (1988) *DCR Technology* Ministry of Economics, Technology and Traffic, Hanover.

Chapter 5

1. Thorburn, S. and Buchanan, N.W. (1987) Building on chemical waste, in: *Building on Marginal and Derelict Land*, Thomas Telford, London, pp. 281–296.
2. Collins, S.P. *et al.* (1987) Rehabilitation of the Old Palace Gasworks site, in: *Building on Marginal and Derelict land*, Thomas Telford, London, pp. 449–496.
3. Mills, G. and Clark, J.C. (1987) The redevelopment of the Wandsworth Gasworks site, in: *Building on Marginal and Derelict land*, Thomas Telford, London, pp. 497–520.
4. House of Commons Select Committee on the Environment (1990) *First Report on Contaminated Land*, HMSO, London.
5. Jones, A.K. (1980) Monitoring of reclaimed contaminated sites, unpublished report to Department of the Environment.
6. Lord, A. (1991) Options available for problem solving, in: *Recycling Derelict Land*, ed. G. Fleming, Thomas Telford, London, pp. 145–190.
7. Cairney, T. (1987) Soil cover reclamations, in: *Reclaiming Contaminated Land*, ed. T. Cairney, Blackie, Glasgow, pp. 144–169.
8. ICRCL (1987) Guidance Note 59/83, *Guidance on the Assessment and Redevelopment of Contaminated Land*, Department of the Environment.
9. Sharrock, T. (1986) Methods of evaluating soil cover materials and quantifying design proposals, *Civil Engineering Technology*, **9**(8), 2–11.
10. Rawls, W.J. and Brakensiek, D.L. (1982) Estimating soil water retention from soil properties, *Journal of the American Society of Civil Engineers*, **108**(IR2), 166–171.
11. Russo, D. (1988) Determining soil hydraulic properties by parameter estimation: on the selection of a model for hydraulic properties. *Water Resources Research*, **24**(3), 453–459.
12. Bloemen, G.W. (1980) Calculation of steady state capillary rise from a groundwater table and through multi-layered soils, *Zeitschrift für Pilanzenern*, **143**, 701–719.
13. Anders, I.J. (1989) Evaluation of the soil cover reclamation method for chemically contaminated land, Ph.D. thesis (unpublished), Liverpool Polytechnic.
14. Bhuiyan, S.I. *et al.* (1971) Dynamic simulation of vertical infiltration into unsaturated soils, *Water Resources Research*, **7**, 1597–1605.

15. Al Saeedi, A. (1992) Irrigation design in arid environments, Ph.D. thesis (unpublished), Liverpool Polytechnic.
16. Driscoll, R. (1983) The influence of vegetation on the swelling and shrinking of clay soils in Britain, *Geotechnique*, **XXXIII**, 93–105.
17. Waters, P. (1980) Comparison of the ceramic plate and the pressure membrane to determine the 15 bar water content of soils, *Journal of Soil Science*, **31**, 443–446.
18. Cairney, T. (1983) Accelerated techniques for predicting the movement of contaminants in soils, Unpublished research report, EC Environmental Programme, ENV/675/UK(H).
19. Department of the Environment. (1986) *Land filling wastes*, *Waste Management Paper No. 26*, HMSO, London.
20. Knox, K. (1991) Water management at landfills: water balance and practical aspects, Lecture notes (unpublished) for NAWDC Course, Coventry.
21. European Communities (1991) Proposal for a Council Directive on the landfill of waste, 91/C190/01, *Official Journal of the European Communities*, C190/33, EC.
22. Bouma, J. Belmans, C.F.M. and Dekker, L.W. (1982) Water infiltration in a silt loam subsoil with vertical worm channels. *Soil Science Society of America*, **46**, 917–921.
23. Hillel, D. (1980) *Application of Soil Physics*, Academic Press, New York.

Chapter 6

1. Chipp, P. (1990) Geotechnical processes for the prevention and control of pollution, in: *Symposium on Management and Control of Waste Fill Sites*, Leamington Spa.
2. European Communities (1991) Proposal for a Council Directive on the landfill of waste, 91/C190/01, *Official Journal of the European Communities*, C190/33, EC.
3. Her Majesty's Inspectorate of Pollution (1989) *Waste Management Paper 27, The Control of Landfill Gas*, HMSO, London.
4. Haxo, H.E. *et al.* (1985) Liner materials for hazardous and toxic wastes and municipal solid waste leachate, *Pollution Technology Review No. 124*, Noyes Publications.
5. Mitchell, J.K. (1991) Conduction Phenomena, 31st Rankine Lecture, *Geotechnique*, September.
6. Anon (1985) First Stent wall installed at Kingston upon Thames, *Ground Engineering*, **October**, 27–31.
7. Xanthakos, P. (1979) *Slurry Walls*, McGraw-Hill, New York.
8. Philipp Holtzman Aktiengesellschaft (1991) *Innovative Glastechnologie für Deponiedicht-wande*.
9. D'Appolonia, D.J. (1980) Soil–bentonite slurry trench cut-offs, *Journal of the Geotechnical Engineering Division*, ASCE, **106**(4), 399–417.
10. Krause, R. (1989) *New Developments and Trends in Ground Water Protection with Flexible Membrane Liners*, LT Lining Technology GmbH.
11. Hass, H.J. and Hitze, R. (1986) All-round encapsulation of hazardous wastes by means of injection gels and cut-off materials resistant to aggressive agents, in: *ESME3 Seminar on Hazardous Waste*, Bergamo, Italy.
12. Oil Companies Materials Association (1973) *Specification No. DFCP 4, Drilling Fluid Materials Bentonite*.
13. Brice, G.J. and Woodward, J.C. (1984) Arab potash solar evaporation system: design and development of a novel membrane cut-off wall, *Proceedings of the Institution of Civil Engineers*, Part 1, **76**, 185–250.
14. Jefferis, S.A. (1985) Discussion on the Arab potash solar evaporation system, *Proceedings of the Institution of Civil Engineers*, Part 1, 641–6464.
15. Jefferis S.A. (1992) Contaminant–grout interaction, in: *ASCE Specialty Conference, Grouting, Soil Improvement and Geosynthetics*, New Orleans.
16. Howsam, P. (ed.) (1990) *Microbiology in Civil Engineering*, E & F.N. Spon, London.
17. Jefferis, S.A. and Mangabhai, R.J. (1989) The divided flow permeameter, in: *Materials Research Society, Symposium on Pore Structure and Permeability of Cementitious Materials*, Vol. 137.

Chapter 7

1. Beever, P.F. (1989) Subterranean fires in the U.K. — the problem, *BRE Information Paper, IP3/89*.
2. Redpath, P.G. (1989) Containment, spread and effect of an industrial site fire, in: *BRE Research Colloquium*.
3. Drake, D. (1987) *Subterranean Heating at Oakthorpe Village*, Institute of Mining Engineering, S. Staffs. Branch.
4. Rainbow, A.K.M. (ed.) (1990) Reclamation, treatment and utilization of coal mining wastes, in: *Proceedings of the Third International Symposium on the Reclamation, Treatment and Utilization of Coal Mining Wastes*, Glasgow.
5. Fardell, P.J. and Lukas, C. (1987) Understanding Fire, *Chemistry in Britain*, **March**.
6. Bowes, P.C. (1984) *Self Heating: Evaluating and Controlling the Hazards*, HMSO, London.
7. Street, P.J., Smalley, J. and Cunningham, A.T.S. (1975) Hydrogen as an indicator of the spontaneous combustion of coal, *Journal of the Institute of Fuel*, **September**.
8. British Standards Institution (1975) BS476, *Fire Tests on Building Materials and Structures*, London.
9. British Standards Institution (1973) BS1016, *Method for the Analysis and Testing of Coal and Coke, Part 3. Proximate Analysis of Coal*, BSI London.
10. Ball, D.F. (1964) Loss-on-ignition as an estimate of organic matter and inorganic carbon in non-calcareous soils, *Journal of Soil Science*, **15**, 84–92.
11. British Standards Institution (1973) BS1016, *Method for the Analysis and Testing of Coal and Coke, Part 5. Gross Calorific Value of Coal and Coke*, London.
12. Davies, C. (1970) *Calculations in Furnace Technology*, Pergamon Press, Oxford.
13. Chigier, N.A. (1981) *Energy, Combustion and Environment*, McGraw-Hill, New York.
14. Rose, J.W. and Cooper, J.R. (eds.) (1977) *Technical Data on Fuel*, 7th edition, The British National Committee World Energy Conference, Scottish Academic Press, Edinburgh.
15. Smith, M.A. (1991) in: *Recycling Derelict Land*, ed. S. Fleming, published for the Institute of Civil Engineers by Thomas Telford, London.
16. Beever, P.F. (1982) Spontaneous combustion, *BRE Information Paper IP 6/82*.
17. Beever, P.F. (1982) Spontaneous Combustion — Isothermal Test Methods, *BRE Information Paper IP 23/82*.
18. Baker, B. (1989) Subterranean Fires, *BRE Research Colloquium*.
19. Sebastian, J.J.S. and Mayers, M.A. (1937) Coke reactivity. Determination by a modified ignition method, *Journal of Industrial and Engineering Chemistry*, **29**(10).
20. British Standards Institution (1975) BS1377, *Methods of Testing for Soils for Civil Engineering Purposes*, London.
21. Nagata, N. (1988) Air permeability of undisturbed soils, *Bulletin of the Faculty of Agriculture*, Mie University, Japan, pp. 35–55.
22. CP3 (1972) *Wind Loads*, HMSO, London, Chap. V, Part 2.

Chapter 8

1. Health and Safety Executive (1991) *Occupational Exposure Limits*, Guidance Note EH 40/91.
2. Department of the Environment (1986) *Landfilling Wastes, Waste Management Paper No. 26*, HMSO, London.
3. Department of the Environment (1985) *Building Regulations, Part C — Site Preparation and Resistance to Moisture*, HMSO, London.
4. Department of the Environment (1987) *Development of Contaminated Land*, Circular 21/87, HMSO, London.
5. ICRCL (1990) Guidance Note 17/78, *Notes on the Redevelopment of Landfill Sites*, 8th edition, Department of the Environment.
6. Carpenter, R.J. (1988) Building redevelopment on disused landfill sites — overcoming the landfill gas problem, in: *Proc. 5th Int. Solid Wastes Conf.*, Copenhagen, Denmark, Vol. 1, Academic Press, London, pp. 153–160.
7. Her Majesty's Inspectorate of Pollution (1989) *Waste Management Paper No. 27, The Control of Landfill Gas Waste*, HMSO, London.

8. Department of the Environment (1987) *Landfill Sites: Development Control*, Circular 17/87, HMSO, London.
9. US Environmental Protection Agency (1980) *Classifying Solid Waste Disposal Facilities*, SW 828.
10. Carpenter, R.J., Goaman, H.F., Lowe. G.W. and Pecksen, G.N. (1985) Guidelines for site investigation of contaminated land, *London Environmental Supplement No. 12*.
11. Pecksen, G.N. (1985) Methane and the development of derelict land, *London Environmental Supplement No. 12*.
12. Crowhurst. D (1987) *Measurement of Gas Emissions from Contaminated Land*, Building Research Establishment, Fire Research Station, Borehamwood, Herts.
13. Institute of Wastes Management (1989) *Monitoring of Landfill Gas*, IWM, Northampton.
14. European Commission (1991) Proposal for a Council Directive on the landfill of waste, 91/C190/01, *Official Journal European Communities*, C190/33, EC.
15. Harries, C. (1991) The application of laboratory methods to the evaluation of methane production, in: *Methane: Facing the Problems, 2nd Symp.*, Nottingham University, Paper 2.2.
16. Smith, M.A. (1991) Data analysis and interpretation, in: *Recycling Derelict Land*, ed. G. Flemming, Thomas Telford, London.
17. Barry, D. (1987) Hazards from methane (and carbon dioxide), in: *Reclaiming Contaminated Land*, ed. T. Cairney, Blackie, Glasgow.
18. Gregory, R.C., Gardner, N. and Lyth, D. (1991) Assessment of gas production and the effects of engineered control measures on gas migration, in: *The Planning and Engineering of Landfill Conference*, Birmingham University, Midlands Geotechnical Society.
19. Needham, C. (1989) Cut off barriers for methane control, in: *Methane: Facing the Problems, 2nd Symp.*, Nottingham University, Paper 3.1.
20. Raybould, J.G. and Anderson, D.J. (1987) Migration of landfill gas and its control by grouting — a case study, *Quarterly Journal of Engineering Geology*, **20**, 75–83.
21. Ingle, J.L. and Kavanagh, S.T. (1991) Case studies of the use of slurry walls to control leachate and gas migration, in: *Methane: Facing the Problems, 2nd Symp.*, Nottingham University, Paper 3.1.
22. Clarke, A.D., Board, N.P. and Griffiths, C.M. (1991) Protection of structures from landfill gas, in: *Methane — Facing the Problems, 2nd Symp.*, Nottingham University, Paper 5.4.
23. Emberton, J.R. and Parker, A. (1987) The problems associated with building on landfill sites, *Waste Management and Research*, **5**, 473–482.
24. Clark, R.G. and Warby, I.S. (1991) Gas protection measures for buildings on or adjacent to landfill sites, in: *The Planning and Engineering of Landfills Conference*, Birmingham University, Midlands Geotechnical Society.
25. Parker, A. (1990) Building on landfill sites, in: *Brunel Landfill Conference — Problems and Solutions*, Brunel University.
26. Carpenter, R.J. (1986) Redevelopment of land contaminated by methane gas. The problems and some remedial techniques, in: *Proc. 1st Int. TNO Conf. on Contaminated Soil*, Utrecht, Netherlands.
27. Sheriff, J.A., Stevenson, D. and Wright, P.A. (1991) From site survey to building design. The safe cost–effective solution? in: *Methane: Facing the Problems, 2nd Symp.*, Nottingham University, Paper 4.4.
28. O'Riordan, N. and Warren, R. (1991) Methane and the design of foundation systems, in: *Methane: Facing the Problems, 2nd Symp.*, Nottingham University, Paper 5.2.
29. Lord, J.A. (1991) Recycling landfill and chemical waste sites, in: *Containment of Pollution and Redevelopment of Closed Landfill Sites, Proc. Leamington Spa Conf.*, Paper 6.1.
30. Crawford, J.F. and Smith, P.G. (1985) *Landfill Technology*, Butterworth, London.
31. Rys, L.J. and Johns, A.F. (1986) The investigation and development of a landfill site, in: *Proc. 1st Int. TNO Conf. on Contaminated Soil*, Utrecht, Netherlands.

Chapter 9

1. Strahler, A.N. (1960) *Physical Geography*, Wiley, New York.
2. Roberts, R.D. and Roberts, J.M. (1986) The selection and management of soils in Landscape Schemes, *Ecology and Design in Landscape*, pp. 99–126.

3. Gasson, P.E. and Cutler, D.F. (1990) Tree root plate morphology, *Arboricultural Journal*, **14**, 193–264.
4. Samuel, P. (1991) Revegetation of reclaimed land, in: *Proceedings of Land Reclamation — An End to Dereliction*, Cardiff, pp. 366–376.
5. Brophy Organic Products (1991) Having your cake and selling it, *Landscape Industry International*, **IX**(5), 24–25.
6. Scullion, J. (1991) Re-establishing earthworm populations on former open-cast coalmining land, in: *Proceedings of Land Reclamation — An End to Dereliction*, Cardiff, pp. 377–386.
7. Gemmell, R.P. (1985) Wildlife habitats created by mining and tipping, *Land and Mineral Surveying*, **August**, 422–431.

Chapter 10

1. British Standards Institution (1981) BS5750, Part 6, HMSO, London.
2. Department of Transport (1986) *Specifications for Highway Works — Part 2*, HMSO, London.
3. Lord, D.W. (1987) Appropriate site investigations, in: *Reclaiming Contaminated Land*, ed. T. Cairney, Blackie, Glasgow.
4. Smith, M. (1991) Data analysis and interpretation, in: *Recycling Derelict Land*, ed. G. Fleming, Thomas Telford, London.

Chapter 12

1. Carson, R. (1963) *Silent Spring*, H. Hamilton.
2. Bellandi, R. (technical editor) (1988) *Hazardous Waste Site Remediation the Engineer's Perspective*, O'Brien Gere Engineers, Inc.
3. PL 91–190 of 1 January 1970, 42 U.S.C. Section 4321 *et seq.*
4. Environmental Regulation Course, Executive Enterprises, Inc., Salt Lake City, 1986.
5. *Ibid.*
6. Environmental Research and Technology, Inc and Sidley & Austin. *Superfund Handbook*, Second Edition, April 1987.
7. *Ibid.*
8. US Environmental Protection Agency (1988), *The Waste System*, Solid Waste and Emergency Response, November 1988.
9. US Environmental Protection Agency (1986). Solving the Hazardous Waste Problem. EPA's RCRA Program, Office of Solid Waste, EPA/530-SW-86-037, November 1986.
10. US Environmental Protection Agency (1988). *The Waste System*.
11. *Ibid.*

Chapter 13

1. ICRCL (1992) Guidance Note 64/85, *Asbestos on Contaminated Sites*, 2nd edition, Department of the Environment, London (in preparation).
2. ICRCL (1987) Guidance Note 59/83, *Guidelines on the Assessment and Redevelopment of Contaminated Land*, 2nd edition, Department of the Environment, London.
3. ICRCL (1986) Guidance Note 61/84, *Notes on the Fire Hazards of Contaminated Land*, 2nd edition, Department of the Environment, London.
4. Building Research Establishment (1987) Fire and explosion hazards associated with the development of contaminated land, BRE Information Paper 2/87.
5. Health and Safety Executive. *Entry into Confined Spaces*, Guidance Note GS5, HMSO, London.

6. Health and Safety Commission (1985) *Approved Code of Practice for the Protection of Persons against Ionising Radiations Arising from any Work Activity*, HMSO, London.
7. Department of the Environment (1982) *Radioactive Substances Act 1960*, A Guide to the Administration of the Act, HMSO, London.
8. Building Advisory Service (latest edition) *Construction Safety Manual*, BAS, London.
9. Health and Safety Executive (1988) *Approved Code of Practice: Control of Substances Hazardous to Health: Control of Substances Hazardous to Health Regulations 1988*, HMSO, London.
10. Health and Safety Executive. *COSHH Assessments: A Step by Step Guide to Assessment and the Skills Needed for it*, HMSO, London.
11. Health and Safety Executive (1985) *Control of Lead at Work: Approved Code of Practice*, revised edition, HMSO, London.
12. Health and Safety Executive (1981) *Control of Lead: Outside Workers*, Guidance Note EH29.
13. British Standards Institution (1988) Draft for Development, *DD175: 1988 Code of Practice for the Identification of Potentially Contaminated Land and its Investigation*, BSI, Milton Keynes.
14. Health and Safety Executive (latest edition) *Occupational Exposure Limits*, Guidance Note EH40.
15. Health and Safety Executive. *Respiratory Protective Equipment (RPE), A Practical Guide for Users*, HS(G)53.
16. Health and Safety Executive (1990) *Approved Code of Practice: First Aid at Work*, HMSO, London.

Index